DANIEL S. LORDAHL, Ph. D., University of Wisconsin, is Associate Professor of Psychology at the University of Miami. He formerly taught at Washington University, St. Louis. Dr. Lordahl is presently a member of the Board of Directors of the University of Miami Computing Center. He frequently contributes papers to the leading journals in psychology and is currently working on a research grant on cognitive processes from the National Institute for Mental Health.

MODERN STATISTICS FOR BEHAVIORAL SCIENCES

DANIEL S. LORDAHL

UNIVERSITY OF MIAMI

THE RONALD PRESS COMPANY • NEW YORK

Library of Congress Catalog Card Number: 67–14485

PRINTED IN THE UNITED STATES OF AMERICA

To

Thelma and *Carl*

Preface

This book is designed for use in an introductory course in statistics for the behavioral sciences. Its emphasis is upon those techniques and concepts most frequently encountered in the current research literature in psychology, sociology, and education.

An attempt is made in this book to include few statistics which are uncommon or irrelevant to the development of the more frequently used methods. In this context, an attempt is made, as well, to reduce the burden of computational techniques imposed upon the student. This is done because of the availability of desk calculators, as well as digital computers, to both students and researchers. For instance, although coded scores provide insight into certain mathematical relationships, their advantages for computational purposes are dated. In such cases, the relationships upon which these techniques are based are discussed, but the actual computational schemes are omitted.

The second important feature of this book is the emphasis on variance tests. Traditionally, analysis of variance has been postponed to advanced courses or merely introduced in beginning texts. In effect, this practice has prepared the student completing a single course in statistics neither to read research literature in a meaningful way nor to analyze results of many experiments having more than two conditions. The solution to this problem attempted here is quite direct: three chapters on analysis of variance are included at the earliest possible time. It should be mentioned that this procedure has been quite successful in the author's experience. In fact, the student of average ability seems to have fewer problems in mastering the analysis-of-variance procedures than a simple *t*-test for differences between means. The advantages inherent in the generality of the analysis-of-variance procedures are taken as sufficient justification for the emphasis on these methods.

A third problem considered in the writing of this book was in the organization of its material. Students often report a failure to see relationships between the various statistical methods presented to them. Although numerous factors may account for this state of affairs, it is hoped that this problem will be overcome through two features of this book: the inclusion of twelve chapters, all of which may be covered in a single semester; and an emphasis upon the communality of purposes among various statistical procedures. This concept of the organizing superstructure of statistical methods is introduced in Chapter 4 and serves as a point of departure (and return) in later work.

Some of the concepts of set theory and notation have been utilized in the presentation of material. These ideas are not difficult, indeed are elementary for many students of the "new" mathematics, and form more of a vocabulary than a formal method of proving theorems.

The chapter on probability is quite extensive for an introductory textbook and has been placed rather late in the sequence of material. As written, however, this chapter could be taken at any point following Chapter 5 without creating a burden for the student.

Formal derivations are not emphasized; on the other hand, the rationale of such derivations is stressed, and thereby the requirements for the appropriate use of different statistical procedures are both indicated and justified for the student.

The exercises provided are divided into two types in the last six chapters, exercises which simply allow application of the procedures presented in the immediately preceding chapter, and applications which provide more meaningful data for analysis and also require the student to select the appropriate method of analysis from among the entire set of those already covered.

The author wishes to acknowledge the help of many people in the formulation and writing of the book. The aid of a number of students in introductory statistics classes, both voluntary and involuntary aid, stimulated the initiation of this text and shaped many of the principles of organization and presentation. I specifically wish to thank Dr. Herbert Harari and Dr. Bruce Dunn for their valuable suggestions about certain portions of the manuscript, and to acknowledge the effort and help provided by George Dauth IV, who read the entire manuscript.

I am indebted to the Literary Executor of the late Sir Ronald A. Fisher, F.R.S., Cambridge, to Dr. Frank Yates, F.R.S., Rothamstead, and to Messrs. Oliver & Boyd, Edinburgh, for permission to reprint Tables Nos. III and IV from their book *Statistical Tables for Biological, Agricultural, and Medical Research.* The material taken from these tables is found in Tables D and F, respectively, in the Appendix.

I offer a special thanks to my wife, Jo Ann, who encouraged me throughout the course of writing and prepared the complete typewritten copy.

DANIEL S. LORDAHL

Coral Gables, Florida
January, 1967

Contents

MODERN STATISTICS FOR BEHAVIORAL SCIENCES

1

Introductory Concepts

Statistics deals mainly with numbers. However, the function of statistics is to promote the understanding, discovery, and application of knowledge. This book is intended to provide information which will allow the reader to use statistics in this manner, particularly in regard to knowledge in the behavioral sciences.

The concept of uncertainty is one important feature of statistical reasoning. Not too many years ago, the weatherman might have made a prediction of rain, only to have this prediction prove incorrect for many parts of the forecast area. Today, the prediction might well be given as, "there is a 70 per cent chance of rain." This current type of prediction is not based, in any sense, on different sorts of information than the earlier type of prediction. The weatherman in prior years was just as aware of the probabilistic nature of his prediction as are current forecasters. However, today there seems to be a realization that the public is able to comprehend the meaning of a probability statement. At least some degree of uncertainty is inherent to all fields of scientific study of the universe, although the layman is not always aware of this fact.

A second basic feature of statistical thinking is the use of numeric information to supplement or even replace the natural language. Although our language permits an almost complete description of certain characteristics in the environment by the use of numbers, there are few abstract terms which summarize quantitative aspects of the world; *all, none, some,* a *majority, few,* etc., are all abstract terms, but the degree of accuracy possible with such terms is not high. Nontechnical words in the language which describe statistical concepts are rare. The concept of the average, for example, is one term used frequently today, but even here many people cannot give a precise definition of the concept.

The limitations of a two-valued logic system, such as that developed by Aristotle, are clearly evident when one attempts to reason from premises which are probabilistic in form. These limitations are overcome, to some extent, by the concept of *degree* which arises from the use of numbers. However, the results of reasoning from probabilistic premises must themselves be stated in terms of probability or uncertainty. This reasoning process and the statement of conclusions require the use of abstractions concerning variability.

In large measure, the study of statistics is the study of variability. However, the characteristic of variability must be attributed to something. We shall first consider the kinds of entities that are important enough to require a study of their variation.

1–1. Sets. Any collection of things grouped together for any reason may be called a *set*. Each entity or thing within such a collection will be termed an **element** or member of that set. An example would be the collection of automobiles in a specific parking lot at a certain time. Each automobile is considered to be an element of the set of automobiles. It will be assumed hereafter that each element of a set can be distinguished from every other element. This is considered true even when two elements are identical in all respects; for instance, two automobiles which appear identical in all observable respects are still seen to be two different objects. Because of this somewhat confusing case of identical but distinguishably different elements, it is common to label the elements (usually by numbering them) for ease in keeping track of each element. Symbolically, let us refer to an element of a set as E_i, where the letter i as a subscript indicates the label of some particular element. For instance, E_{57} would refer to the element numbered 57 within some set.

The usual type of set dealt with in statistics is one in which the elements are numbers. Such sets are ordinarily obtained from the characteristics of sets of objects or people, but aspects of the objects or people other than those used to obtain the set of numbers are ignored. In the example used, one might be interested in the weights of the automobiles in the collection of automobiles considered. The resulting numbers (weights in some appropriate unit) form the elements or scores of this new set. When a set under consideration has numbers as elements, the symbol X will be used in place of the symbol E. Once again it will prove convenient to number the elements or scores in order to keep track of each and to refer to any one of the scores in an unambiguous fashion. Letting the general notation for an element in a set of numbers be X_i, it is usual to let the value of i start at 1 and increase by 1 for each additional element labelled. In this case, the last value of i (call it N) will be equal to the number of elements in the set. Care must be taken not to confuse the value of the *index* (the subscript i) with the number or score value represented. For example, X_{25} is used to indicate the 25th score in a set of numbers. That score might be 2,647,209, and it may also include some unit of measure, such as pounds or inches.

1–2. Population Sets. In statistics, any set of interest may be designated as a *population*. Usually a population is a set of numbers. These scores (numbers) may all be known to the investigator or they may be largely or wholly unknown. This aspect of the set is not important in defining the set as a population so long as any particular score can be distinguished from other scores *and* can theoretically be investigated and its value discovered.

The number of scores in a population is ordinarily quite large. This is not a requirement; even a single score could be defined as a population, but rarely is so small a set of interest.

Some sets are defined in such a way that they contain an "uncountable" or infinite number of elements. These infinite sets will not be considered in a mathematically rigorous fashion in this book, but the reader will obtain correct solutions to most problems involving infinite sets if he simply considers them as containing an extremely large but "countable" number of scores.

Sets which have a definite number of elements are called *finite*. Most of the population sets dealt with in research in the social sciences are finite.

1–3. Sample Sets. A type of set called a *sample* is always defined with respect to a population. If every element of a set is also an element (or member) of a population set, then the first of these sets is called a sample of, or from, the population.

Two special characteristics of samples should be noted. The first of these arises directly from the above definition of a sample and is the fact that if an element of the sample is not also a member of the population, then the so-called sample is not really a sample from that population being considered. A second aspect of some samples is that they may have two or more elements which are in fact the same element in the population. For example, a group of people entering a lottery might be considered as a population, and the set of people winning prizes might then be considered a sample from that population. If a single person wins two prizes (or more) then that person is a single element in the population but is two or more elements in the sample of winners. Such a condition need not be feared in statistics but may be avoided if desired by appropriate selection of population elements when obtaining a sample.

1–4. Random Samples. A type of sample called a *random sample* is one which is constructed from a population of scores in a very special way. Roughly speaking, we can define a *random sample* as a set of elements selected from a population in such a way that every element of the population has an equal chance or likelihood of being included in the sample.

The terms *chance, likelihood,* and *probability* are typically used in the definition of random selection or sampling. They will not be defined at this point, but the reader should have already formed a good intuitive grasp of such concepts.

The concept of a random sample is of utmost importance to the logic and interpretation of a multitude of statistical procedures. The concept will be used frequently, explored more fully, and illustrated concretely in later chapters.

Sampling procedures. Although there are other useful methods of obtaining a sample from a population, only random samples will be discussed in this book. There are, however, two important types of random sampling procedures which have subtle but important differences, called (a) *sampling with replacement,* and (b) *sampling without replacement.* In (a), the selection of elements for the sample can be considered as taking place one at a time in a consecutive fashion. The selection of element E_i for a sample element does not prevent its later selection as another sample element. It is as if the element just chosen for the sample is placed back in the population (although it also stays in the sample) before the next selection is made. Hence the term replacement is used to describe this procedure. This sort of sampling is the one which could have given two or more prizes to the same person in the lottery illustration used earlier.

The selection of a sample without replacement (b) can be thought of as a series of selections involving one element each time, in the same manner as (a). The basic distinction is that in (b) an element selected for a sample element is not placed back in the population. Another way of saying the same thing is that the population element selected is immediately removed from the population (and placed in the sample). Since the population element chosen is not replaced, such an element can only be used one time in the selection of a sample.

Duplicate scores. A point of confusion can now be avoided in dealing with populations and samples of numbers. Consider the following small set of numbers as a population: 3, 4, 4, 5, 8. These scores will be labelled consecutively as X_1, X_2, X_3, X_4, and X_5. If a random sampling procedure were used to select a sample of two elements from this population, the two scores chosen could be the same number with or without the replacement feature being used. When replacement is used, any of the *four values* (3, 4, 5, and 8) could occur twice in the sample of two numbers chosen. If sampling were without replacement, the result might still be a sample consisting of the numbers 4 and 4. This is possible because one might have selected the elements X_2 and X_3. Although these are two distinctly different scores, their values are the same.

1–5. Definition of a Statistic. The fundamental part of all statistical tests, inferences, and conclusions is the statistic. The capital letter S will be used to denote any statistic in general. A statistic, S, may be defined as a function of a set of numbers. Symbolically, this could be stated as

$$S = f(X_i), \quad i = 1, N \tag{1–5.1}$$

In equation (1–5.1), the f denotes a function or rule, the X_i indicates numbers in a set, and the notation $i = 1, N$ (read as i equals one through N) implies that all of the N numbers in the set are considered when applying the function.

It should be noted that the term function is used in its broad mathematical sense; i.e., a function is any rule, table, equation, procedure, or combination of these that allows one to obtain a single number (usually) when applied to a set of scores.

The term *statistic* is actually used in two ways. The first one refers to whichever particular function has been employed. In this use, the student will find that the reason for different names for statistics is that different rules are used in each case. The second use of the term is for the particular value or answer obtained when a certain rule has been applied to a given set of numbers.

The name of a statistic, then, refers to the rule used to obtain an answer for any set of numbers, and the value of the statistic is the number obtained when that rule is applied to a particular set of numbers.

Examples of a statistic. In order to illustrate the definition of a statistic, let us consider a very simple but useful statistic which has been encountered in earlier discussion. This is the number of scores in a set, N. The rule defining this statistic can be stated in a precise but somewhat cumbersome way if desired, but let us simply say that the rule applied to the set is "count the number of numbers in the set, counting each score only once." Recall that two or more scores may have the same value. Care should be taken to count the number of scores and not simply the number of different values occurring in the set. In the set of the numbers 3, 4, 4, 5, 8, there are 5 different scores (hence $N = 5$) even though only 4 different score values are present.

Let us use this same set of 5 scores to illustrate another frequently used statistic, the **total** or **sum** of a set of numbers. If we have N numbers in a set, the definition of the total, T, is

$$T = X_1 + X_2 + X_3 + \cdots + X_N \tag{1-5.2}$$

The series of dots $(\cdots)$ in this equation merely implies that all the values of the scores labelled with subscripts between 3 and N are also added to those shown explicitly. In the present example, the calculation of the total (or sum) of the set of scores may be written out as

$$T = 3 + 4 + 4 + 5 + 8 = 24$$

Note that two statistics, N and T, have been computed on the same set of numbers. This set of scores can be described, incompletely, of course, by giving the results of these statistics:

$$N = 5 \qquad \text{and} \qquad T = 24$$

A statistic can be viewed as a type of abstraction from a set of numbers. As such, it should be clear that the value of a statistic is only a partial description of the set of numbers from which it is obtained. Since the meaningfulness and usefulness of any abstraction depend upon one's purposes, there are a large number of different statistics used in different areas by statisticians and researchers.

1–6. Misuses of Numbers. The use of numbers to describe aspects or characteristics of objects or events is common in our culture. Common too is the use of operations on numbers, such as addition, subtraction, multiplication, and division. In fact, the teaching of arithmetic by traditional methods has so well combined these two different uses of the number system that many students find it difficult to imagine that some uses of numbers to describe aspects of the world do not mix well with the highly practiced operations on numbers.

If numbers are to be anything but toys and games, then they must represent something. In this sense, meaningful operations performed on numbers must also represent meaningful operations on the characteristics for which the numbers stand. If the planets of our solar system were numbered from 1 to 9, what is represented by the addition of the identification numbers 2 and 8?

As will be discussed shortly, certain uses of numbers permit only a limited set of operations to be performed while still retaining some meaning in the results. Later chapters will point out that certain statistical procedures assume, or are designed for, particular types of representational uses of numbers. The major point here is that any student, professor, or computer can take nearly any set of numbers and perform any statistical procedure on these numbers. The statistical test or procedure has no way of knowing that it cannot give meaningful results on this set of numbers. The serious use of statistics requires that one keep in mind the meaning of numbers and of operations on numbers in terms of what the numbers represent.

In this context, it may be appropriate to answer the charges that the use of statistics can mislead or aid actual falsehoods. Certainly the concept that statistics are abstractions indicates that incomplete description is involved. However, because of man's limited abilities, complete description of many sets of data (scores) leads to an incomprehensible mass of numbers. It is also true that inappropriate uses of statistics (as discussed in this section) may give erroneous impressions to many people. By selecting certain statistics, ignoring others, and misusing numbers, it is possible to give false impressions to many persons. However, consider the phrase, "etudinus refratis cum gesundheitus." To readers without a reasonable background in Latin, such an insert in a textbook may well give the impression that something meaningful is being said. The phrase is actually without coherent meaning in any language known to the present author. The competent student of statistics will not be fooled easily by misuses of the subject matter; moreover, he will gain valuable information when it is presented in appropriate statistical form.

1–7. Scales of Numbers. When possible, it is usual to order objects along a hypothetical line or dimension with respect to their relative amounts of some characteristic. This is fairly easy when speaking about magnitudes, frequency, or even abilities or preferences. If one now assigns numbers to the points along such a dimension, we may speak of the combination as a scale of numbers, representing the characteristics of the elements being considered. A few such scales and their differences will now be given.

Nominal scales. The most restrictive use of numbers to be discussed is called the *nominal* scale. The term *scale* is actually somewhat misleading in this case, because the objects or entities to which the numbers are assigned are not actually arranged along any dimension related to their characteristics. The assignment of nominal numbers is absolutely arbitrary. This use of numbers is merely one of giving labels or names, i.e., the numbers, to each element in a set of elements. The previous use of index numbers is a good illustration. If such numbers are assigned to the elements without regard for their characteristics, then the resulting scale of numbered elements is a nominal one.

It is meaningless to add, subtract, divide, or multiply the nominal scale numbers. Indeed, the fact that E_2 is called by the number 2 and E_5 by the number 5, gives no information about their relative size, magnitude, or other pertinent aspects. The only fact to be inferred from their two numbers (2 and 5) is that they are separate elements.

In a more refined use of the nominal scale of numbers, one can use the same numbers to describe elements with identical or very similar characteristics; in this more general case, the fact that two elements have different numbers tells us that they have different characteristics, and that two elements have the same scale number tells us they have the same characteristics. One can readily see that the manipulation of nominal scale numbers is difficult if meaning is to be given to the results.

Ordinal scales. A somewhat less restrictive and more powerful use of numbers is called an *ordinal* scale. In this and scales to be discussed later, the notion of dimensionalized characteristics is essential. On an ordinal scale, if two elements have numbers which are different, their characteristics are different on the dimension considered, just as in a nominal scale. In addition to this, if one element has a higher number than a second element, this means that the first is in a position on the scaled dimension closer to the high-numbered end than is the second element. For instance, if we assign numbers on the scale so that high numbers mean more of a certain attribute than low numbers, then the element with the higher number has more of that attribute than the lower numbered element. We may symbolize this relationship as $E_i < E_j$.

This statement describes the fact that the scale number for E_i is a smaller number than for E_j, and it indicates that E_i possesses less of the scaled characteristic than does E_j, assuming an ordinal scale.

A slightly stronger variation of the ordinal scale, called a *strictly ordinal* or *strictly ordered* scale, allows certain mathematical manipulations and logical transformations to be performed with inequalities. For instance, if we are given as true relationships

$$E_i < E_j \qquad \text{and} \qquad E_j < E_k$$

then the algebra of inequalities allows the conclusion that

$$E_i < E_k$$

And this conclusion is meaningful and true in terms of the scale of attributes under consideration.

The addition, subtraction, multiplication, and division of ordinal scale numbers do not give meaningful results, in general. For instance, a common ordinal scale is one in which baseball teams are *ranked* in terms of the percentage of games won during a season's play. By convention, the highest or winning team is given a rank of 1 (first place), the second-best team a rank number of 2, etc. This system of assigning numbers does not indicate the important feature of how close together two given teams may have been. If the first-place team won by a great deal, but the second- and third-place teams were very close in terms of percentage of games won, this feature is ignored by the assigned scale numbers. The difference between the *numbers* 1 and 2 is the same difference as between the *numbers* 2 and 3, but these equal differences in the numbers do not reflect equal differences in the characteristics being described. For this reason, adding the ranks 2 and 7 together may give the same numerical result as adding the ranks 1 and 8, but we cannot meaningfully interpret the equality of the two totals as showing equal totals of percentages of games won.

The student is encouraged to consider other examples of this type and to note the strange concepts which arise from attempts even to add certain types of characteristics together in meaningful ways.

A commonly used example of an ordinal scale from the area of psychology is the intelligence quotient (I.Q.). The addition of two I.Q. scores together has rather ambiguous implications if one means to represent a person's ability by such a score. The corresponding addition of the abilities of two persons presents many questions of interpretation. Even if a meaningful definition of adding abilities together were accepted, the present I.Q. scores would probably still be only an ordinal scale.

Interval scales. A still more powerful scale of numbers is called an *equal interval* scale (or simply an interval scale). In this type of scale, all the properties of the nominal and ordinal scales apply with some very important additions. Specifically, the distances between points within the dimension considered are exactly represented by the differences in scale numbers assigned to those points. The way in which such distances are meaningfully defined is beyond the scope of this text. One can grasp the notion of defining these distances by considering the physical measurement of short distances along a straight line with a standard ruler. The concept of equal distances might involve the *operation* of placing two segments side by side and noting that their beginning and end points coincide. Addition operations might involve the placing of two segments end to end and noting that the resulting sum of the numbers representing the added distances is equal to the number representing the distance of a segment which "matched" the two added segments.

Such fundamental aspects of measurements are indeed crucial for the proper use of number scales, but the reader is asked to assume the meaningfulness of most measurements used in this text.

When a scale of numbers may be considered an interval scale, it becomes possible to meaningfully add, subtract, multiply, and divide the scale values assigned to different elements. However, division in this case has limitations, as illustrated below. Constants or *pure* numbers (those without units) may be used as divisors, but care is required in other cases. Equations involving the interval scale numbers may be manipulated according to the usual rules of algebra, with the important exception that proportionalities or ratios may not be employed easily. To illustrate this restriction, consider the situation in which the temperature in Houston is 64° Fahrenheit while Chicago enjoys 32° Fahrenheit. While it is true that the number 64 is twice as large as the number 32, it is not really meaningful to say that it is twice as warm in Houston as it is in Chicago. If one asserts such a condition, it is interesting to note that a Centigrade thermometer would show 0° Centigrade in Chicago and 17.8° Centigrade in Houston. Is it infinitely warmer in Houston? The problem arising from such a use of an equal interval scale is caused by the lack of a meaningful zero point on such scales. The statement that New York is twice as far away as San Francisco illustrates the deficiency. "Twice as far away" from what starting point? Once a starting point can be agreed upon, a value of zero can be meaningfully assigned to that point and other numbers assigned accordingly.

Ratio scales. When a zero point is chosen which has no meaning of zero with regard to the aspect being scaled, it is called an arbitrary zero point. When a meaningful zero point (or absolute zero point) is added to a scale of the interval type, the resulting scale is called a *ratio* scale. In this case, all the usual arithmetic operations on the numbers and equations can be performed in a meaningful way with respect to the characteristics represented by the numbers. Specifically, equality of ratios of numbers now means some type of equality in the "ratios" of the characteristics referred to by the numbers. One commonly used scale which can almost always be considered a ratio scale arises from the simple operation of counting the number of events of the same type. Such a scale will be referred to hereafter as a *frequency scale*.

Continuous and discrete scales. Another sort of characterization of number scales concerns the distinction between *continuous* and *discrete* sets of numbers. A good example of a discrete number scale is the frequency scale. In a discrete scale, two numbers on the scale may be selected in such a way that no other number of the scale exists between the two chosen values. For instance, in counting the number of deaths in a certain community for a given year, only integer values result. There is no value between 16 and 17 (for this purpose, 16.7 deaths is impossible). By contrast, a scale of all real numbers is said to be continuous. If the reader suggests any two numbers which are exceedingly close together, it will always be possible to find some other number which has a value between the two selected. Expressing two numbers in decimal form such as 2.67888999 and 2.67889000 permits one to obtain a number between these two by carrying out the number to another decimal place: 2.678889995.

This type of distinction is of great importance in mathematics and certain areas of statistics as well. Where the distinction is relevant to certain statistical problems, further discussion of continuous and discrete scales will be presented in later chapters.

1–8. Algebraic Manipulations.

Although a working knowledge of algebra on the part of the reader is assumed, it may be valuable to review a few elementary operations which may be applied to expressions and equations. The coverage which follows is selective. In view of the frequency of occurrence in statistics, the operations reviewed here are considered essential for the understanding and application of subsequent material.

In manipulating expressions and equations, keep in mind the distinction between *terms* and *factors*. Study the expression

$$4XY - Y + (A - 2X) + \frac{(X - Y)}{2A} \qquad (1\text{–}8.1)$$

Consider all quantities within parentheses to be single entities, at least at first. In this case, items separated from other items by addition or subtraction symbols are called *terms* of the expression. The four terms present in the illustration are: $4XY$, Y, $(A - 2X)$, and

$(X - Y)/2A$. Note that further analysis would permit the subdivision of the terms within parentheses into two more terms in each case. But for the moment, just consider the four listed. Any term may be composed of *factors*, this is true when the numbers, symbols, or groups of symbols are connected (related) by multiplication or division operations. The first term in the above expression may be viewed as being composed of three factors: 4, X, and Y. The last term above has two factors related by division and one of these, $(2A)$, may be considered as two factors itself, 2 and A. One convenient method for viewing a term involving division of factors is to consider the reciprocals of the divisor factors and relate all factors by multiplication. In this case, the fourth term above may be expressed as $(X - Y)(\frac{1}{2})(1/A)$.

The expression in (1–8.1) may be rewritten by clearing parentheses, hence (1–8.1) may be expressed as

$$4XY - Y + A - 2X + \frac{X}{2A} - \frac{Y}{2A}$$

In this case, there are six terms, and four of these have more than one factor.

The simplification of expressions is accomplished through two fundamental operations. (a) Two *entire* terms may be eliminated if their sum is equal to zero (subtraction is viewed as addition of a negative term), or two terms may be combined into one term by addition or subtraction. For example,

$$X + 2Y - (X + Y) = X + 2Y - X - Y = 2Y - Y = Y$$

The second type of operation on expressions involves factors. (b) Factors are "cancelled" by dividing one factor by an equal factor, and obtaining a result of 1. For instance,

$$\frac{(A + B)(C + D)}{(A + B)} = (1)(C + D) = (C + D)$$

But take careful note of the following expression:

$$\frac{(A + B) + C}{(A + B)}$$

This expression is *not* equal to $1 + C$, or C. The perceptual tendency to divide the numerator quantity of $(A + B)$ by the equal denominator is incorrect because the first of these is not a factor! The expression may be modified, if wished, by the following correct manipulations:

$$\frac{(A + B) + C}{(A + B)} = \frac{(A + B)}{(A + B)} + \frac{C}{(A + B)} = 1 + \frac{C}{(A + B)}$$

Any term of an expression may also be modified without changing its value by multiplying *and* dividing that term by a non-zero quantity or expression; thus,

$$\frac{(A + B + C)}{(A + B)} = \frac{(X + Y)(A + B + C)}{(X + Y)(A + B)} \quad \text{where } (X + Y) \neq 0$$

When two expressions are equal to one another, the resulting algebraic statement is called an equation. Equations may be modified so as to preserve the equality of the expressions on the left and right sides of the equals sign by the following rules (simple examples are also given):

1. Either the left- or right-hand expression may be simplified.
2. If two expressions are equal in value (usually identical), one of them may be added to one side of the equation and the other to the other side of the equation:

$$X = A + B$$
$$X + 5 = A + B + 5$$

3. If two expressions are equal in value, one of them may be subtracted from one side of the equation and the other subtracted from the other side of the equation:

$$X = A + B$$
$$X - Y = A + B - Y$$

4. If two expressions are equal in value, one side of an equation may be multiplied by one of the expressions if the other side of the equation is multiplied by the other expression:

$$X = A + B$$
$$CX = C(A + B)$$

5. If two expressions are equal in value (but not zero), one side of an equation may be divided by one expression if the other side of the equation is divided by the other expression:

$$X = A + B$$
$$\frac{X}{C} = \frac{A + B}{C}$$

where $C \neq 0$.

The expressions used in applying rules 2 through 5 are usually identical ones, such as multiplying both sides of an equation by the number 10 or adding a constant to both sides of an equation.

The consistent use of the above rules will give correct solutions for nearly all elementary statistical problems. The use of "short cut" methods will save time but increase the probability of error for most students who are not expert in mathematics.

The exercises at the end of this chapter will give opportunities to apply these elementary operations to expressions involving statistical notations for variables and constants.

A number of statistical procedures call for the squaring of numbers and expressions. The reason for this is somewhat difficult to explain to the beginning student in statistics although some limited reasons will be suggested in later chapters.

In order to eliminate certain common forms of error in the use of squared terms, the following relations are given without further comment:

$$X^2 = (X)(X) \tag{1–8.2}$$

$$X^2 + Y^2 = (X)(X) + (Y)(Y) \tag{1–8.3}$$

$$(X + Y)^2 = (X + Y)(X + Y) = X^2 + 2XY + Y^2 \tag{1–8.4}$$

$$(X - Y)^2 = (X - Y)(X - Y) = X^2 - 2XY + Y^2 \tag{1–8.5}$$

$$(X^2 - Y^2) = (X - Y)(X + Y) \tag{1–8.6}$$

$$(XY)^2 = (XY)(XY) = X^2Y^2 \tag{1–8.7}$$

$$\frac{(XY)^2}{X} = \frac{X^2Y^2}{X} = XY^2 \tag{1–8.8}$$

$$\left(\frac{XY}{X}\right)^2 = \frac{X^2Y^2}{X^2} = Y^2 \tag{1–8.9}$$

Further examples involving statistical notations will be given shortly.

1–9. Statistical Notation. It has been observed by many teachers of statistics that there is no system of symbolic notation that is not ambiguous to the introductory student. The lack of standardized notation among textbooks in this field suggests that authors add a corollary to the above observation. This seems to be something like, "my notation is less ambiguous than yours."

The student is cautioned that the notation used in this text, or indeed any text, may well differ from the notation used in any other text or reference. If one takes the time to find the definitions related to the notation system, quite different equations often are found to be identical in meaning.

The use of symbolic notation stems from at least two important considerations. The first is that operations and procedures may be stated in a much more general form than is possible by just giving specific examples. The second is that space and time are saved by using symbolic notations. Two other possible considerations are avoidance of ambiguity inherent in the English language and relative ease of manipulating symbols as opposed to corresponding language statements.

When studying equations and symbolic notations, keep in mind that a few lines of such information may represent procedures which could take pages of description in prose. Do not expect to read a statistical equation at the same rate you would read an English sentence.

Statistics based on samples and populations. It is frequently necessary to make a distinction between a statistic calculated on a set of population scores and the same statistic calculated from a sample set of scores. Two methods of denoting a statistic based on a population set of scores will be used. The first method is to place the symbol $(_*)$ as a subscript to the letter used to denote the statistic. In our general scheme, S_* would represent any statistic computed from a population of scores. The second method, one widely used in statistics, is to denote this population value of a statistic, called a **parameter,** by use of a Greek letter. Such letters as σ and μ, will be used in later work to denote statistics based on a population.

Whenever a sample set of scores is used to compute the value of a statistic, a Latin letter will be used to symbolize the statistic. When the population value is denoted by a Greek letter, the Latin letter equivalent will be employed for the sample value. Later work will substitute s for σ and M for μ when samples rather than populations are involved. The same statistic may be computed, then, on either a population set or a sample set. In terms of our existing definitions, this distinction may be stated and illustrated as follows:

S_* is a statistic computed by using a set of population scores. (1–9.1)

T_* is the sum of all the scores in a population.

N_* is the number of scores in a population.

S is a statistic computed by using a set of sample scores. (1–9.2)

T is the sum of all the scores in a sample.

N is the number of scores in a sample.

This notational difference affects only the set of scores used, not the rule applied.

In later work we will often use Greek letters for a statistic based on a population. When a symbol for a statistic is given, the related symbols for scores from the set are defined by implication and no further symbolic distinction will be used. The score X_i, then, could be either a population score or a sample score depending upon which of the above notations for the related statistic is used.

Summation notation. The Greek letter Σ will be used to indicate the addition of a number of terms. The symbol or symbols following such summation signs often will have subscripts on them. In these cases it is necessary to give the reader information about just which scores are to be added. This is done by indicating, below and above the Σ, the relevant subscript (such as i), the starting value of the subscript (often 1) and the ending value of the subscript. This ending value sometimes takes on an explicit value, for example, 15; however, it often takes on a general value denoted by some symbol such as N. The scores to be added together are those which have the appropriate subscript equal to all values between the initial and ending values, inclusive. The following definitions and illustrations should make the elementary uses of summation notation clear:

$$\sum_{i=1}^{N} X_i = X_1 + X_2 + X_3 + \cdots + X_N \qquad (1\text{–}9.3)$$

The method of interpreting summation notation may be described in a series of steps. First, we must determine which symbols in an expression are related to the summation sign. The convention here is that the summation sign refers to a single *term* of any expression. Any symbols beyond the first addition or subtraction sign which are not within parentheses are not a part of the summation notation. Once the term to which the summation sign refers has been found, the next step is to write out this term, replacing the appropriate subscript letter with the first value given in the summation sign. We then place an addition sign in the list and again write out the term being used, this time replacing the index, subscript, with a value one greater than the last time. This process is repeated until the last value of the subscript has been used. Other terms of the original expression are then added as given.

Examples of this theme follow; note carefully the conventions indicated in the use of parentheses:

$$\sum_{j=1}^{N} Y_j = Y_1 + Y_2 + \cdots + Y_N$$

$$\sum_{i=3}^{7} X_i = X_3 + X_4 + X_5 + X_6 + X_7$$

$$\sum_{i=2}^{5} (X_i - Y_i) = (X_2 - Y_2) + (X_3 - Y_3) + (X_4 - Y_4) + (X_5 - Y_5)$$

$$\sum_{i=1}^{N} X_i^2 = X_1^2 + X_2^2 + X_3^2 + \cdots + X_N^2$$

$$\left(\sum_{i=1}^{N} X_i\right)^2 = (X_1 + X_2 + \cdots + X_N)^2$$

$$\sum_{i=2}^{7} (Y_i + 4) = (Y_2 + 4) + (Y_3 + 4) + (Y_4 + 4) + (Y_5 + 4) + (Y_6 + 4) + (Y_7 + 4)$$

$$\sum_{k=1}^{N} X_k + 5 = (X_1 + X_2 + \cdots + X_N) + 5$$

$$\sum_{i=1}^{3} \frac{X_i}{Y_i} = \frac{X_1}{Y_1} + \frac{X_2}{Y_2} + \frac{X_3}{Y_3}$$

$$\sum_{j=3}^{N} 7X_j = 7X_3 + 7X_4 + \cdots + 7X_N$$

All of the examples follow from (1–9.3), the discussion of the use of the summation sign, and conventions concerning parentheses.

Manipulations of summation notations. The use of algebraic manipulations on expressions involving summation signs requires careful handling of the subscripts, and a clear awareness of which symbols represent scores from a set and which represent constant numbers or values. Those symbols referring to scores are called *variables* in a given equation, while those symbols which indicate a single value throughout a given equation (and the resulting series of summations) are regarded as *constants*.

Two valuable theorems will now be given. Use of these theorems will save time and many steps in later derivations and manipulations where summation notation is used.

Theorem 1–1. If X_i is a variable and c a constant, then

$$\sum_{i=1}^{N} cX_i = c \sum_{i=1}^{N} X_i$$

The proof of this theorem and the next requires only the definition of the summation sign and a little algebra:

$$\begin{aligned} \sum_{i=1}^{N} cX_i &= cX_1 + cX_2 + cX_3 + \cdots + cX_N && \text{(from (1–9.3))} \\ &= c(X_1 + X_2 + X_3 + \cdots + X_N) && \text{(by factoring out } c\text{)} \\ &= c \sum_{i=1}^{N} X_i && \text{(from (1–9.3))} \end{aligned}$$

Theorem 1–2. If X_i is a variable and c a constant, then

$$\sum_{i=1}^{N} (X_i + c) = \sum_{i=1}^{N} X_i + \sum_{i=1}^{N} c = \sum_{i=1}^{N} X_i + Nc$$

In proving this theorem, it will be useful to place a subscript on c simply in order to count up the number of c's used; keep in mind that c has the same value no matter what subscript it may have:

$$\begin{aligned} \sum_{i=1}^{N} (X_i + c) &= (X_1 + c_1) + (X_2 + c_2) + \cdots + (X_N + c_N) && \text{(from (1–9.3))} \\ &= (X_1 + X_2 + \cdots + X_N) + (c_1 + c_2 + \cdots + c_N) && \text{(by regrouping terms)} \\ &= \sum_{i=1}^{N} X_i + \sum_{i=1}^{N} c_i && \text{(from (1–9.3))} \\ &= \sum_{i=1}^{N} X_i + Nc && \text{(from the basic notion of counting, i.e., } c + c = 2c,\ c + c + c = 3c, \text{ and so forth, up to } Nc\text{)} \end{aligned}$$

Let us now apply these theorems and definition (1–9.3) to a kind of manipulation common to many statistical derivations:

$$\sum_{i=1}^{N} (X_i + c)^2 = ?$$

$$= \sum_{i=1}^{N} (X_i^2 + 2X_ic + c^2)$$ (by the general expansion of $(a + b)^2$)

$$= \sum_{i=1}^{N} (X_i^2 + 2cX_i) + Nc^2$$ (by use of Theorem 1–2)

$$= \sum_{i=1}^{N} X_i^2 + \sum_{i=1}^{N} 2cX_i + Nc^2$$ (by expanding the summed terms, regrouping and using (1–9.3))

$$= \sum_{i=1}^{N} X_i^2 + 2c \sum_{i=1}^{N} X_i + Nc^2 \qquad (1\text{–}9.4)$$ (from Theorem 1–1)

Example 1–1. Given the set of scores 0, 2, 1, 7, 14, 10, 7, and 8, use a value of $c = -2$ and show directly that the values of the expressions on the right and left of Equation (1–9.4) are actually equal.

The work may proceed in many fashions, but the following tabular presentation is a way of organizing the steps involved. Note that the labelling of the scores is arbitrary in the sense that any one of them could have been called X_1, X_2, etc. Also notice that there are eight scores, so that $N = 8$.

(a)	(b)	(c)	(d)	(e)	(f)
i	X_i	$X_i + c$	$(X_i + c)^2$	X_i^2	$2cX_i$
1	0	−2	4	0	0
2	2	0	0	4	−8
3	1	−1	1	1	−4
4	7	5	25	49	−28
5	14	12	144	196	−56
6	10	8	64	100	−40
7	7	5	25	49	−28
8	8	6	36	64	−32
$\sum_{i=1}^{N} =$	49	–	299	463	−196

The column totals (sums) shown are those related to (1–9.4). The total (summation) of column (d) is the value represented by the left side of (1–9.4) and is 299. We now wish to show that the same value is obtained by calculating the sum of the three terms on the right side of the equation. It can be seen that the total of column (e) corresponds to the first of these three terms, and is 463. The second term may be evaluated by multiplying the total of column (b) by $2c$, giving $2(-2)(49) = -196$. This value is the same as the total of column (f), as application of Theorem 1–1 would suggest. The last term of (1–9.4) is Nc^2 which gives a value of 32. Adding the values of these three terms we obtain: $463 - 196 + 32 = 299$.

This numerical illustration, Example 1–1, has been given to acquaint the student with the manipulation of the numbers represented by the notations so far encountered. If the student has difficulties with the tabular summary of the work in Example 1–1, it is suggested that he write out the terms of (1–9.4) with $N = 8$ and then substitute the values of the variable and constant c.

EXERCISES

1. If the following set of numbers is considered as the population, what is the maximum possible value for a sample value of T, when sample size N is four and (a) sampling is without replacement; (b) sampling is with replacement?

   1, 5, 5, 6, 7, 10, 11, 14, 14

2. For each of the following suggested uses of numbers, indicate which type of scale (nominal, ordinal, interval, or ratio) seems most appropriate:
   (a) Measuring the weights of people in pounds.
   (b) Numerically recording the month of the year in which each of a population of people was born.
   (c) Numerically describing the favorite food preferences of a group of persons.
   (d) Describing the order of preference for five television programs based upon the votes of a population of viewers.
   (e) Measuring time in seconds, above or below a national average time, needed by seventh-grade boys to run 100 yd.
   (f) Using a grade point average to describe the scholastic achievement of college seniors.

3. Simplify each of the following expressions:
   (a) $\dfrac{(A + B)(A - B)}{A^2}$
   (b) $\sum_{i=1}^{10} X_i^2 - X_1^2 - X_{10}^2$
   (c) $\sum_{i=1}^{N} (X_i + c)^2 - Nc^2$
   (d) $\sum_{j=1}^{N} kX_j + \sum_{j=1}^{N} cX_j$
   (e) $\sum_{i=4}^{6} X_i - \sum_{i=1}^{100} X_i$

4. Use the following set of scores to evaluate the expressions indicated below:

i	X_i	i	X_i
1	3	6	10
2	3	7	10
3	5	8	11
4	6	9	12
5	7	10	13

(a) T and N

(b) $\sum_{i=2}^{N} X_i$

(c) $\sum_{i=1}^{N} X_i^2$

(d) $\sum_{i=1}^{N} \left(X_i - \frac{T}{10}\right)$ (use Theorem 1–2)

(e) $\sum_{i=1}^{N} \left(X_i - \frac{T}{10}\right)^2$ (hint, Theorem 1–2 does not apply directly)

(f) Express the following calculation in summation notation:

$$3 + 5 + 6 - (11 + 12 + 13) + 7^2 + 10^2 + 10^2 + 11^2$$

5. For $N = 5$, show that

$$\sum_{i=1}^{N} X_i^3 - \sum_{i=1}^{N} X_i^2 = \sum_{i=1}^{N} X_i^2(X_i - 1)$$

6. Give a numerical illustration which shows that

$$\frac{\sum_{i=1}^{N} X_i^2}{\sum_{i=1}^{N} X_i} \neq \sum_{i=1}^{N} X_i$$

2

Descriptive Statistics

2–1. Purpose and Limitations. The description of any set of scores may be complete; i.e., every single score may be given. Such complete description is called *enumeration*. The description of a set of scores by enumeration is time-consuming, confusing to most people, often without direction or purpose, and almost useless *per se* for answering statistical questions.

It is possible to describe a set of scores in an incomplete fashion as compared to the complete-description enumeration method. When such an incomplete characterization of a set of scores is given in statistical terms, the descriptions used are called *descriptive statistics*.

The use of descriptive statistics, instead of enumeration, is an attempt at reduction of each of the above problems. The fact that a statistic is an abstraction makes the use of a descriptive statistic less time-consuming, less confusing, and specific to certain aspects of the scores. Statistical questions are framed in terms of statistics or require the use of statistics as intermediate steps even when they refer to a complete set of scores.

Once a set of scores has been described to some degree by the use of a statistic or group of statistics, it is of value to ponder what has been accomplished. A set of scores has been characterized to some limited degree. If the intent of such description is merely to perform manipulations on a set of numbers or to illustrate purely mathematical relationships among numbers, then one may feel satisfied at this point and stop. On the other hand, if the numbers represent something about the world, and one's *intent* is to describe and *predict* certain aspects of the world, then descriptive statistics alone are a dead end. As representations of the world, these statistics can only be classed as history.

This statement is made because of the fact that all real things are in a constant state of change. To be sure, some things change at a less rapid rate than others: our sun, a table, and a mountain range all seem to change at a very slow rate relative to our life span. Each of these does change, and at quite different rates. A description of the locations and intensities of sun spots is valid only for the particular time at which the set of needed measurements was taken. Certainly a few days later conditions on the sun may have changed drastically. Descriptive statistics then represent aspects of the world at some time already passed. Even the historian may find this state of affairs undesirable. The scientist must go beyond such a temporally isolated set of descriptions of the world.

Prediction is essential to the practice of relating descriptive statistics to descriptions of the current state of the world. In one of its simplest forms, prediction is actually used when one expects to get the same results by measuring the "same" thing at two different times. This concept of *reliability* can be viewed as a prediction that what once was, is still the same. More sophisticated types of predictions are those in which descriptions at a given time and under certain conditions are used to guess (predict) what the descriptions will be at a later time and under slightly different conditions.

When descriptive statistics are used to make predictions in either of these two general ways, the procedure is called *inferential* statistics. It should be clear from this discussion that descriptive statistics are of little use to the scientist without the tools of inferential statistics.

However, descriptive statistics without reference to inferential uses will be studied for two reasons: (a) they are used as stepping stones or basic components for inferential techniques, and (b) they illustrate the basic characterizations or abstractions of sets of numbers which are of primary interest.

2–2. Classification and Counting. The use of nominal, and often ordinal, scales permits only a limited set of descriptive statistics to be meaningfully employed. The number of elements in a set (N) is one such elementary but useful statistic. The statement that there are 37 persons in a given class room at time t, is a descriptive statement which can be thought of as statistical in form, but this information by itself is of no great importance. The *comparison* of this statistic ($N = 37$) with the number of persons in another class room at the same time may be of interest. Let us say that the other class room contained 63 persons. Even this simple comparison requires the concept of *classification*. Note that we are now considering 100 different people, or elements of a population set, in making the comparison. The comparison is possible only when we *subdivide* this group into two parts. By considering certain aspects of each element, it is possible to classify the people, in this case, into two groups. Let us say "room A" and "room B" are the names given to the two classes. Once this system of classification has been established, it is possible to assign each element to one of the two classes. Let us now reconsider our statistical description of the situation. The total number of elements in the set is 100; hence $N_* = 100$. We can still describe the number of people in the two classes as additional information. We might say $N_A = 37$, to indicate

TABLE 2–1

Selected Characteristics of a Group of People

Name	Color of Eyes	Color of Hair
Joe	Blue	Black
Mary	Brown	Blond
Tom	Brown	Brown
Peter	Blue	Brown
Ellen	Blue	Brown
Mike	Brown	Brown
Sue	Brown	Blond
Sally	Brown	Blond
David	Brown	Black
John	Brown	Brown
Ralph	Brown	Blond
Helen	Green	Red

that 37 elements are classified as belonging in class A, and $N_B = 63$, to indicate the number of elements in the other class.

The classification system just used is called a **partitioning** of a set. The *partitioning* of a set of elements is the assignment of each element of a set to a class or *subset* on the basis of one or more characteristics of the elements, with the restrictions that: (a) each element must be assigned to some subset, and (b) no element can be assigned to more than one subset on the basis of the classification system being used.

When a set of elements having many aspects is to be partitioned, one must first decide which characteristics are appropriate to the definition given above, and also which characteristics seem useful or are interesting to the investigator. Table 2–1 contains brief descriptions of 12 hypothetical people. The names are given to indicate the person's gender, and the eye and hair coloring of each are indicated. Three possible means of partitioning the people are readily evident; sex, hair color, and eye color.

When using sex as the relevant characteristic for the partitioning, we need only two classes or subsets: men and women. Once this grouping has been accomplished, the resulting number in each class can be given as a sort of description. Letting $N_* = 12$, we might say $N_M = 7$, and $N_W = 5$, to give the number of men and women, respectively.

A separate and distinctly different partitioning of the same 12 elements on the basis of eye color requires at least three classes to be formed, call them blue, brown, and green. When completed, this partitioning gives the following results when the numbers in each class are counted: $N_{Bl} = 3$, $N_{Br} = 8$, and $N_G = 1$.

Using hair color as the defining characteristic for the partitioning procedures requires at least four classes or subsets. The resulting N's, using abbreviated subscripts, are $N_{Br} = 5$, $N_{Blk} = 2$, $N_{Bld} = 4$, and $N_R = 1$.

It is possible to use combinations of the aspects given to form other types of partitions for this set of persons. For instance, we might wish to consider the sex *and* hair color for these people. As the set is now described, six classes will suffice to form the partitioning. However, note that classifying and counting blond men and blond women, no matter how interesting this may be, does not constitute a partitioning of the population since not all elements may be placed into one of these two classes. This type of incomplete partitioning must be carefully avoided in making certain statistical tests to be considered later; special ways of handling such incomplete subsets must be used. Also of interest is the possibility of establishing classes which are in fact not used. In the above combination of sex and hair color, eight classes are suggested from the existence of four hair colors and two sexes. Each of the four hair-color classes is, in a sense, partitioned once again into two sex groupings for the total of eight classes. The fact that no women have black hair is then described by saying that the N_i involved equals zero. The same count of zero is also found for the class of red-haired men. Technically, such sets (subsets here) are called *null* or *empty* sets. When using two or more aspects to form a partitioning, it is well to use all possible combinations, for the fact that a possible set is empty is often as useful or interesting as the fact that one finds a large number of elements in a class.

The basic statistics for nominal scale data are the number of elements in the population and the number of elements in each subset of a partitioning of that population. Two related statistics are readily formed from the above N's. These are *proportions* and *percentages*.

First, let us consider proportions. This term implies a comparison of two N's or *frequencies,* and is therefore a *relative* measurement of frequencies rather than an *absolute* measurement.

Letting N indicate the number of elements in some set, and N_i the number of elements in a subset, the proportion (P_i) of elements in the subset, relative to the total set, is defined as:

$$P_i = \frac{N_i}{N} \tag{2–2.1}$$

If a set is partitioned into K subsets, then

$$\sum_{i=1}^{K} P_i = 1.00 \tag{2–2.2}$$

The proof of (2–2.2) is left as a simple exercise for the reader.

The notion of a percentage is again one of relativity. If we simply compare two numbers N_j and N_k, we say that N_j is Z_j per cent of N_k when

$$Z_j = \frac{N_j}{N_k}(100) \tag{2–2.3}$$

Restricting the use of percentages to sets and their partitions, the simple relationship between proportions and percentages may be expressed as

$$Z_i = (100)(P_i) \tag{2–2.4}$$

The use of proportions and percentages to describe the relative number of elements in subsets is quite common and useful. Note that these statistics do not give as much information as the use of N's for the same data (unless the N_* is also given). If N_* is given, with the percentages or the proportions, one can readily compute any of the other statistics being discussed. In fact, if percentages or proportions for each subset are given, just a single N_i will allow all of the other N_i's, including N_* to be computed, simply by manipulation of equation (2–2.1) or (2–2.4).

In summarizing the description of nominal scale data, we may say that (a) classification is the procedure which leads to meaningful description, (b) counting is the procedure which yields the appropriate descriptions, and (c) the resulting statistics are frequencies (N's) and relative frequencies (percentages or proportions). For nominal data, one needs no other descriptive statistics for this type of elementary course.

The description of ordinal scale data may be handled in the same way as nominal data. However, some additional descriptive statistics are useful for ordinal scales but will be introduced in later chapters when they will be more meaningful to the student.

2–3. Frequency Distributions. When dealing with sets of scores, a convenient method of partitioning the set leads to what is called a *frequency distribution.* The basis of partitioning is the value of the scores. It is usual to name each class or subset by using the score value it represents. After the assignment of the scores to classes, one counts the number of scores in each class. This is equivalent to counting the number of scores having each possible value. The results of this counting procedure may be presented in the form of a table or graph.

The scores in Table 2–2 will be used to illustrate this procedure and its usefulness in describing a set of scores completely. In Part A of this table, one may observe the lack of perceptual organization inherent to the enumeration of scores without regard for their values. One

TABLE 2–2

A Set of Thirty Scores Presented in Three Different Forms

PART A: Scores listed in order of selection from the population

18	8	25	21	14	4
5	8	12	8	11	19
21	13	9	15	15	6
9	21	15	12	11	18
10	15	7	14	17	16

PART B: The same scores given in order of decreasing value

25	18	15	13	10	8
21	18	15	12	9	7
21	17	15	12	9	6
21	16	14	11	8	5
19	15	14	11	8	4

PART C: The same scores in the form of a frequency distribution

Score Value (X)	Frequency (f_i)	Score Value (X)	Frequency (f_i)
25	1	14	2
24	0	13	1
23	0	12	2
22	0	11	2
21	3	10	1
20	0	9	2
19	1	8	3
18	2	7	1
17	1	6	1
16	1	5	1
15	4	4	1

relatively easy way to organize a set of scores is illustrated in Part B. A quick examination of the scores as shown in Part B will allow discovery of the rule of organization being applied, i.e., the listing of the scores in the order of their values. In the present case, a decreasing order of values has been used, although, in general, there is no reason to avoid an increasing order of values. Further examination of the scores as shown in Part B reveals certain interesting features of the set: (a) the highest score has the value 25, the lowest is 4; (b) there are a number of scores with duplicative values; and (c) there are some score values between the highest and lowest scores in the set which did not occur in this sample. These characteristics of the sample are rather obvious when examining Part B of the table, but are much less obvious from the examination of Part A, even though the two parts contain the same scores.

The presentation of this set of scores in the form of a frequency distribution is illustrated by Part C of Table 2–2. This frequency distribution is formed, formally, by using the value of the scores as the relevant characteristic for the partitioning of the set, and by naming each subset by the corresponding score value. Examination of the set reveals that only 18 different subsets need be used, because there are only 18 different score values. The use of some empty sets is again found convenient, in this case to create a continuity along the *dimension* of score values. Specifically, the classes with values of 24, 23, 22, and 20 have been included in the partitioning procedure in spite of the fact that no scores are placed in these classes. Continuing in this formal construction of the frequency distribution, each score is now assigned to its appropriate subset by virtue of its value. Then the number of scores within each subset is counted. The results of this entire procedure are given in summary form in Part C. It may be noted that the organizing principle of decreasing score values is used in the presentation of the results.

The less formal manner of constructing a frequency distribution can be described in much simpler terms: Write down the possible score values and place a mark beside the corresponding value for each score in the set. Then count the number of marks related to each score value.

In future uses of a frequency distribution, the notation X will be used without a subscript to indicate score values or classes, and the notation f_i, or simply f, will be used to indicate the corresponding frequencies. If the reader will keep in mind the formal procedures for obtaining frequency distributions, he should be aware that the label X in this usage does not really indicate a score but rather a score value.

Lest this distinction seem trivial, consider the computation of the sum of the scores shown in Table 2–2.

The equation for the sum or total of a set of scores (1–5.2) is usually written in summation notation. Hence Equation (1–5.2) becomes

$$T = \sum_{i=1}^{N} X_i \tag{2-3.1}$$

If the student were given the data as shown in Table 2–2, Part C, a *misapplication* of (2–3.1) might lead to the following computations: $T = 25 + 24 + 23 + 22 + \cdots + 4 = 319$.

This summation is *not* indicated by the equation applied. What has been done above is to add the score values together, which is quite a different thing from adding the scores together.

The correct application of (2–3.1) may proceed directly from the table by using either Part A or Part B. The addition of all thirty scores gives a sum of 397.

In the *incorrect* procedure given first, the confusion might have been avoided by calculation of N. The f_i's shown in Part C of the table represent what we have called N_i's in our discussion of partitioning. It should then be clear that

$$\sum_{i=1}^{K} f_i = N \tag{2-3.2}$$

where K is the number of classes in the partitioning. The summation of the f_i's will give the correct N of 30, as will a simple counting of the scores as shown in Part A or Part B of Table 2–2. The incorrect calculations shown above involved only 22 numbers.

A frequency distribution gives the same information as enumeration when given as in Part C of the table. But notice that the frequency distribution actually does *not list even a single score.* By giving the score value for each class and the related frequencies, the frequency distribution allows the reader to recreate the scores represented if so desired. For that reason, we may say that all parts of Table 2–2 give precisely the same information, but in somewhat different form (especially Part C).

The application of (2–3.1) to Part C of the table requires the following line of reasoning: "The value 25 is represented by one score so I add 25 (to zero); the values 24, 23, and 22 are not present in the set, so nothing is added; the value 21 is represented by three scores so I add 21 + 21 + 21 to the old total of 25. This gives 88. The value 20 is not present, so nothing is added there. One score of 19 is present so I add 19 to 88 . . . ." This reasoning is continued until the scores have all been added together.

In this computation, it is clear that short-cut methods are possible. Many textbooks in statistics offer special formulas for this type of computation. All of these equations are based on the simple reality that the sum of j numbers, all of which are the same, is obtained by multiplying j by the value of the numbers. In the description of finding the sum of the above scores, the operation of adding $21 + 21 + 21$ equals 63 is more easily accomplished by using the operation 3×21 equals 63. The sum of scores in a frequency distribution may be gotten fairly rapidly, then, by multiplying each score value by its corresponding frequency, and then adding these products together. The development of a formula for this procedure is given as an exercise at the end of this chapter.

Other common forms of the frequency distribution involve the use of proportions or percentages. Equation (2–3.2) points out the slight change in notation between the numbers of scores in a frequency distribution (f_i's) and the numbers of scores in different classes (N_i's) which were discussed with regard to nominal scale data. The equations for obtaining proportions and percentages can be easily applied to the data shown in a frequency distribution. When this is done, the frequencies for each score value are simply changed to proportions (or percentages if desired). In Table 2–3 the resulting proportions are shown for the data from Table 2–2.

The computations involve the division of each f_i in Table 2–2 by N (which is 30 in the present case).

TABLE 2–3

A Relative Frequency Distribution Using Proportions

X	Proportion of cases	X	Proportion of cases
25	.033	14	.067
24	.000	13	.033
23	.000	12	.067
22	.000	11	.067
21	.100	10	.033
20	.000	9	.067
19	.033	8	.100
18	.067	7	.033
17	.033	6	.033
16	.033	5	.033
15	.133	4	.033

The use of proportions is a great aid to the presentation of sets of scores when the number of scores is large. Also, the concept of probability or chance is more easily demonstrated with proportions than with raw frequencies. For these reasons, frequent use of proportions to describe distributions of scores will be made in later discussions. Again keep in mind that proportions alone give less information than the raw scores or the basic frequency distribution. One needs a corresponding frequency (f_i) for at least one of the proportions in order to reconstruct the complete set of scores.

Graphic forms of the frequency distribution. The information given in a frequency distribution is very often presented in a graph or figure. This method gives a better picture of a set of scores than a tabular presentation. Global aspects of the distribution are obtained at a glance from such a figure, but computations of statistics are difficult because reading the values from most figures is imprecise.

The formation of a frequency distribution in graphic form usually begins with a set of scores in a frequency distribution table. The graph or figure (see Fig. 2–1) has two axes, the x-axis (horizontal) and the y-axis (vertical). By convention, the score values are arranged in order from low to high (reading left to right) along the x-axis. The frequencies are denoted by placing a number scale along the y-axis, with values increasing as one goes vertically higher and higher.

Each score value and its corresponding frequency (or proportion if the y-axis is so labelled) is represented by a single point in the graph. This point is found by going directly up from the appropriate score value and directly over from the corresponding frequency. Where these two courses cross is the desired point. Figure 2–1 illustrates this procedure with the data from Table 2–2. The set of points obtained when this procedure has been carried out for all score values gives a somewhat disjointed appearance. Various means are used to make the picture easier to view. Two such methods are shown in Fig. 2–1: in Part B of this figure, a bar graph or *histogram* is shown. Each bar in the histogram represents the frequency of one score value and is centered over the corresponding point on the x-axis. Comparison of Part A of this figure, in which the points alone are given, with Part B of the figure should make the method of constructing the histogram apparent. Also, such comparison will reveal the perceptual advantage in the histogram.

The second method of showing the frequency distribution in clear graphic form is illustrated in Part C of Fig. 2–1 and is called a *frequency polygon.* In this case, the points in Part A of the figure have been con-

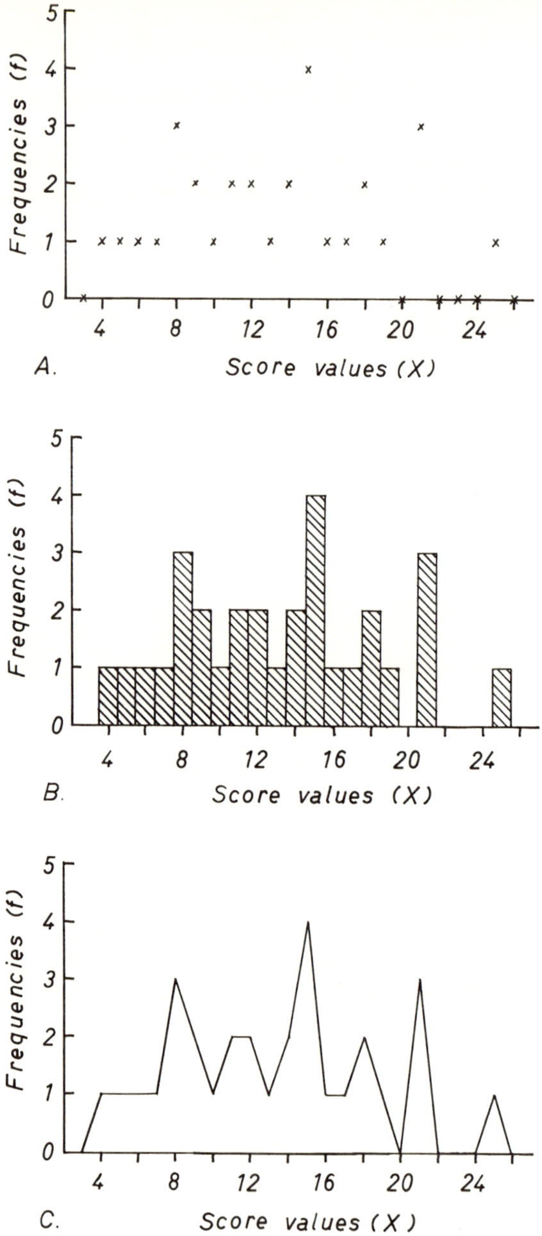

Fig. 2–1. Part *A* shows a frequency distribution in graphic form, part *B* shows the same distribution in the form of a *histogram*, and part *C* shows this distribution in the form of a *polygon*.

nected by straight lines, going from each point to its adjacent points. Note that all frequencies are used including those which are zero.

Both the frequency histogram and polygon may be modified to show relative frequency distributions by means of proportions (percentages are rarely employed). The y-axis is simply scaled from 0 to 1.00, and proportions (rather than absolute frequencies) are indicated for each score value. Figure 2–2 shows a relative frequency polygon based on

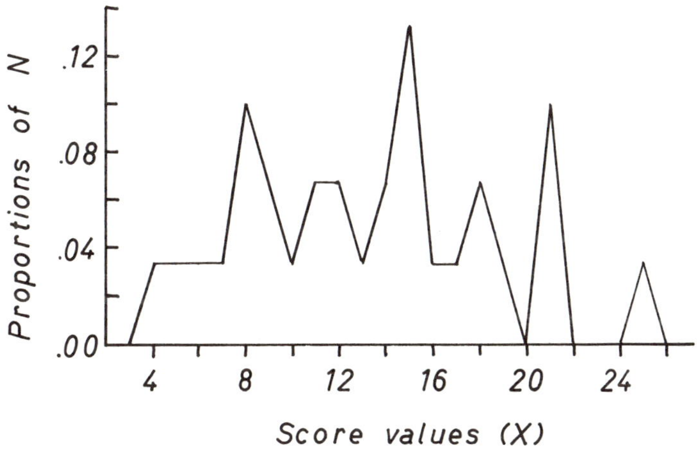

Fig. 2–2. A relative-frequency distribution in polygon form.

the proportions given in Table 2–3. Notice that the shape of the curve is the same as in Fig. 2–1, Part C.

2–4. Measures of Central Tendency. The description of a set of scores by means of statistics requires that one decide which aspect or aspects of the set are to be abstracted. The concept of the middle of a distribution of scores is an exceedingly useful and common characteristic to describe. The difficulty arises in the definition of the term *middle*. The concept of the middle of a set of scores is referred to as the *central tendency* of the scores. This term implies a characteristic observed in many sets of scores obtained in the course of statistical inquiry. The characteristic is that a large number of scores have values quite close to one another and relatively few scores are found with much higher or lower values. The idea behind measures of central tendency is to describe about where this middle of the distribution is in terms of a score value.

Although many statistics are possible which reflect central tendencies of a set of scores, only three are often used by statisticians. These statistics are called **mean, median,** and **mode.** Of these three, the mean is by far the most useful for interval and ratio scale data.

The mean. The precise name for the mean is *arithmetic mean.* But it is common to simply use the term *mean* or *average.* The mean of any set of scores, which will be denoted as M or μ, is defined as

$$M = \frac{\sum_{i=1}^{N} X_i}{N} \quad \text{where } M \text{ is based on a sample} \tag{2–4.1}$$

$$\mu = \frac{\sum_{i=1}^{N_*} X_i}{N_*} \quad \text{where } \mu \text{ is based on a population} \tag{2–4.2}$$

The numerator of these equations is itself a statistic, the one we have called T, the sum or total of a set of scores. It is easy to describe the mean, then, as equal to the sum of the scores divided by the number of scores.

The technique already discussed for finding the total of a set of scores presented in a frequency distribution table may be used to find the mean of the set. For example, the total of the scores in Table 2–2 was found to be 397. The mean of the set is then obtained by dividing this total by the number of scores ($N = 30$). The result, rounded off at the second decimal place, is a value for the mean of 13.23. Locating this value of 13.23 on the x-axis of Fig. 2–1 or 2–2, the reader should note that it is certainly in the "middle" of the distribution of scores.

The mean has many interesting properties. One of these is the physical analogy of a balancing point. If each of the scores represented in Table 2–1 stood for objects of equal weight, and the value of each of these scores stood for the location of the object on a ruler (numbered as is the x-axis in the figure), then the mean value indicates the balancing point of the set of objects. It is necessary to ignore the weight of the ruler in this illustration, but the notion of the mean as a measure of central tendency is apparent from this relation to the real world.

The mean has another property which makes it different from the median and the mode. This is the fact that the mean takes into account the value of every score in the set. If the value of any single score is changed, then the value of the mean is changed. This is not true for the median and the mode.

The median. The *median* may be defined as that score value which partitions the set into two subsets, with an equal number of scores in each, so that the number of scores above the median value is equal to the number of scores below the median.

The actual computation of the median is done through inspection of a set of scores and counting. Usually this is most easily done when the scores are in the form of a frequency table. Often some problem is presented by duplicate scores at or near the approximate median value or by the fact that some sets of scores have an odd number of scores. In the latter case, it is impossible to find a value which will divide the scores into two equal sets. When an odd number of scores

TABLE 2–4

A Frequency Distribution of 16 Scores

X	f
10	1
9	1
8	0
7	4
6	5
5	3
4	1
3	1
	$\Sigma f = 16$

is present in a set, the one score which could go either way is used as the value of the median. For instance, if a set has the scores 1, 1, 3, 5, 6, 8, and 9, then the median score is the value 5. Here, one score is exactly at the median, think of it as the odd score; and the other scores are split into two groups of equal numbers of scores.

Examine once again the scores represented in Table 2–2, Part C. Since there are 30 scores, the median of this set is a value which separates the set into the higher 15 scores and the lower 15 scores. Counting from either end of the X values, we find that the desired point is somewhere between 13 and 14. In this case, the value of the median is taken as being half way between these two values; i.e., 13.5.

The more difficult application of the definition of a median to a set of scores will be illustrated by using the set of scores shown in Table 2–4.

Since there are 16 scores in the set ($\Sigma f = 16$), we wish to find such a score value that 8 scores are above that value and 8 are below it; this will be the median. The problem arises when we note that the value 6 will not quite meet the requirements. If 6 were taken as the median, 6 scores would fall above this value but only 5 scores would be lower (5 scores would be exactly at this point). The solution of the problem involves some refinement of the five scores with the value 6. In order to proceed, we must assume that these five scores are not really all the same value after all. The score value of 6 is therefore assumed to be a rounded approximation of each of the five scores in question. We merely assume that the score value of 6 really represents all true scores from 5.5 up to 6.49999 . . . . This score interval contains the five scores distributed in some unknown way. For simplicity, we assume that the five scores are spread out at equal intervals in the range 5.5 to 6.5. A diagram of this hypothetical analysis is given in Fig. 2–3. In

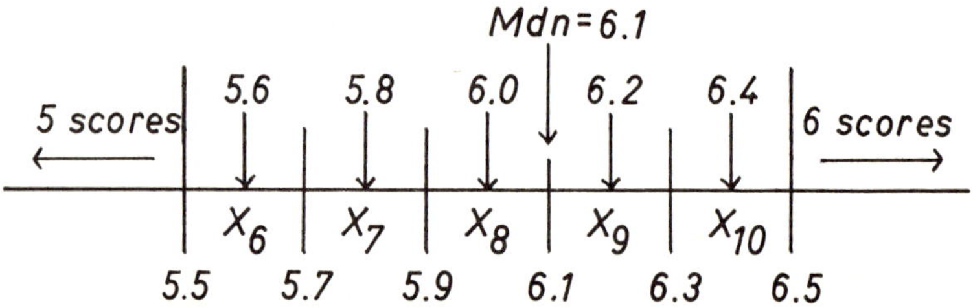

Fig. 2–3. The hypothetical distribution of five scores with values of 6, for computation of the median.

the figure, the five scores are labelled X_i and their values are shown to indicate the type of spacing that is assumed. The desired median value is now found as before. Since there are an even number of scores, 16, we wish the value midway between the eighth and ninth scores, counting from either end. The point desired is 6.10, half way between 6.0 and 6.2.

Although many authors give a formula for the computation of the median in such difficult cases, it is the opinion of the present author that this tends to confuse the student needlessly. Certainly the formula obscures the logic behind the computations, which is illustrated in Fig. 2–3. Moreover, the calculation of a median will rarely be encountered in the remaining chapters.

The mode. The *mode* of a set of scores is defined as the value of the most frequent score. This value is readily obtained by inspection of a frequency distribution. It is possible that ties may exist. When two values occur equally often and more than any other single score value, then there are two modes for the set. Such a distribution is called a *bimodal* distribution. The term *multimodal* distribution is used to describe sets of scores in which more than two values are all tied for highest frequency honors.

2–5. Comparisons of the Measures of Central Tendency. The three measures of central tendency may each be computed on the same set of scores. It is possible, but not very likely, that the values would be exactly equal for each of them. Usually the mean, median, and mode will be different values even though they describe the same set of scores. Clearly, if they give different results in such a case, they must be showing slightly different sorts of central tendencies.

The mode may result in an answer which is near one end or **tail** of a distribution of scores. When this happens, it is somewhat erroneous to interpret the modal value as a measure of central tendency of the scores. This misinterpretation can be avoided easily enough by recalling the definition of the mode and the fact that this single most frequent score could be the highest or lowest value in any set of numbers. It is true that a great many types of distributions have their highest frequencies near the middle of the set of score values, but this is not always the case.

On the other hand, the median value in a set of scores always has the democratic property of dividing the set into halves. There are always as many scores above the median as below. This is certainly one readily acceptable meaning of the center or middle of a group of scores. However, the median has one limitation also possessed by the mode: many of the scores in the set can be changed in value without changing the value of the median.

Consider the small set of scores 1, 2, 3, 4, and 5, and compare it with the set 1, 2, 3, 40, and 50. The median is 3 in each case. The fact that the second set has two large numbers above the median and the first has two small numbers is not reflected at all by the median. The comparison of the mean and median in this respect yields the following statement: *the mean is sensitive to the precise values of each score in a set while the median is relatively insensitive to the exact values of many of the scores in a set.*

Another comparison of these measures of central tendency is based upon the type of scale numbers used. In the case of interval or ratio-scale scores, any or all of the three measures can be meaningfully employed. But if the data have only ordinal-scale properties, the mean value no longer gives meaningful information about aspects represented by the numbers. The computation of the mean requires the operation of addition of scores, and this is not a meaningful operation for ordinal-scale numbers. The median and mode both involve the operation of counting, hence these measures are meaningful when applied to ordinal data. In the case of nominal scales, the mode alone remains as a meaningful measure of "central tendency." In such a case, the numbers do not reflect properties which are scaled in magnitude, hence the median is meaningless. The concept of central tendency is even strained to the breaking point for such nominal data. The interpretation of the mode in this case may be likened to the results of an election with many candidates; candidate number 11 won because he got the most votes.

2–6. Uses of the Measures of Central Tendency. Let us restrict our attention to interval and ratio scale numbers, in which case any or all of the measures of central tendency can be used. Why would one measure be used in preference to the others? The intent of the description given will usually lead to the selection of either the mean, median, or mode, or will require that more than one of them be used. If one's intention is to be complete in the description, but without giving all of the scores in the set, then all three of the measures of central tendency must be reported. Often the intent of description allows a single measure to be used. Our question then becomes, what kinds of uses of sets of numbers call for one of the three measures in preference to the others?

The uses of numbers which allow or require the mean value as a statistical description of central tendency are so numerous and varied that one might well answer the above question in a non-specific fashion: "When in doubt, use the mean."

There are some specific conditions to keep in mind that require the use of the mode or median over the mean, even with interval or ratio-scale data. For example, if one wishes to describe the most frequent score, the mode is preferred by definition. A similar case arises when one wishes to use a set of scores for predictive purposes involving a right or wrong type of outcome. For instance, if you were offered a prize for guessing the age of the person sitting in row 3, seat 12, of a nearby classroom, and you were given a frequency distribution of the ages of all the people in the room, then it should be clear that your best guess would be the mode of the distribution. In this case the term "best" takes into account that close doesn't count. You are either exactly right or you are wrong. The magnitude of your error is irrelevant. By contrast, if you are to be penalized one dollar for every year by which your guess misses the person's age, the problem is quite different. You would wish to consider other measures of central tendency quite carefully. Actually, the best decision in this case involves the value of the prize and other statistical factors not yet discussed.

The purposes which require the use of the median are those in which the counting of the number of scores is somewhat more important than the exact value of each score. For example, if a student wishes to compare his numerical grade on the first test in statistics with the rest of the class, then the median score in the class may be of primary interest. If the student's score is at or above the median he knows he has done better than half of the other students. The mean value cannot be used in this way without additional information about the distribution of scores. Another example of this counting notion of central tendency may be given by considering two small sets of scores. Let us say that the scores show the number of correct items on a sixth grade arithmetic test for 10 different students in each of two classrooms. These hypothetical scores are given in Table 2–5. In this table, the mean (M) and the median (denoted as Mdn) are given for each set of scores. The two sets of scores are amazingly similar (an advantage of using hypothetical data) with the exception of the last two scores in each group. Notice that the mean for Room B is affected by the two students with exceptionally high scores, whereas the median is unchanged. The use of the mean as a measure of central tendency for Room B would give the impression that the scores tend to cluster around the value of 31.0. In fact, there are no scores at all very close to a value of 31.0. If one is interested in a "representative" score here, then the median value of 21.5 might well be selected.

TABLE 2–5

Scores on an Arithmetic Test for Two Groups of Ten Students

Room A	Room B
18	18
18	18
19	19
20	20
21	21
22	22
22	22
24	24
25	66
31	80
$M = 22.0$	$M = 31.0$
$Mdn = 21.5$	$Mdn = 21.5$

Notice that in Table 2–5 the mean and median are very close together for Room *A*, while they are quite far apart for the Room *B* scores. The reason for the differences in the degree of correspondence between the two measures of central tendency is best explained by considering the shapes of frequency distributions.

2–7. Common Types of Distributions. The use of graphic distributions of scores will aid in the discussion of certain common types of sets of scores. The aspects of the distributions to be considered are easily seen from the shapes of these graphic presentations.

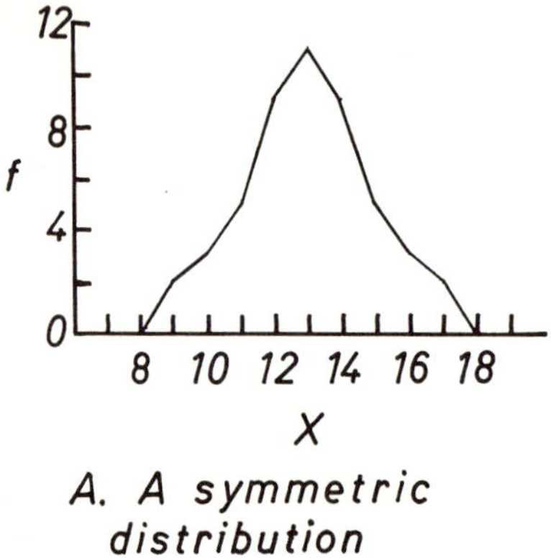

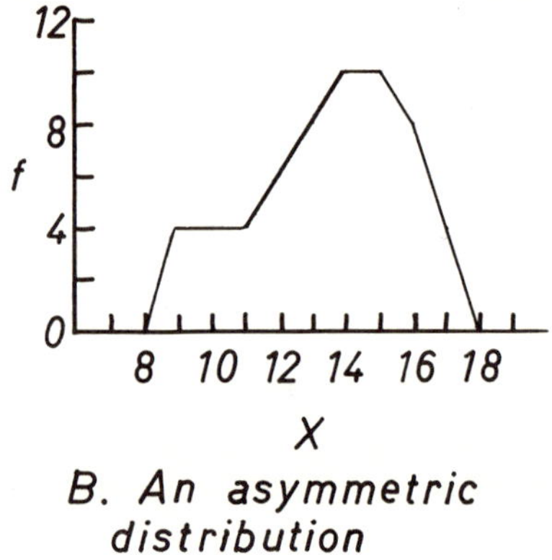

Fig. 2–4. Frequency polygons showing symmetric and asymmetric distributions of scores.

Figure 2–4 gives an example of a **symmetrical** distribution and an example of a distribution which is not symmetrical, called **asymmetrical.** Symmetry involves the concept of a midpoint or center. In the case of distributions of scores, either the mean or median will serve this purpose, because if a distribution is symmetrical, the mean and median will have the same value. In a symmetrical distribution, there are an equal number of scores at all values which lie an equal distance in either direction from the mean (or median). If one were to cut out the paper in the shape of the frequency polygon for a symmetrical distribution, the paper could be folded along a vertical line at the mean and the two halves would coincide perfectly. This, of course, would not be true for an asymmetrical distribution.

For any symmetrical distribution of scores, the mean and the median will be equal. It may also be stated that, if this distribution has a single mode, the value of all three measures of central tendency will be the same. It is possible for multimodal distributions to be symmetric too. One interesting case of this sort is given in Fig. 2–5 and is called a **rectangular** distribution. All score values which occur have equal frequencies and there are no "gaps" between the highest and lowest scores.

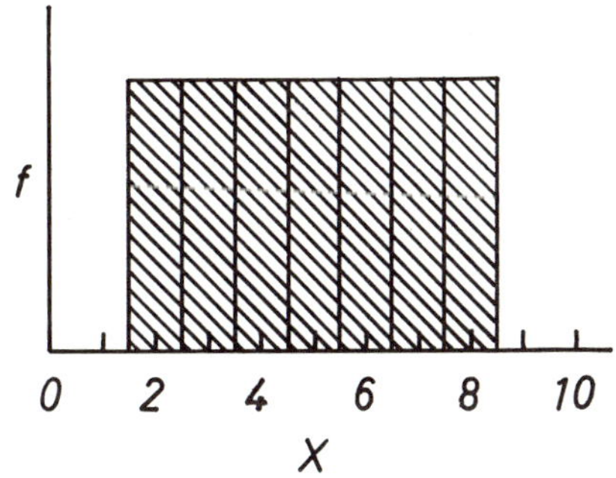

Fig. 2–5. A histogram showing a rectangular distribution.

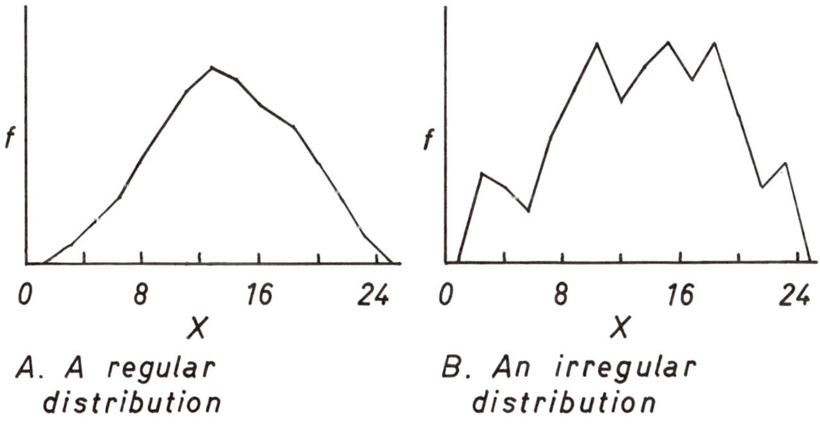

Fig. 2–6. Frequency distributions of the regular and irregular types.

Although a symmetrical distribution will always have a mean value equal to the median, the converse of this is not true; i.e., a set of scores may have equal mean and median values but not be symmetrical.

Of the infinitely many shapes of distribution possible, those which have increasing frequencies going from one end of the score values up to some point and then decreasing frequencies from there to the other end of the scale are called **regular** distributions. Figure 2–6 illustrates a regular distribution of scores and one which would be considered

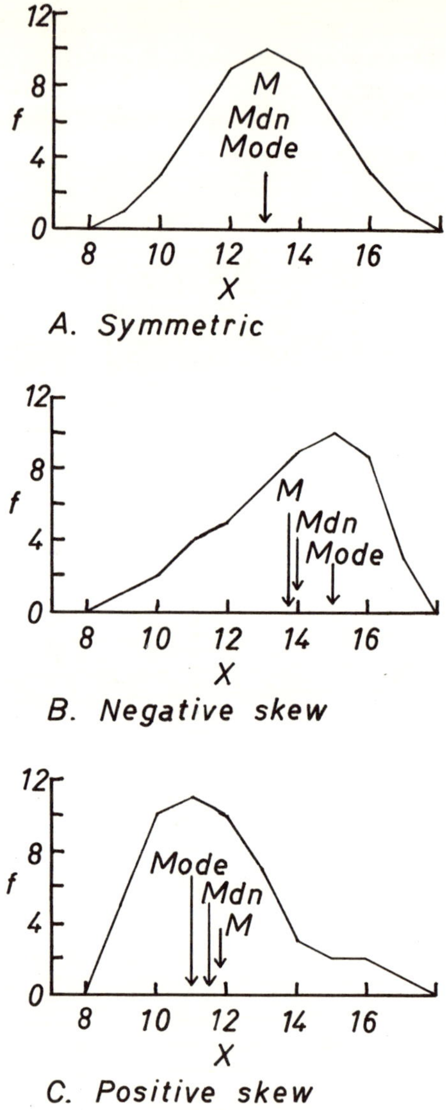

Fig. 2–7. Symmetric and skewed distributions showing the relative positions of the measures of central tendency.

irregular. It is of interest to look at certain forms of these regular distributions and to examine how the measures of central tendency are affected by the shape of the curve or polygon.

Examination of the three frequency polygons shown in Fig. 2–7 will show the differences between regular distributions called **skewed** distributions and the regular symmetrical type. The skewed distribution has a relatively long tail on one end which is not present (to the same degree) at the other end. The direction in which the tail "points," so to speak, is described by the terms *positive* and *negative* skewness. The distribution is said to be negatively skewed when the tail is at, or points to, the lower score values (Part B of the figure) and is said to be positively skewed when the tail is at the higher-valued end of the scale (Part C).

As already noted, the mean, median, and mode all have the same value for the regular symmetrical distribution. Figure 2–7 illustrates an interesting relationship between the measures of central tendency

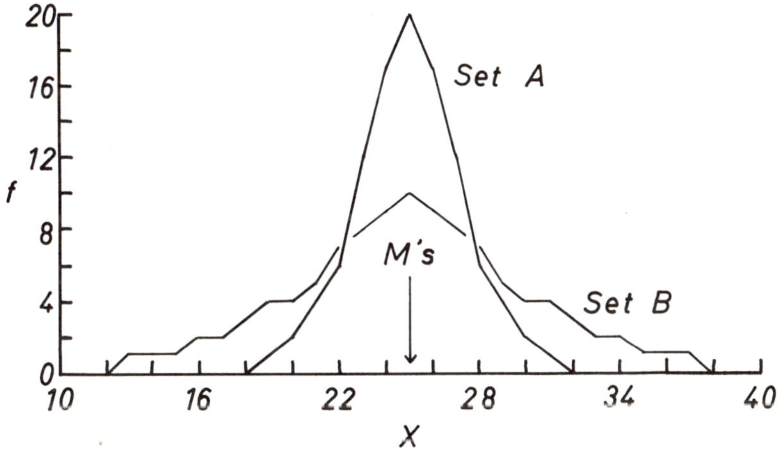

Fig. 2–8. Two frequency distributions with the same means but with different degrees of dispersion.

for skewed distributions of this regular type. The mean, median, and mode have each been indicated for the two skewed distributions shown. The relationships shown are fairly general ones for all regular skewed distributions: (a) The median lies between the mean and the mode. (b) The mean is toward the tail end of the distribution. (c) The mode is toward the end of the distribution having the shorter tail.

2–8. Measures of Dispersion. Two sets of scores may yield identical values for the corresponding measures of central tendency, but the tendencies for the scores to cluster around the central values may be quite different. This characteristic of the closeness of the scores to one another is called the **dispersion** or **variability** of the scores. Figure 2–8 illustrates this concept by representing two sets of scores on the same frequency polygon. Since the distributions are symmetrical, we need consider only the means as measures of central tendency. The mean value is the same for each distribution, as is the total frequency (N) for each set. The figure shows that the two distributions are markedly different with regard to the spread or dispersion of the scores around their mean value.

The concept of *variability* or *dispersion* is probably the most important one in all of statistics. Indeed, inferential statistics becomes nearly impossible without this concept. Therefore, the measures of dispersion to be discussed have a much more important function than mere description. The reasons for computing the measures of dispersion in the manner to be presented may remain somewhat mysterious until later chapters.

The requirements for an adequate statistical measure of dispersion may be illustrated to some degree by considering the smallest set of scores which can have variability of scores, i.e., a set of two numbers. If the two numbers in such a set have the same value, it is clear that their separation or dispersion is as small as possible. It is natural to set the value of a measure of dispersion equal to zero in such a case. As the difference between the two scores becomes greater and greater, we would then require that the value of the measure of dispersion become larger. All of the measures of dispersion to be discussed will follow this simple pattern.

The range. The easiest measure of the dispersion of a set of scores to compute is the *range*. The range is defined as the difference between the highest and lowest scores in a set. In general, the range suffers from the same sort of limitation that the median and mode do, that is, the range is insensitive to the values of scores between the highest and lowest in the set. Various measures of dispersion are based upon the notion of the range between two score values. The differences between these other measures and the range is that the two score values used are not the highest and lowest ones. The use of range-type measures of dispersion is presently quite limited, and no further examples of these measures will be given at this point.

Deviation scores. In order to define three very useful measures of dispersion, it is necessary to first define a new type of score called a *deviation score*. The term deviation is used in this context to speak about the difference between the value of a score and the mean value of the set of scores under study. A deviation score, x_i, is defined as

$$x_i = X_i - M \qquad \text{or} \qquad x_i = X_i - \mu \tag{2–8.1}$$

A deviation score may be computed for each score in a set of scores.

This is done by subtracting the value of the mean of the set from each score. We may speak of the scores of a set, X_i's, as *raw* or *untransformed* scores to distinguish them from the deviation scores.

Since a deviation score is based upon the mean of a set of scores, we may say that each set of raw scores has a corresponding set of deviation scores. The deviation score, x_i, gives information about its corresponding raw score, X_i. The sign, positive or negative, of the deviation score indicates whether the corresponding raw score is above the mean of the set (a positive deviation) or is below the mean of the set (a negative deviation). The value of the deviation score, ignoring the sign, indicates just how far from the mean value the raw score happens to be.

TABLE 2–6

A Set of Raw Scores with the Corresponding Deviation Scores

X_i	x_i
4	−5
7	−2
7	−2
8	−1
8	−1
10	+1
11	+2
17	+8
$\sum_{i=1}^{N} X_i = 72$	$\sum_{i=1}^{N} x_i = 0$

$$N = 8$$
$$M = 9.0$$

Table 2–6 gives a set of eight raw scores in the first column. The calculation of deviation scores requires the computation of the mean, M, which is 9.0 in the example given. This value of 9.0 is now subtracted from each of the raw scores in turn and this procedure yields the set of deviation scores found in column two. Note that some of the deviation scores are negative values and some are positive. Unless all the raw scores in a set have the same value, there will always be some negative and some positive deviations from the mean. An interesting feature of deviation scores is shown by the summation of all eight deviations. This total, which is zero, is shown at the bottom of the second column of Table 2–6. The fact that this sum or total is zero is no coincidence. The following simple theorem indicates that the sum of deviation scores will always be zero for any set of scores.

Theorem 2–1. For any set of N scores, X_i's, the sum of all N of the corresponding deviation scores is zero. Put into notation form,

$$\sum_{i=1}^{N} x_i = 0$$

The proof of Theorem 2–1 employs only the algebra of summation notation, definition (2–8.1), and the definition of the mean (2–4.1). The following manipulations are given without further comment:

$$\sum_{i=1}^{N} x_i = \sum_{i=1}^{N} (X_i - M) = \sum_{i=1}^{N} X_i - \sum_{i=1}^{N} M = \sum_{i=1}^{N} X_i - NM$$
$$= \sum_{i=1}^{N} X_i - N\left(\frac{\sum_{i=1}^{N} X_i}{N}\right) = \sum_{i=1}^{N} X_i - \sum_{i=1}^{N} X_i = 0$$

The sum of squares. An important measure of dispersion is called the *sum of squares*, denoted as SS. This term is a bit misleading, for what is really meant is the *sum of squared deviations from the mean.* We will define this measure as

$$SS = \sum_{i=1}^{N} x_i^2 \qquad (2\text{–}8.2)$$

The computation of the sum of squares from (2–8.2) requires the use of deviation scores. Using the same scores as given in Table 2–6, the computation of the sum of squares is illustrated in the following example.

Example 2–1. The computation of the sum of squares of a set of numbers. The raw scores given in the first column below are transformed to deviation

X_i	x_i	x_i^2
4	−5	25
7	−2	4
7	−2	4
8	−1	1
8	−1	1
10	+1	1
11	+2	4
17	+8	64
Sum 72	0	104

scores, x_i's. This procedure requires the calculation of M, which is found to be 9.0, and then the subtraction of the mean from each raw score.

Each of the deviation scores is now squared. These values are given in the third column. The addition of the entries in the third column gives a total of 104, which is the sum of squares for this set of numbers.

Students often misinterpret (2–8.2) as indicating the summation of the deviation scores before squaring. This is obviously seen to be incorrect when Theorem 2–1 is recalled. If one summed the deviation scores before each was squared, the total would always be zero, which is still zero when squared. Each deviation score must be individually squared before the process of addition is applied, in order to obtain the correct value for the sum of squares.

Properties of the sum of squares. One characteristic of the sum of squares is that its value will be zero if all the scores in a set have the same value. If all the scores are identical, the mean is equal to that value also. Then all of the corresponding deviations from the mean are zero. Squaring and adding zeros will always give a total of zero.

As the scores in a set become more and more dispersed from the mean, the deviation scores get larger and larger. The squaring of these deviation scores gives larger and larger values as well. Hence, the sum of squares increases as the scores in a set become more widely dispersed from the mean.

Another characteristic of the sum of squares statistic is that the value of the SS depends in part upon the number of scores in the set. For example, if each score in a set of 10 numbers is duplicated so as to form a new set of 20 numbers, the sum of squares would be doubled in value. This is a reasonable result in one respect since the new set has twice as many scores as the original one. However, the two sets in question have identical degrees of dispersion if viewed from the standpoint of *relative* frequency distributions. For instance, if the score value 57 occurred twice in the set of 10 scores, its relative frequency would be equal to .20. In the set of 20 scores, the score value 57 would now occur 4 times and would still yield a relative frequency of .20. This equality of relative frequency would be true for each score value. In this sense, the tendency for the scores to be dispersed from the mean can be thought of as equal for the two sets of scores. We often wish to speak of the dispersion of relative frequency distributions rather than the sum of squares of a set.

The variance. The *variance* is a measure of dispersion which reflects the variability of scores as represented by a relative frequency distribution. Using the notation s^2 or σ^2 to indicate the variance, we define this measure as:

$$s^2 = \frac{SS}{N} = \frac{\left(\sum_{i=1}^{N} x_i^2\right)}{N} \qquad \text{or} \qquad \sigma^2 = \frac{SS_*}{N_*} \tag{2–8.3}$$

The variance is simply the sum of squares of a set divided by the number of scores in the set. This is a very similar procedure to that used in obtaining the mean. That is, N numbers are summed and then divided by N. In the case of the mean, we add up the raw scores before dividing by N, while in the case of the variance, it is squared deviation scores which are added up before the division. This similarity permits one to speak of the variance as the *average squared deviation.*

Once the procedure for finding a sum of squares has been mastered, the variance is quickly obtained by dividing the SS by the number of scores in the set.

The relationship between the SS and s^2 of any set of numbers can also be stated in a different form by manipulating (2–8.3). The result is

$$SS = Ns^2 \tag{2–8.4}$$

which permits one to obtain the SS if given the variance and N for a set of numbers.

For some purposes, both the SS and the s^2 have a technical disadvantage compared to a third important measure of dispersion. This problem arises when one wishes to consider the meaning of the SS or s^2 value for a set of scores. For example, consider again the scores given in Table 2–6 and in Example 2–1. Suppose that the raw score values represent the number of errors that each of eight persons made on a test in German. The mean value can be expressed as equal to 9.0 *errors*. The units of the scale of numbers are errors and the mean can be represented by pointing to the number 9 on this scale. However, both the SS and the s^2 may be expressed only in units of *errors squared.* The analogy to length and area illustrates this point. If the length of the side of a square is measured and turns out to be 6 inches, squaring the number 6 will give the area of the square. This value is 36 inches squared (or square inches). An area is not expressed in the same terms as is length. It is not possible to point out the distance along the yard stick and say "this distance is equal to the area of the square." In the same way, the SS of 104 (errors squared) and the s^2, which is 13 errors squared, can not meaningfully be located on the same scale as the raw scores, mean, and even deviation scores. All of the latter values are in terms of errors. The *standard deviation* is a measure of dispersion which can be expressed in the same scale units as the raw scores it describes.

The standard deviation. The *standard deviation* will be referred to symbolically as s or σ and is defined here as the *positive* square root of the variance of a set of scores. Using our earlier definitions for measures of dispersion, this definition may be expressed in a variety of ways,

$$s = \sqrt{s^2} = \frac{\sqrt{SS}}{\sqrt{N}} = \sqrt{\frac{\sum_{i=1}^{N} x_i^2}{N}} \qquad \text{or} \qquad \sigma = \sqrt{\sigma^2} \qquad (2\text{–}8.5)$$

As in previous cases, the Greek letter, σ, refers to a population and the Latin letter, s, to a sample.

Although it is valuable to be able to find the square root of a number by one of the methods ordinarily taught in the elementary or high school grades, a great deal of time is saved by the use of tables. Table A in the Appendix gives the squares and the square roots of the numbers from 1 to 1,000.

The use of Table A to find a square root is straightforward when the exact number is an integer between 1 and 1,000. For example, the s^2 found from Example 2–1 data is 13; this number is found in the first column of Table A and its square root is read from the corresponding line in the third column as 3.6056. To find the square root of a number not given in the table, care must be taken in the movement of the decimal point. For example, suppose we wish to find the square root of 1.64, which is not given in the table. It is permissible to move the decimal point in the number *two* places right or left in order to obtain a tabled number. In this case, moving the decimal place two places to the right yields the number 164. The square root of 164 is read to be 12.806. In order to change this answer to the square root of 1.64, we move the decimal place in the answer *half* as many places as in the number and in the *opposite* direction. This results in the change of 12.806 to 1.2806, which is the desired square root. Notice that the square root of the number 16.4 cannot be found in the $\sqrt{N}$ column. Movement of the decimal point two places to the right or left still gives a value not in this column. However, the square root of 16.4 may be found by using the column headed $\sqrt{10N}$ and the number 164. The tabled value of 40.497 is the square root of 1,640. To obtain the square root of 16.4, we simply move the decimal point one place to the left and find the answer to be 4.0497.

Although the standard deviation is widely used and will be discussed in great detail in later chapters, it should be pointed out that the variance gives the same information without involving the computation of a square root. Given the s^2 of a set one can obtain the s by taking the square root. Given the s of the set one can obtain the s^2 by squaring the value of the standard deviation.

The next two examples illustrate the computations involved in obtaining all three measures of dispersion on the same set of scores. In Example 2–2 the entire set of scores is presented and the defining equations for the measures of dispersion are applied in a direct fashion. In Example 2–3 the scores from Example 2–2 are given in the form of a frequency table. The computations of the mean and sum of squares from the frequency distribution is presented in a systematic form. Care must be taken in these computations based on a frequency distribution. The student is urged to think in terms of the basic definitions of the mean and sum of squares. This will make the procedures in Example 2–3 much clearer and more meaningful than reliance on special formulas for the frequency distribution table.

Example 2–2. The computation of the SS, s^2, and s is performed below on a set of 25 scores. Note that the mean is first calculated in order to obtain the set of 25 deviations scores.

X_i	x_i	x_i^2	X_i	x_i	x_i^2
8	−8	64	16	0	0
9	−7	49	16	0	0
9	−7	49	16	0	0
11	−5	25	18	2	4
12	−4	16	18	2	4
12	−4	16	18	2	4
12	−4	16	19	3	9
14	−2	4	20	4	16
14	−2	4	21	5	25
15	−1	1	23	7	49
15	−1	1	24	8	64
15	−1	1	29	13	169
16	0	0			

$$\sum_{i=1}^{N} X_i = 400 \qquad SS = \sum_{i=1}^{N} x_i^2 = 590$$

$$N = 25$$

$$M = 16$$

$$s^2 = \frac{SS}{N} = \frac{590}{25} = 23.60$$

$$s = \sqrt{23.6} = 4.86$$

Example 2–3. The set of 25 scores used in Example 2–2 is given below in the form of a frequency distribution in columns (*a*) and (*b*). The computations involved in obtaining the *SS*, s^2, and s from this type of distribution are in the last four columns. As explained beneath the data, the symbols X, x, and x^2 which are used without subscripts do not refer to individual scores in the same way as do the symbols X_i, x_i, and x_i^2 in Example 2–2. For instance, in column (e), headed x^2, the numbers give the value of *any single* squared deviation score found in the corresponding score class. The value of 36 found in column (e), row three, indicates that any score with the value of 10 will have a squared deviation value of 36. Column (b) shows that there actually are no such raw scores, or deviation scores, in this set of numbers. The meanings and appropriate use of the columns can be understood by thinking about the procedures used in Example 2–2 on the same data.

(a) X	(b) f_j	(c) Xf_j	(d) x	(e) x^2	(f) x^2f_j
8	1	8	−8	64	64
9	2	18	−7	49	98
10	0	0	−6	36	0
11	1	11	−5	25	25
12	3	36[a]	−4	16	48[b]
13	0	0	−3	9	0
14	2	28	−2	4	8
15	3	45	−1	1	3
16	4	64	0	0	0
17	0	0	1	1	0
18	3	54	2	4	12
19	1	19	3	9	9
20	1	20	4	16	16
21	1	21	5	25	25
22	0	0	6	36	0
23	1	23	7	49	49
24	1	24	8	64	64
25	0	0	9	81	0
26	0	0	10	100	0
27	0	0	11	121	0
28	0	0	12	144	0
29	1	29	13	169	169
Totals	25	400	–	–	590

$$M = \frac{\sum_{i=1}^{N} X_i}{N} = \frac{\sum_{j=1}^{K} (Xf_j)}{\sum_{j=1}^{K} f_j} = \frac{400}{25} = 16$$

[a] The value 36, for example, equals $12 + 12 + 12 = (3)(12)$.
[b] This total of 48 is obtained by adding $(-4)^2 + (-4)^2 + (-4)^2$, or $(3)(-4)^2$.

$$SS = \sum_{i=1}^{N} x_i^2 = \sum_{j=1}^{K} (x^2 f_j) = 590$$

$$s^2 = \frac{SS}{N} = \frac{590}{25} = 23.60$$

$$s = \sqrt{23.6} = 4.86$$

Column (a): Entries show the score value of each class of number.

Column (b): f_j's indicate the number of scores in a class.

Column (c): The products Xf_j give the sum of the scores within each class.

Column (d): Entries show the deviation score for each score within each class.

Column (e): The value of each squared deviation score is given for each class.

Column (f): The products $x^2 f_j$ equal the sum of the x_i^2's within each class of numbers.

Note that the summations involved are for K classes rather than for N scores directly. The totals of the appropriate terms summed over K classes of scores will, of course, equal the sums found directly from the N scores as in the previous example.

2–9. Degrees of Freedom. An exceedingly important part of inferential statistics involves the comparison of the amounts of variability found in two or more sets of numbers. This comparison process will be discussed at length in later chapters, but a basic concept related to such discussion is that of *degrees of freedom*. We may speak of the degrees of freedom associated with the variability in a set of scores as a number equal to N (the number of scores) *minus* the number of restrictions placed on the set. For example, the SS based on N scores involves deviations from the mean. In this case, the mean must be computed from the set of scores and we are restricting the potential variability of the set by using the particular value of the mean found in the set. This may be illustrated further by the following equation for the sum of three scores:

$$T = X_1 + X_2 + X_3 \tag{2–9.1}$$

Here, if we know the value of T, and hence the mean of the set, only two of the three numbers, X_i, are *free to vary*. Once two of the scores are assigned values, the value of the third score is automatically determined by the restriction that the sum is equal to its given value. If we have a set of three scores which total 32, for instance, two of the three scores may take any value, such as 25 and -3. The third score now becomes fixed with the value 10 so as to give the total of 32. The degrees of freedom associated with the SS based on these three scores is then equal to two, the N minus the one restriction involving the total of the scores.

The meaningfulness of this concept may be partly indicated by examining the variability possible in a set consisting of a single number. Clearly, the mean of the set is equal to the value of the one score, and the measures of dispersion all equal zero. This fact is better described by saying that the associated degrees of freedom equals zero, 1 minus 1 for the restriction on the mean, than by saying the variability is based on one score. The latter seems to imply that some variability is possible, whereas the degrees-of-freedom measure gives the correct impression that no variation is even possible.

The SS, s^2, and s all have $N - 1$ degrees of freedom. This is true because each of these measures is based on deviation scores which require the calculation of the mean of the set. This is equivalent to placing a restriction on the total of the N scores in the set.

EXERCISES

1. Use the population of persons given in Table 2–1 to form a partitioning based on:
   (a) Sex *and* hair color.
   (b) Hair color *and* eye color.
   (c) Sex *and* hair color *and* eye color.
   Give the frequencies for each resulting subset.
2. Define and use your own notation to develop a formula which describes the calculation of T, the sum of a set of scores, from a frequency distribution of the scores.
3. Use the result of exercise 2 to express the calculation of the mean from a frequency distribution of scores.
4. Check your formula from exercise 3 by computing the mean for the following set of scores. First use your formula for this frequency distribution, then reconstruct the entire set of raw scores from this distribution and calculate the mean directly.

X	f	X	f
16	2	9	4
15	0	8	3
14	1	7	4
13	4	6	1
12	3	5	0
11	5	4	0
10	2	3	1

5. Use the distribution given in exercise 4 and construct a frequency polygon from the set of scores.

Find the mean, median, and mode for this distribution and compare the three measures of central tendency by indicating the value of each on your figure.

6. Convert the following raw scores to deviation scores:
   (a) Verify that the sum of deviation scores is zero.
   (b) Compute the *SS*, s^2, and s for this set.

11	7	5	3
9	6	5	2
9	6	5	1
8	6	4	0
8	5	4	−4

7. Calculate the *SS* and s^2 from the following frequency distribution:

X	f	X	f
140	1	135	4
139	2	134	3
138	2	133	2
137	3	132	2
136	4	131	1

8. Find the *SS* and s for the following set: 1, 3, 4, 5, 7.
   Now add 10 to each of these scores and find the *SS* and s of the resulting set of scores.
9. Find the *SS* and s for the following set:

6	9
6	10
8	10
8	12
9	12

Now multiply each raw score by 10 and find the *SS* and s of the new set of scores.

3

Scale Transformations

3–1. Linear Transformations. It is often convenient and even necessary to change a set of scores measured in one type of unit to a different type of unit. This is equivalent to changing the measurement scale. Although there are an infinite number of methods for transforming a set of numbers into a different set, the method called a *linear transformation* is of great importance in statistics. This transformation uses the equation for a straight line. We shall define a linear transformation of a set of scores as

$$X_i' = aX_i + b \tag{3–1.1}$$

where a and b are constants and X_i' stands for the value of the transformed (new) score. It is intended that (3–1.1) be applied to all of the scores in the set to be transformed. The constants a and b may be any finite value, including numbers smaller than 1 and negative values. However we shall restrict a to a non-zero number, i.e., $a \neq 0$.

A classic illustration of a linear transformation is the change of temperature scales. For instance, to change scores which are measured in degrees Fahrenheit to scores on the Centigrade scale, we may apply the rule

$$°\text{F} = \tfrac{9}{5}°\text{C} + 32° \tag{3–1.2}$$

It may be readily seen that this rule is a linear transformation with a equal to $\frac{9}{5}$ and b equal to 32°.

We have already encountered another example of a linear transformation. Examine once again (2–8.1), the definition of a deviation score. In this definition, the value of a is 1 and the value of b is $-M$, whatever that may be for the set. Deviation scores are, then, linear transformations of the set of corresponding raw scores.

3–2. Raw Score Formulas. Once the two constant values in a linear transformation are known, it is possible to convert transformed scores back into their corresponding raw scores. We now wish to use this property, and the knowledge that deviation scores are linear transformations, to obtain some convenient variations in the expressions used to obtain the measures of dispersion.

Recall that SS, s^2, and s are all based on deviation scores (see (2–8.3)). Since any deviation score, x_i, is equal to $X_i - M$, it is possible to express the SS as follows:

$$SS = \sum_{i=1}^{N} X_i^2 - M \sum_{i=1}^{N} X_i = \sum_{i=1}^{N} X_i^2 - \frac{T^2}{N} \qquad (3\text{–}2.1)$$

The derivation of this equation, dropping the notation of the obvious limits of the summation sign, starts with (2–8.2) and is

$$\begin{aligned} SS &= \Sigma\, x_i^2 = \Sigma\,(X_i - M)^2 = \Sigma\,(X_i^2 - 2MX_i + M^2) \\ &= \Sigma\, X_i^2 - 2M\,\Sigma\, X_i + NM^2 = \Sigma\, X_i^2 - 2M\,\Sigma\, X_i + \frac{N(\Sigma\, X_i)(\Sigma\, X_i)}{NN} \\ &= \Sigma\, X_i^2 - 2M\,\Sigma\, X_i + M\,\Sigma\, X_i = \Sigma\, X_i^2 - M\,\Sigma\, X_i = \Sigma\, X_i^2 - \frac{(\Sigma\, X_i)(\Sigma\, X_i)}{N} \\ &= \Sigma\, X_i^2 - \frac{T^2}{N} \end{aligned}$$

The definitions of the s^2 and s and (3–2.1) permit the following raw score forms of the equations for the variance and standard deviation:

$$s^2 = \frac{\sum_{i=1}^{N} X_i^2}{N} - M^2 \qquad (3\text{–}2.2)$$

$$s = \sqrt{\frac{\sum_{i=1}^{N} X_i^2}{N} - M^2} \qquad (3\text{–}2.3)$$

Equation (3–2.2) is obtained by dividing each term in (3–2.1) by N. The square root of the variance is, of course, the standard deviation. These raw score equations for computing the measures of dispersion yield the same answers as their corresponding deviation score equations, but the raw score computations are usually easier to use with real data. This is because the raw scores are often whole numbers, but the mean is usually a decimal number. In that case, the use of the deviation score equations requires obtaining and squaring many decimal deviations. The following example will illustrate this point and also show the use of the raw score equations for computing the SS, s^2, and s.

Example 3–1. The problem is to compute the SS, s^2, and s for the set of 10 numbers given in the first column below. The calculation of the sum of the scores is the first step, as it is used no matter which method of calculation is chosen. The second and third columns give the intermediate calculations required in the deviation score formula for the SS, and the fourth column entries show the squared raw scores for use in (3–2.1) through (3–2.3).

X_i	x_i	x_i^2	X_i^2
8	−6.2	38.44	64
9	−5.2	27.04	81
9	−5.2	27.04	81
11	−3.2	10.24	121
14	−0.2	.04	196
15	.8	.64	225
17	2.8	7.84	289
17	2.8	7.84	289
19	4.8	23.04	361
23	8.8	77.44	529
Σ 142	0	219.60	2,236

$$M = \frac{142}{10} = 14.2$$

From (2–8.2), the $SS = 219.60$. Then $s^2 = \dfrac{219.6}{10} = 21.96$.

Equation (3–2.1) may be used to obtain the same answer by subtracting the value of T^2/N from 2,236, the sum of the squared raw scores; $SS = 2{,}236 - \dfrac{(142)^2}{10} = 2{,}236 - 2{,}016.4 = 219.6$.

Division by N will again give the desired s^2 of 21.96. If one wishes to compute the variance directly without obtaining the SS, (3–2.2) may be used. Here the sum of squared raw scores is divided by N and M^2 is subtracted from the result:

$$s^2 = \frac{2{,}236}{10} - (14.2)^2 = 223.60 - 201.64 = 21.96.$$

The standard deviation is obtained in either method by finding the square root of the variance. The value here is 4.69, approximately.

When data are presented in the form of a frequency table, care must be used to obtain the proper sum of squared raw scores. The next example illustrates the use of (3–2.1) for obtaining the SS from a frequency-distribution table.

Example 3–2. Find the SS of the scores represented in the distribution given below in the first two columns.

X	f	fX	X^2	fX^2
5	1	5	25	25
6	0	0	36	0
7	2	14	49	98
8	5	40	64	320
9	4	36	81	324
10	2	20	100	200
11	2	22	121	242
Σ	16	137	–	1,209

Since the $SS = \Sigma X_i^2 - (T)^2/N$, we may proceed by recalling that

$$T = \sum_{j=1}^{K} f_j X_j$$

and noting that

$$\sum_{i=1}^{N} X_i^2 = \sum_{j=1}^{K} f_j X_j^2$$

(where K is the number of classes). This yields

$$SS = 1{,}209 - \frac{(137)^2}{16} = 1{,}209 - 1{,}173.0625 = 35.9375$$

3–3. Central Tendency and Dispersion in Transformed Scores. The measures of central tendency and dispersion describing a set of transformed scores may be readily calculated from knowledge of the corresponding statistical values from the raw scores when the transformation rule is of the linear type. This represents a great saving in labor from the procedure of recalculation of these values from the transformed set.

Let us first consider the effect on the mean value when a linear transformation is applied. The following results for the mean are also true for the median and mode.

Equation (3–1.1) may be thought of as representing two separate operations: (a) the multiplication of each score in a set by a constant, a; and (b) the addition of a constant b, to each score in a set. When both of these are applied, unless $a = 1$ or $b = 0$, the steps must be followed in the order given, i.e., each score is multiplied by a and then each *new* score is changed by adding b. The result of this general case may be given as a theorem.

Theorem 3–1. If a set of N scores with mean M is transformed in a linear fashion (a as a multiplier and b as the constant of addition), then the mean of the transformed set of scores, M', will be equal to $aM + b$.

The derivation of this theorem follows quickly when (3–1.1) is substituted into the formula for the mean of the transformed scores:

$$M' = \frac{\Sigma X_i'}{N} = \frac{\Sigma (aX_i + b)}{N} = \frac{a \Sigma X_i + Nb}{N} = aM + b$$

It is of interest to examine the result shown in this theorem in the special cases where a is 1 and secondly where b is 0. The first case is equivalent to adding a constant, b, to each score; the result is that the mean of the new set of scores is found by adding b to the mean of the original scores. In the case where b is zero, all scores are multiplied by a constant, but no addition (or subtraction) takes place; then it is seen that the transformed mean is found by multiplying the original mean by this same constant.

The first of these two special cases has already been partially illustrated for the case of deviation scores. Recall that the formation of deviation scores requires that the mean of the set be subtracted from each raw score. This is a case where $a = 1$ and $b = -M$. Use of Theorem 3–1 then shows $M' = M - M = 0$. This has been expressed in effect earlier when it was shown that the sum of deviation scores is always zero. As the sum is zero, division by N to obtain the mean also results in the above value of zero.

Since the s^2 and s are both based on the SS measure of dispersion, let us find the value of the SS', the sum of squares of the transformed scores, in terms of the SS of the raw scores and the values of the constants. Then division by N will yield a similar expression for s'^2 and finally s'. The results will be given first as a theorem, the derivation of which shows one method of obtaining these relationships.

Theorem 3–2. If a linear transformation is applied to a set of scores with measures of dispersion SS, s^2, and s, then the corresponding measures of dispersion in the set of transformed scores, SS', s'^2, and s', may be expressed as

1. $SS' = a^2(SS)$
2. $s'^2 = a^2s^2$
3. $s' = |a|s$, where a is the constant of multiplication.

The following derivation for SS' uses, mainly, substitutions of definitions and Theorem 3–1. The reasons for each step are left out to provide an exercise for the reader.

$$\begin{aligned} SS' &= \Sigma (x_i')^2 = \Sigma (X_i' - M')^2 \\ &= \Sigma [aX_i + b - (aM + b)]^2 = \Sigma (aX_i + b - aM - b)^2 \\ &= \Sigma (aX_i - aM)^2 = \Sigma [a(X_i - M)]^2 \\ &= \Sigma (ax_i)^2 = \Sigma (a^2x_i^2) = a^2 \Sigma x_i^2 \\ &= a^2SS \end{aligned}$$

Parts 2 and 3 follow directly:

$$s'^2 = \frac{SS'}{N} = \frac{(a^2SS)}{N} = a^2s^2$$

$$s' = \sqrt{s'^2} = \sqrt{a^2s^2} = |a|s$$

It should be carefully noted that Theorem 3–2 clearly implies that the constant of addition, b, in a linear transformation has no effect at all on the change in the dispersion of the scores. This means that one can freely add or subtract any constant value for each score in a set and still find that the sum of squares, variance, and standard deviation of the transformed scores are equal in value to their corresponding raw score values.

In this sense, it can be stated that the mean can be changed *independently* of the variance when applying linear transformations. This point will be of some importance in later discussions and may be taken here to illustrate that the measures of central tendency and dispersion actually do reflect different (independent) characteristics of sets of scores.

The results of the previous two theorems are summarized in Table 3–1. In this table, three types of manipulations are considered: (a) multiplication of the scores by a constant, (b) addition (or subtraction) of a constant to each score, and (c) both (a) and (b) in that order.

TABLE 3–1

The Mean and Measures of Dispersion of Linearly Transformed Sets of Scores in Terms of Original Score Values

Transformed Values	Transformation Applied		
	$X' = aX$	$X' = X + b$	$X' = aX + b$
$M' =$	aM	$M + b$	$aM + b$
$SS' =$	a^2SS	(unchanged) SS	a^2SS
$s'^2 =$	a^2s^2	(unchanged) s^2	a^2s^2
$s' =$	$\|a\|s$	(unchanged) s	$\|a\|s$

3–4. The z-Score Transformation. In the comparison of two distributions of scores, we have already commented in Chapter 2 on the need to equate the groups of scores in terms of the numbers of scores.

This was done when comparing dispersions by using the average squared deviation, s^2, instead of the SS measure. However, two sets of scores to be compared may be measured in different units. In that case, differences between the two sets in terms of their means and dispersions may arise from the differences in units used. This is true even when the two sets of scores are based on identical observations reported in both units. For instance, consider once again the measurement of temperature. In Fig. 3–1, 60 hypothetical measurements of temperature are reported in two forms, degrees Fahrenheit and degrees Centigrade. Although the shapes of the two distributions are similar, the

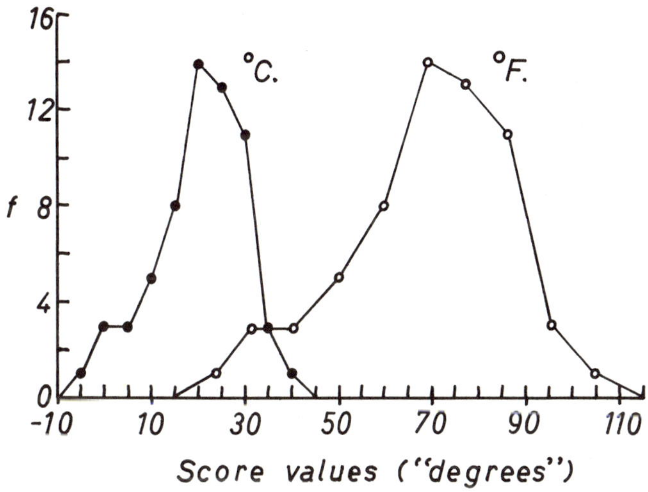

Fig. 3–1. A single set of temperatures displayed in two forms, degrees Centigrade (°C.) and Fahrenheit (°F.).

identity of the two sets of measurements is obscured by the different units used. One attempt at displaying the identity of the two sets of scores is illustrated in Fig. 3–2, in which each set of scores has been converted to deviation-score form. This procedure removes the differences in the mean values between the two sets, because the mean deviation is zero for any set of scores. It is clear from Fig. 3–2 that the distributions are not yet identical, on account of their different degrees of dispersion.

One way of showing that the two sets of temperature scores are identical (when their meanings are considered) would be to change one scale of values to the same units as the other. Another method, one with considerable advantages over the first, in general, is to convert each scale of values to a kind of standard scale. This is the intent of the *z-score transformation.*

We define a *z-score,* with respect to a set of raw scores, as

$$z_i = \frac{x_i}{s} \tag{3–4.1}$$

It is clear from (3–4.1) that any set of raw scores can be converted to a corresponding set of z-scores. This is done by finding the M and s

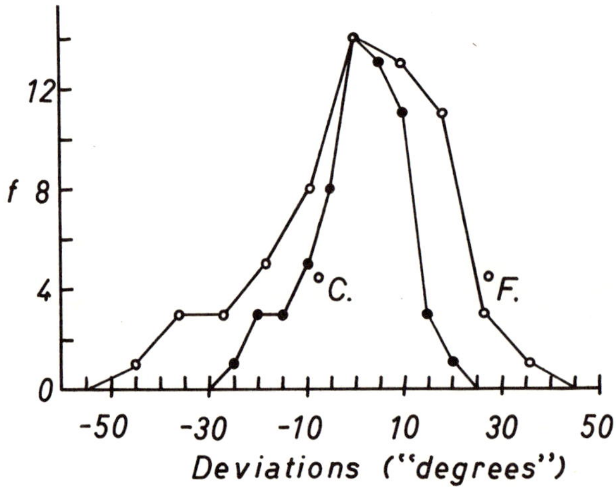

Fig. 3–2. A single set of temperatures presented as deviations from the mean in terms of Centigrade (°C.) and Fahrenheit (°F.) units.

of the raw scores, transforming each score to deviation-score form, and then dividing each deviation score by s.

Before examining the properties of this sort of transformation, let us convert the temperature measurements in Figs. 3–1 and 3–2 to z-scores. This may be accomplished easily with the data as shown in Fig. 3–2, where deviation scores are shown for each set. The value of the standard deviation of the Centigrade scores, s_c, is calculated to be 9.574. The similar value on the Fahrenheit scores is 17.234. When each deviation score value is divided by its appropriate standard deviation value, as shown in Table 3–2, one may clearly see the identity of the two distributions, in terms of z-scores.

TABLE 3–2

The Conversion of the Two Sets of Temperature Scores to Their z-Score Equivalents

x(°C)	x(°F)	f	z (from °C)	z (from °F)
−25	−45	1	−2.611	−2.611
−20	−36	3	−2.089	−2.089
−15	−27	3	−1.567	−1.567
−10	−18	5	−1.044	−1.044
−5	−9	8	−0.522	−0.522
0	0	14	0	0
5	9	13	0.522	0.522
10	18	9	1.044	1.044
15	27	3	1.567	1.567
20	36	1	2.089	2.089

$z_i = \dfrac{x_i}{s}$ $\quad s_c = 9.574 \quad s_f = 17.234$

3–5. Properties of the z-Score Transformation. First let it be noted that the z-score transformation is a linear transformation of the raw scores. This may be shown by manipulating the right side of (3–4.1) in the following ways:

$$z_i = \frac{x_i}{s} = \frac{X_i - M}{s} = \frac{X_i}{s} - \frac{M}{s} = \frac{1}{s}X_i - \frac{M}{s}$$

The final expression is in the form of a linear transformation in which the constant of multiplication, a, is equal to $1/s$ and the constant of addition, b, is equal to $-M/s$.

Since the z-score transformation is a linear one, we may apply Theorems 3–1 and 3–2 to obtain the following results.

Theorem 3–3. If any set of N scores is transformed to a set of corresponding z-scores, (1) the mean of the set of z-scores, M_z, is zero, and (2) the variance and standard deviation, s_z^2 and s_z, of the set of z-scores are equal to 1.

Using (3–1.1), with $a = 1/s$ and $b = -M/s$, we may apply Theorem 3–1 to obtain the M_z:

$$M_z = aM + b = \frac{M}{s} - \frac{M}{s} = 0$$

In the same way, application of Theorem 3–2 gives the s_z^2 and s_z:

$$s_z^2 = a^2 s^2 = \left(\frac{1}{s}\right)^2 s^2 = 1$$

and

$$s_z = |a|s = \frac{1}{s} s = 1$$

Another interesting property of z-scores arises from the observation that the value of a z_i also gives the value of that score's deviation from the mean of zero. This shows that the SS_z, the sum of squared z-score deviations from the mean, is equal to the sum of the squared z-scores. The value of this computation is always the same, for any set of z-scores. Recalling from (2–8.4) that $SS = Ns^2$ we find for z-scores that

$$SS_z = \sum_{i=1}^{N} (z_i - M_z)^2 = \sum_{i=1}^{N} z_i^2 = Ns_z^2 = N \qquad (3\text{–}5.1)$$

The sum of squared z-scores is always equal to N.

The z-score transformation gives a sort of standardized distribution to which any set of raw scores may be converted. This distribution will always have a mean of zero and a standard deviation of 1. The comparison of two or more sets of raw scores, in terms of the shapes of their distributions, is then feasible in spite of differences in means and dispersions in the raw-score sets. Also, differences in the units of measurement are eliminated, in a sense, when z-score distributions are used. This last point will become clear if the following transformation of a Centigrade temperature from Fig. 3–1 to its z-score equivalent is examined.

$$z_i = \frac{X_i - M}{s} \qquad \text{(where } X_i = 30° \text{ C, } M = 20° \text{ C, and } s = 9.574° \text{ C)}$$

$$= \frac{30° \text{ C} - 20° \text{ C}}{9.574° \text{ C}} = \frac{10° \text{ C}}{9.574° \text{ C}} = 1.044$$

Note that the z of 1.044 is a pure number; the units have been cancelled in the formation of the ratio of two numbers with the same units. The z-score transformation is often called a scale measured in standard deviation units. This merely refers to the fact that a z-score of 1 reflects a raw score which is 1 standard deviation from the raw score mean.

If a distribution of scores is given in terms of the z-scores, and the mean and s of the raw scores are given, it is possible to convert each z_i to its corresponding X_i by using (3–4.1) and (2–8.1).

The full usefulness of the z-score transformation will become increasingly clear in subsequent chapters.

EXERCISES

1. Calculate the SS and s^2 for the following set, using both the deviation and raw score formulas:

   2, 3, 3, 4, 5, 5, 5, 6, 7, 8, 8, 10, 10, 14

2. Use the raw score formula to calculate the SS and s^2 from this frequency distribution.

X	f
12	10
13	14
14	16
15	17
16	11
17	8

3. If each raw score in exercise 2 were multiplied by 14 and a constant of 3 were then subtracted from each, what would be the M, SS, s^2, and s of the transformed set of scores?
4. Transform the following set of raw scores to the corresponding set of z-scores:

   2, 3, 3, 4, 4, 5, 5, 5, 7, 10, 17

   Verify that

$$\sum_{i=1}^{N} z_i^2 = N$$

5. Prove that the set of z-scores from any set of numbers will be identical to the set of z-scores corresponding to a set of numbers which is a linear transformation of the first set.

4

Random Sampling Distributions

Having examined basic terminology, measures of central tendency, measures of dispersion, and linear transformations, the student should be in a position to gain an insight into the purposes and general methods of inferential statistics. The present chapter presents the major types of statistical questions asked by the scientist, and indicates the general theoretical tool which is employed in answering these questions. The majority of the statistical tests and operations discussed in the remainder of this text are really just special cases of the more general procedures outlined in this chapter. Attempts will be made to point out the type of question asked in all later statistical procedures and frequent reference will be made to the generality of the theoretical basis used in obtaining answers to these questions.

4–1. Types of Statistical Questions. The vast majority of statistical procedures and tests are designed to answer one or more of three basic questions. These three questions may be stated as:

Question I. Given knowledge of the population, what is the probability of selecting a sample of scores which has a specified value or values for a given statistic?

Question II. Given a sample of scores, what is the best estimate of the population value (parameter) of a specified statistic?

Question III. Given one or more samples of scores, what is the probability that they were selected from a population or populations with certain assumed parameter values?

Let us discuss each of these questions briefly in order to make them clear.

4–2. The Probability of a Certain Sample Value. In Question I, a sample of scores is selected from some defined population. The question to be answered concerns the chances (probability) of finding that the value of a certain statistic computed from the sample is equal to some particular number or numbers. For example, we may wish to know the probability of obtaining a sample of 10 scores with a mean value of 156.50 when the population consists of the measured weights (in pounds) of all adults in a certain hospital. An exact answer to this question is not possible unless we know something of the population scores. Usually, this question is asked only when one knows at least the *relative frequency distribution* of the population scores.

It is possible to ask this question in a less exacting way, when desired. For example, rather than specifying the exact value of 156.50, one might ask about the probability that the sample mean will be any value between 150 and 160.

4–3. Estimates of Population Parameters. Information at one's disposal is often less complete than desired. Question II is directly oriented to such cases as it asks a question which is to be answered with incomplete data. For instance, if we have a sample of 10 scores from the above mentioned population of weights, but do not know the mean value in the population, we may still use the information in the sample to make some kind of estimate (or guess) about the population mean. The problem is now to find the best single estimate, given the sample scores. To answer this question, it is usually necessary to know or assume certain things about the population scores, such as the shape of the distribution.

When a set of sample scores is used to estimate the population value of a statistic (called a *parameter*), the symbol $\widehat{S}_*$ will be used to denote the estimate. For example, $\hat{\mu}$ indicates an estimate of a population mean, μ, based on a sample from that population.

Two slight variations on this question may be asked. The first of these requires that the estimate of the population value be a range of scores instead of an exact value. The second refinement involves the calculation of the probability that the true population value is actually within the estimated range of values. For instance, based on a sample mean of 148.0 lb. in the above illustrations, appropriate calculations might lead to the answer that the population mean is between 142.0 and 154.0, and that this estimate has a 50 per cent "chance" of being correct.

4–4. Hypothesis Testing. Question III is a general statement of what is often called *hypothesis testing* in statistics. The question is directed at the population value of a statistic, as is Question II, but is usually intended to rule out or eliminate one or more possible population values rather than to estimate their true values. Quite frequently, Question III is concerned with two or more populations and the possibility that they do not have equal values for a particular statistic.

This question arises when one assumes particular values and characteristics about a population or populations and proceeds by calculating the probability that an actual sample or samples with obtained statistical values could have been selected from the assumed populations. The object of this reasoning is commonly to reject or disprove the assumed population values.

Two brief illustrations will be given here prior to a later detailed development of these questions. Let us assume that a population consists of all white marbles. If we obtain a sample of 9 white marbles and 1 red marble, it is clear that our assumption about the population cannot be true. Put another way, if we assume that the population contains all white marbles, the probability that our sample was drawn from this population is zero, i.e., it is impossible. This example is not typical of statistics, however, for some degree of uncertainty is characteristic of statistical answers.

The second illustration is based on the earlier ones dealing with weights of persons. Let us make the assumption that the mean weight of the adults in one hospital is equal to the mean weight in a second hospital. If we now select a random sample of 10 scores from each of these populations and find that the sample means are 113 lb. and 187 lb., respectively, we have reason to question the truth of our assumption. By asking the question, "What is the probability that two sample means will differ by as much or more than the two obtained do?" we may arrive at an answer which is a very low probability. If this answer is quite close to a probability of zero, which indicates an impossibility, we may reasonably reject the assumption of equal population means, upon which our calculations have been based.

4–5. Chance and Probability. Before an outline of the methods of answering the above questions is attempted, the notions of *chance* and *probability* must be more completely examined. The concept of chance, as used in statistics, does *not* imply lack of causation. For example, in rolling a pair of dice, the multitude of physical forces operating on the dice may be considered as causing whichever faces to turn upward as the dice come to rest. We may still say that the outcome (an event) is determined by chance. This use of the term chance implies that the factors influencing the outcome event are not under strict control. Another way of saying this is that on the next *trial*, circumstances do not permit a duplication of the causal factors. Because of the variation in the values of such causal factors, variation in the results may occur from one trial to the next. This is not meant to imply that each possible outcome has an *equal* chance of taking place. What is required is that more than one result is possible and that one does not have sufficient control over the governing factors to predict the exact result beforehand.

The concept of *probability* may be defined in a number of ways. Only two such definitions will be given here, although a more formal and slightly different definition will be given in the chapter on probability. The probability, P, that an event, E, will occur will be expressed as a number, $P(E)$, such that

$$0 \leq P(E) \leq 1.00 \tag{4–5.1}$$

Two ways of obtaining this number, $P(E)$, for any event E will be considered: (1) the *relative frequency* method, and (2) the *mathematical limits* method.

The first of these methods uses the relative frequency of occurrence of the event E over some number, N, of occasions on which the event could have occurred. This relative frequency, expressed as a proportion of N, is said to be equal to the $P(E)$. Clearly, this is simply a way of expressing past history. To use the resulting value to *predict* what may happen on future trials involves a good deal of risk, but sometimes such a procedure is the best one available.

The definition of probability in terms of the mathematical concept of a *limit* requires that one make certain mathematical assumptions (or postulates) which allow one to describe the relative frequency of the event in question by means of an equation. If it can be shown that this relative frequency value approaches some stable value (limit) as N, the number of trials, approaches infinity, then this limiting value of the relative frequency of E is defined as the probability of the event.

The concept of a limit may be explained in mathematical terms as well as by analogy to physical events. If an expression has a limit as N approaches infinity, this means that the value of the expression can be made as close to the limiting value as desired, simply by selecting a large value of N. For instance, if $Y = .50 + 1/N$, we may say that

$$\lim_{N \to \infty} Y = .50 \tag{4–5.2}$$

The statement in (4–5.2) is true because the value of Y may be made as close to .50 as desired merely by selecting a large enough value for N.

An analogy to this concept might be the water level at some point in a large container as a function of the time since a rock is thrown into the water. Here time corresponds to N and, assuming no further disturbances, one can see that the effect of the stone gets less and less as time passes. The limiting value corresponds to the "final" level of the water.

Using the notation $f_n(E)$ to indicate the relative frequency of event E, we may then define the $P(E)$ as

$$P(E) = \operatorname*{limit}_{n \to \infty} f_n(E) \tag{4–5.3}$$

Equation (4–5.1) illustrates a basic feature of the probability of an event; this probability can be no lower than zero and no higher than 1.00 for any event. The value of $P(E)$ may be more clearly interpreted by noting its correspondence to the values of proportions, which have the same upper and lower limits in the case of partitions of a set of elements.

Unless otherwise stated, we will use the mathematical limits concept of probability in preference to the relative-frequency definition.

4–6. Random Sampling Distributions. The fundamental theoretical tool used to answer each of the three types of statistical questions discussed in this chapter involves the concept of the *random sampling distribution of a statistic*. Let us refer to such a distribution as the $RSD(S)$. As the name implies, a random sampling distribution arises from the selection of random samples from some population. The $RSD(S)$ is a type of relative frequency distribution which differs in two ways from frequency distributions encountered up to this point. In the first place, a $RSD(S)$ describes the relative frequency of the sample value of a statistic, S, rather than a single score, X. Secondly, the relative frequency of the value of the statistic is expressed as a probability, in the mathematical limits sense.

An illustration of a random sampling distribution is given in graphic form in Fig. 4–1. The statistic under study in Fig. 4–1 is the sum, T, of a sample of four scores. Each sample has been selected from a population consisting of the digits 0, 1, 2, 3, 4, 5, 6, 7, 8, and 9. It is assumed that this selection has been made so as to satisfy the definition of random sampling *with replacement*. Under these assumptions, there are 37 possible sums, T, which may result from any single random sample. These possible values of the statistic are given along the x-axis in Fig. 4–1.

On the y-axis of the $RSD(T)$ are found the probability values. Each point on the curve indicates the probability of obtaining a random sample which has a sum of scores equal to the corresponding value of T on the x-axis. These probability values have been calcu-

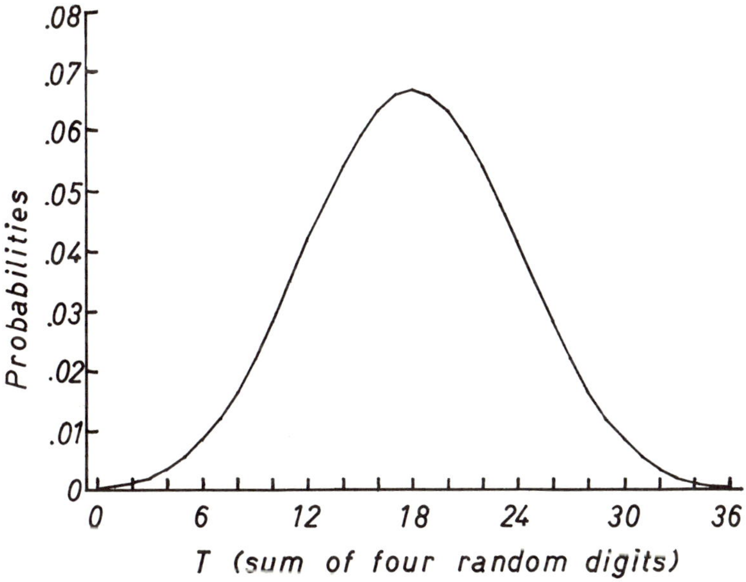

Fig. 4–1. A random sampling distribution of T. Each point on the figure indicates the theoretical probability of a single random sample of four digits having a total (T) of the value shown on the x-axis.

lated by means of a mathematical analysis which yields the limiting relative frequencies for each value of T as sampling proceeds indefinitely; i.e., an infinite number of samples is drawn. In terms of (4–5.3), the n indicates that the number of samples approaches infinity, although each sample has only four raw scores.

Although a $RSD(S)$ is ordinarily obtained by mathematical analysis, an approximate $RSD(S)$ may be obtained for any statistic by actually drawing a large number of random samples from an appropriate population of scores. It is of value to generate such an *empirical* distribution in order to illustrate the notion of random sampling and the meaning of a $RSD(S)$.

Once an appropriate population of scores has been defined, the major difficulty in the construction of an empirical sampling distribution is a selection procedure which is truly random, a procedure in which each population score does have an equal chance of being an element in a given sample. Most mechanical means of selection, such as shuffling cards or drawing numbers from a hat, do not actually produce a truly random sample. For experimental and theoretical purposes, a table of random numbers is a better means of obtaining random samples. Table B in the Appendix is such a table of random numbers. In this table, the digits 0 through 9 have been generated randomly so that each number had an equal probability of being selected at each position in the table. Use of a random number table requires that each population element to be sampled be assigned a unique number. Then the table of random numbers is entered at some arbitrary point, and the desired number of random numbers is taken in some predetermined order. When the number of a population element is encountered in the series of random numbers, that element is placed into the sample.

In the generation of an actual distribution of sample T's corresponding to those shown in Fig. 4–1, one would most likely number the scores of the population 0 through 9 so that the value of each score is the same as its label number. Then the first sample of 4 scores is selected from the random number table and the value of the statistic, T, is computed for that sample. This procedure is continued over and over, and from each sample, a single value of T is recorded. The construction of the distribution of the statistic is now accomplished by computing the relative frequency with which each possible sample value of T has occurred. For example, if one had obtained a total of 100 different samples and the sum of the four scores in a sample had been 26 in each of 8 samples, then the relative frequency of $T = 26$ would be $\frac{8}{100}$ or .08.

In practice, such a sampling distribution would be very time consuming if a large number of samples were drawn. However, a high-speed computer may be instructed to perform such labor and will do so in a fraction of the time required by a person. Figure 4–2 illustrates the correspondence between the theoretical distribution of T, the $RSD(T)$, and an empirical distribution of 1,000 sample values of T as selected by a computer. Clearly, the general shapes of the two curves are the same, but the empirical proportions are not as uniform from point to point as are the theoretical proportions (probabilities). Note that the probability definition requires that an infinite number of samples be drawn. The empirical proportions based on 1,000 samples are not equal to the theoretical values because 1,000 is less than infinity. Of course, any finite number of samples is also less than infinity but the correspondence between the $RSD(S)$ and an actual distribution of S

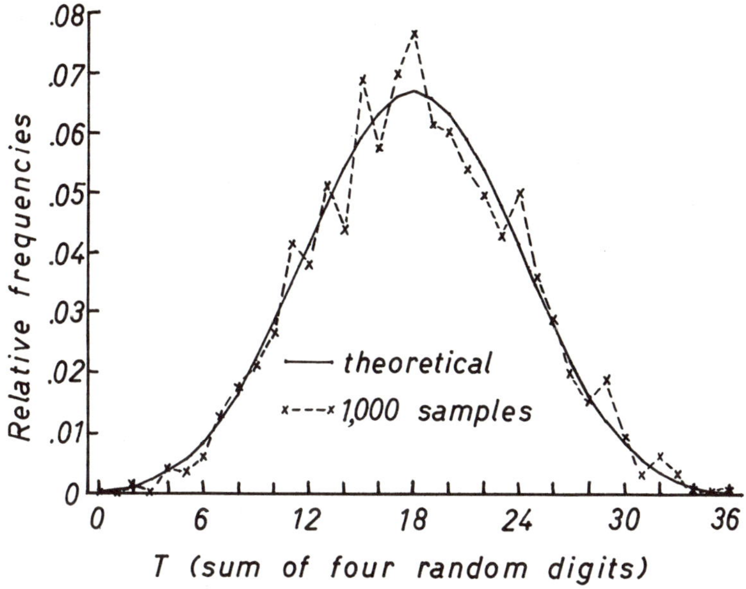

Fig. 4–2. The random sampling distribution of T (theoretical) compared to the empirical distribution of 1,000 sample values.

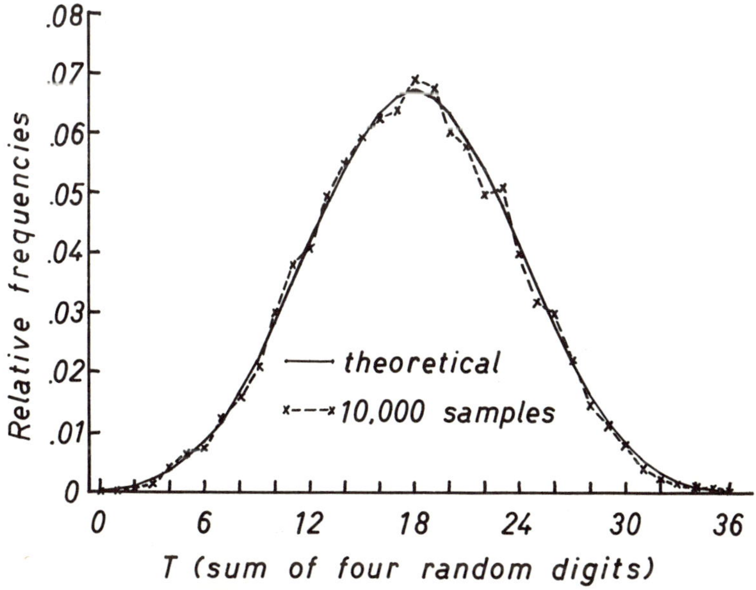

Fig. 4–3. The random sampling distribution of T (theoretical) compared to the empirical distribution of 10,000 sample values.

will become quite close when a very large number of samples is drawn. In Fig. 4–3, a distribution based on 10,000 samples is compared to the $RSD(T)$. Discrepancies still exist but are now much smaller than in Fig. 4–2.

The results of still larger numbers of actual sample values of T are given in Table 4–1. Here the obtained proportions are so close to the

TABLE 4–1

Comparison of the $RSD(T)$ with Empirical Sampling Distributions of T

T	Theoretical $P(T)$	Obtained Proportions		
		50,000 Samples	100,000 Samples	250,000 Samples
0	.0001	.0002	.0002	.0001
1	.0004	.0004	.0004	.0005
2	.0010	.0007	.0009	.0010
3	.0020	.0021	.0018	.0019
4	.0035	.0031	.0034	.0035
5	.0056	.0051	.0054	.0056
6	.0084	.0089	.0089	.0085
7	.0120	.0122	.0117	.0119
8	.0165	.0159	.0161	.0161
9	.0220	.0215	.0220	.0223
10	.0282	.0284	.0276	.0279
11	.0348	.0344	.0347	.0347
12	.0415	.0421	.0421	.0416
13	.0480	.0490	.0494	.0479
14	.0540	.0533	.0540	.0538
15	.0592	.0593	.0596	.0591
16	.0633	.0625	.0637	.0622
17	.0660	.0656	.0648	.0657
18	.0670	.0687	.0677	.0676
19	.0660	.0652	.0644	.0666
20	.0633	.0639	.0637	.0636
21	.0592	.0610	.0600	.0594
22	.0540	.0530	.0539	.0537
23	.0480	.0484	.0483	.0484
24	.0415	.0416	.0415	.0420
25	.0348	.0353	.0355	.0349
26	.0282	.0293	.0290	.0281
27	.0220	.0222	.0223	.0220
28	.0165	.0160	.0157	.0161
29	.0120	.0123	.0119	.0121
30	.0084	.0079	.0083	.0082
31	.0056	.0048	.0052	.0056
32	.0035	.0030	.0030	.0036
33	.0020	.0015	.0018	.0021
34	.0010	.0008	.0009	.0010
35	.0004	.0004	.0004	.0004
36	.0001	.0001	.0001	.0001

probability values that a very large figure would be needed to show the discrepancies. In this table, the $RSD(T)$ values for each possible sum are given as well as the results of three different sets of samples with 50,000, 100,000, and 250,000 samples, respectively. It is of interest to note that the values from the first set of 50,000 samples appear reasonably close to the corresponding probabilities, and yet statistical calculation of a type to be discussed later indicates that this particular set of sample values is more discrepant from the theoretical values than 95 per cent of such sample sets which might have been obtained. Notice the closer correspondence to the theoretical values which accompanies larger and larger sets of samples.

All of the statistics discussed in this book have random sampling distributions, as do other statistics in common use. However, the $RSD(S)$ is not theoretically known for all statistics which have been defined or for those which might be invented for a particular problem. An approximation to an unknown $RSD(S)$ is possible today through the use of high-speed computers, but most common statistics are used for one reason, because their random sampling distributions have been studied and are available in the form of tables for use in statistical problems.

4–7. Uses of Random Sampling Distributions. The procedures used to answer the three major types of statistical questions outlined earlier depend directly upon the $RSD(S)$ for the statistic being studied. This may be illustrated most directly in the case of Question I, in which the probability of a particular sample value or values for a statistic is desired. In this problem, the answer may be obtained immediately from the appropriate random sampling distribution.

For example, let us ask the following question: "What is the probability of randomly selecting a set of four digits from a table of random numbers and finding that the sum, T, of the sample is 18?"

This question is typical of Question I in the following important respects: (1) we are given enough information to describe the relative frequency distribution of the raw scores in the population (a rectangular distribution of the digits 0 through 9 is implied); (2) a single random sample from the population is to be considered; and (3) the question concerns the probability that the sample will have a particular value (or values) for some statistic, T in this case.

The answer to our specific question may be found directly by examining the $RSD(T)$ found in Table 4–1. This distribution has been computed with the exact requirements stated in our question, i.e., selection of random samples with $N = 4$ from a population which has equal proportions of the digits 0 through 9. The answer is obtained merely by taking the probability value from the table which corresponds to the T value of 18. This is a $P(E)$ equal to .0670.

In general, the solution of any problem of the Question I type involves the calculation of, or reference to, the appropriate $RSD(S)$ for the statistic in question. Often, only a desired part or parts of such a $RSD(S)$ need be calculated or examined. In each case, the steps needed in answering a question of this type are mainly those involved in obtaining the appropriate $RSD(S)$. As might be expected, the particular steps needed to generate a $RSD(S)$ are somewhat different for different statistics. The specific methods given in later chapters, especially in the chapter on probability, sometimes seem to bear little relationship to one another, but the object of each method is really the same, the calculation of part or all of a particular $RSD(S)$.

It will be noted here, with the reasons postponed, that the answers to Question I which require more than one value of a statistic may be found by adding the probability values found in the appropriate $RSD(S)$. For example, one might wish to know the probability that the sum of the four digits in our earlier problem will turn out to be an even number. This may be rephrased as asking the probability that the sample T will be 0, 2, 4, 6, . . . , or 36. The desired $P(E)$ will be equal to $P(0) + P(2) + \cdots + P(36)$. Again the values may be obtained from the appropriate $RSD(T)$ found in Table 4–1. The addition of the indicated $P(T)$ values yields an answer of .5000. The interpretation of this answer may be made in terms of proportions or percentages if one considers drawing repeated samples. In this case, one may say the *expectation* is that .5, or 50 per cent, of the samples drawn will have totals which are even numbers. The remaining 50 per cent are expected to have totals which are odd numbers.

The general procedures followed in obtaining answers to Question II also depend in very large measure upon the $RSD(S)$. To illustrate these procedures, let us assume some characteristics of a population of scores and then *estimate* other unknown parameters of that population on the basis of a random sample of scores. Assume that the population consists of any 10 numbers which are spaced one unit apart. If the lowest of these scores is called X_1', then the second lowest score, X_2', will be equal to the first score plus 1; i.e., $X_2' = X_1' + 1$, and so forth.

Let us now ask the following question: Based on a single random sample of 4 scores, what is the best estimate of the population mean, μ?

The answer to this question is that the best estimate of μ is the value of the mean of the sample, M. In order to show that this is the answer desired, we must consider a $RSD(S)$. The statistic in question is a mean, and the appropriate $RSD(S)$ will be a random sampling distribution of means, $RSD(M)$.

In order to illustrate a $RSD(M)$, we need only consider the random sampling distribution of T shown in Table 4–1. Since each of the totals indicated in this table was based on 4 raw scores, the corresponding mean values may be calculated by dividing each T value by 4. The results of this procedure are given in Table 4–2 with the corresponding probability figures. Table 4–2, then, gives the $RSD(M)$ for samples of size 4 drawn from a population of the ten digits 0 through 9.

One interesting feature of the distribution of means shown in Table 4–2 is that the distribution is symmetrical and has a mean of 4.5. This value of 4.5 may be obtained by inspection since the distribution is symmetrical and has only one mode. Recall that the mean, median, and mode are all equal in such cases.

The population mean, μ, is also found to be equal to 4.5. We find, in this example, that the mean of the distribution of means (called the *mean of the means*) is equal to the mean of the population.

Before returning directly to the question posed earlier, consider the possibility that our population of X_i' scores might actually consist of the digits 0 through 9. In that case, the $RSD(M)$ shown in Table 4–2 would be appropriate to the present problem of estimating a population

TABLE 4–2

The Random Sampling Distribution of Sample Means, Where Each Sample Consists of Four Random Digits

Mean	Probability	Mean	Probability
.00	.0001	4.75	.0660
.25	.0004	5.00	.0633
.50	.0010	5.25	.0592
.75	.0020	5.50	.0540
1.00	.0035	5.75	.0480
1.25	.0056	6.00	.0415
1.50	.0084	6.25	.0348
1.75	.0120	6.50	.0282
2.00	.0165	6.75	.0220
2.25	.0220	7.00	.0165
2.50	.0282	7.25	.0120
2.75	.0348	7.50	.0084
3.00	.0415	7.75	.0056
3.25	.0480	8.00	.0035
3.50	.0540	8.25	.0020
3.75	.0592	8.50	.0010
4.00	.0633	8.75	.0004
4.25	.0660	9.00	.0001
4.50	.0670 (mean of means)		

mean. The selection of one random sample would result in some particular mean value, one of those shown in Table 4–2. Now assume that we know that the appropriate $RSD(M)$ is symmetric as in Table 4–2, with the mean of the distribution of means equal to the population mean. If we do not know the mean of the $RSD(M)$ or the population mean, we have no basis for using any value for the estimate, $\hat{\mu}$, except the sample mean, M. This becomes clear if one considers a possible adjustment or correction of the value of M obtained. Since the distribution of sampling means is symmetric, there is exactly as much chance of a single sample mean's being too high as too low! In this case, we call the sample mean an *unbiased estimate* of the population mean. There is no consistent tendency for a sample mean to be either higher or lower than the population mean.

That the sample mean is the best estimate of the population mean will now be demonstrated as correct if we can show that the $RSD(M)$ will have the characteristics of the one shown in Table 4–2, no matter what the mean of the population might happen to be.

The correspondence between the $RSD(M)$ shown in Table 4–2 and the $RSD(M)$ for any set of 10 numbers spaced one unit apart can be illustrated by use of a linear transformation of one set of scores to the other. If we denote the ten digits 0 through 9 as X_i and the other set of ten numbers as X_i', the X_i' scores can be transformed into X_i values by the addition of a constant, c (which may be negative). This will transform the mean of the X_i' set, μ', into the mean of the X_i set, μ, therefore $\mu = \mu' + c$ or

$$\mu' = \mu - c \tag{4–7.1}$$

Since the $RSD(M)$ is known for the X_i population, the $RSD(M)$ for the X_i' may be inferred from (4–7.1). The distribution of M_i is symmetric about its mean, μ, so the distribution of $M_i - c$ (which is equal to M_i') will be symmetric around a mean of $\mu - c$ (which is μ').

This use of a linear transformation illustrates that the $RSD(M)$ will be symmetric around the population mean no matter what the population mean may be, given our assumptions about the type of population. Therefore, the answer to our problem of the Question II type is that the sample mean will be an unbiased estimate of the mean of the population.

Although the answer to this Question II type of problem required extensive computations and reasoning, the time will have been well spent if the reader is convinced of the importance of the $RSD(S)$ in answering such questions. Later problems of this type will be solved in a more efficient manner after other theoretical foundations have been introduced.

A further example of the uses of the $RSD(S)$ may be made by considering a problem of the Question III type. Let us again use a population of ten consecutive digits. We will proceed by making assumptions about the population, developing a random sampling distribution for a statistic, and then deciding whether or not a particular obtained sample value for the statistic is a reasonable outcome. If the result of this analysis indicates that an unlikely sample value for the statistic has been obtained, this will be taken as evidence that our assumptions about the population may not have been true.

First, assume that the population consists of the ten consecutive digits 0 through 9. If this assumption about the population is true, then the random sampling distribution of the total (of samples of size 4) will be found again in Table 4–1. One can quickly see that the most likely total for a single sample of four scores drawn from the assumed population will be near 18. The totals of 0, 1, 2, 34, 35, and 36, for instance, are possible outcomes but not very likely. Values for the total of a single sample which are negative or above 36 are impossible if our assumptions about the population are correct.

As will be discussed later, some prior decision should be made as to what constitutes an event which is so unlikely as to make the assumptions about the population unacceptable.

However, let us now select a sample of four scores from the population, which is supposed to consist of the ten digits 0 through 9, but which might actually be quite different. As an example, let us assume that we select the scores 1, 0, 0, and 0. The total of this sample is 1, which is quite far from the *expected* mean of the distribution of totals, which is 18. Let us find the probability that a sample total would be drawn which is this far or further from the expected value of 18. In answering this question, it is usual to consider deviations in *either direction* from the mean of the expected distribution. In the present case, this would imply that the totals of 0, 1, 35, and 36 are all as far or further from the distribution mean of 18 as is our sample value. To find the probability of such unexpected outcomes, we add the probability values for each of these totals together. Using the probabilities shown in Table 4–1, we have

$$\begin{aligned} P(T = 0, 1, 35, \text{ or } 36) &= P(0) + P(1) + P(35) + P(36) \\ &= .0001 + .0004 + .0004 + .0001 \\ &= .0010 \end{aligned}$$

The answer indicates that a result as unexpected (or worse) than our sample value should occur only once in 1,000 selections, *if* our population assumptions are correct. It is possible that the population is as supposed, and that we have simply had the misfortune to select an atypical sample. However, a probability value as low as .001 is usually taken to be ample evidence that the assumptions made about the population are not true. Note that the *rejection* of the assumptions about the population may be in error, but that the evidence at hand strongly suggests that the population is different than originally supposed.

In the present case, we could have rejected the assumptions about the population with absolute certainty if certain kinds of samples had been selected. For instance, if our sample total had been greater than 36 (one of the scores would have been greater than 9), it is obvious that our assumptions about the population were not true. In the usual application in statistics, any outcome will be possible, but many of them will be exceedingly unlikely, i.e., have very small probabilities. The statistician must learn to live with some degree of uncertainty in nearly all of his decisions.

4–8. A Preview. The majority of statistical procedures and tests to be introduced in later chapters follow the general procedures discussed in this chapter. The particular statistics to be encountered and their random sampling distributions vary in different types of problems, but the reader will find a communality among all of them. The questions asked will usually be one or more of three types just discussed, and the procedures used in answering these questions will depend in large part upon the development or knowledge of an appropriate random sampling distribution.

A particular statistical procedure often seems unrelated to the body of statistics. It is hoped that the present chapter will serve as an organizing force in the study of subsequent material.

EXERCISES

1. The $RSD(T)$, where T is the sum on the faces of a pair of dice, is given below. Develop an empirical random sampling distribution by rolling a pair of dice 200 times (or use a table of random numbers to obtain 200 pairs of the numbers 1 through 6). Compare the $RSD(T)$ with your empirical distribution by constructing a relative frequency polygon showing both distributions.

T	Probabilities	T	Probabilities
2	.028	8	.139
3	.056	9	.111
4	.083	10	.083
5	.111	11	.056
6	.139	12	.028
7	.167		

2. For each of the following problems, indicate which one of the three basic types of statistical questions is appropriate:
   (a) Trying to find out whether teen-aged men have more automobile accidents than teen-aged women.
   (b) Assessing the chances that a hand of five playing cards will have four aces in it.
   (c) Making tests to determine the average number of defects per day to be expected in a production line at a certain factory.
   (d) Testing the hypothesis that college men and women have the same mean grade-point average.
   (e) Trying to find the probability that three people selected at random from census tables will have a mean age between 25 and 30 years.
   (f) Trying to estimate the number of words in a theme by counting the number of words on a single page.
3. Using Table 4–1 and the concept of linear transformation, find the probability of selecting a random sample of four scores which has a total (T) of 28, when the population consists of the integers 5 through 14, inclusive.
4. What is the most likely value of T in exercise 3? Why?

5

The Normal Distribution

Certain distributions of scores are of greater importance in the field of statistics than are other distributions. Frequency of occurrence of distributions and mathematical knowledge about the properties of distributions are two factors which tend to cause this preference for some distributions over others.

5–1. Normal Curve Equation. A very common and well understood distribution of scores in the area of statistics is the *normal distribution*, or *normal curve*. The normal distribution is also called by other names, including such as, "the bell-shaped curve," "the Gaussian curve," and "the normal probability error function." These names, and others, all refer to the same basic function. One form of this function is given in (5–1.1), and examples of the normal curve are shown in Fig. 5–1.

$$Y = \frac{1}{\sigma\sqrt{2\pi}} e^{-[(X-\mu)^2/2\sigma^2]} \qquad (5\text{–}1.1)$$

Equation (5–1.1) gives a value of Y, which may be likened to the relative frequency of a score value, for any X, a score, if one is also given the value of certain constants and *parameters* of the function. Two well known mathematical constants, π and e, are used in the equation but these are always known and stay fixed in value. The σ and μ are called the parameters of the normal curve, and once values for these two have been specified, the value of Y may be calculated for any given value of X.

In the same way that any possible straight line may be described by giving the general equation for a straight line and the values of the two parameters in the equation (called a and b in earlier discussions), any particular normal distribution may be completely described by reference to (5–1.1) and by designating the values of only two parameters, σ and μ. These two quantities are the standard deviation and the mean of the resulting curve, if X is taken as the score value and Y as proportional to the relative frequency of the X's.

If a set of scores forms a normal distribution, one needs only the values of the mean and standard deviation (or variance) in order to

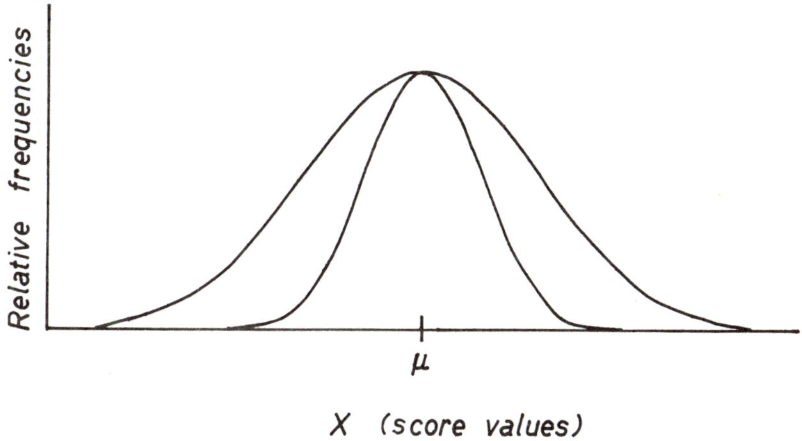

Fig. 5–1. Two examples of a normal distribution. The distributions have equal means but different standard deviations.

describe the relative frequency distribution of the set. All other characteristics of the set of scores which are of interest may be discovered from examining the distribution directly, because use of (5–1.1) and the values of μ and σ allow the entire relative frequency distribution to be constructed.

5–2. Characteristics of the Normal Curve. One important feature of the normal distribution which must concern us at once is that the values of X are assumed to form a *continuous scale* of numbers without any restrictions on the range of values possible. This means that X may be any value at all on the scale of real numbers.

A set of scores which is allowed to contain only discrete values, such as the sum of four random digits considered in the last chapter, can never form a normal distribution. However, such a discrete distribution is often very close in shape to a normal curve. In such cases, the normal curve and its properties are used as an approximation to the discrete distribution.

Another property of the normal curve is that it is a *regular* and *symmetrical distribution.* As such, the measures of central tendency are all equal. Figure 5–1 illustrates this property and as well shows the general bell-shaped nature of the curve.

Further examination of the two curves shown in Fig. 5–1 will show that the vast majority of the scores in a normal distribution are relatively close to the mean. The truth of this statement is apparent when one recalls that the range of scores is from minus to plus infinity! To be more specific, one may state that in a normal distribution 99.994 per cent of the scores lie within four standard deviations of the mean, and about 99.7 per cent of the scores lie within three standard deviations of the mean.

These characteristics indicate the reason that a normal distribution may be presented in the form of a graph by using a range of only five to six standard deviations along the x-axis. Only a small proportion of the scores are omitted in such a graph.

5–3. The Frequency of Normal Distributions. The reasons that the normal curve is important to the study of modern, and the more classical, statistical methods are: (1) the normal curve seems to appear again and again in common types of statistical problems, and (2) the characteristics of the normal curve were discovered quite early in the history of statistical studies. We shall make no attempt to study the history of statistics, but the variety of ways in which a normal curve, or a close approximation to a normal curve, may arise should be clearly understood.

Assume that a type of score, X, is a measurement of some aspect of the world. If the characteristic being measured rises from a number of relatively unrelated causes or factors, then X will tend to so vary as to form a normal distribution. For example, certain types of athletic abilities are related to a variety of inherited characteristics which in turn are controlled by the particular combination of genes found in an individual. If one were to make measurements of the athletic ability so determined, a likely result would be a normal distribution of scores, or a very close approximation.

A normal population of scores does not always result in such cases, but the number of times that a normal distribution is obtained is quite remarkable.

The central limit theorem. An important theorem from mathematical statistics will be offered at this point for two purposes. The first is to indicate another reason why the normal curve is a very frequent distribution in statistical procedures. The second is that the results of this theorem, called the *central limit theorem*, will prove useful in later derivations.

Unfortunately, the proof of the central limit theorem requires concepts well beyond the scope of this text and its assumed prerequisites. The theorem is given, therefore, without proof.

Theorem 5–1. If any population of scores, X_i's, has a finite mean (μ) and variance (σ^2), then the random sampling distribution of means ($RSD(M)$) based on this population will approach a normal distribution as the sample size (N) approaches infinity. The mean of the random sampling distribution of means (μ_M) will be equal to μ, and the variance of the random sampling distribution of means ($\sigma_M{}^2$) will be equal to σ^2/N.

To understand Theorem 5–1, simply consider each sample mean in the $RSD(M)$ distribution to be a score; call it Y. The set of Y's which forms the $RSD(M)$ has a mean and a variance as any set of scores does. The theorem indicates that, as the samples contain more and more raw scores, the corresponding distribution of Y's, resulting from random sampling, will tend to be a normal distribution.

In addition to specifying the shape of the $RSD(M)$, the theorem also indicates what the mean and variance of the set of Y's will be, in terms of the mean and variance of the population of raw scores.

Applications of Theorem 5–1 will be considered later in this chapter. The important point to consider at the moment is that whenever a $RSD(M)$ is needed in answering one of the three types of statistical questions, and the sample size considered is large, then the normal curve is the expected shape of the $RSD(M)$. This will be the case every time Theorem 5–1 is appropriate. It can be noted here that the theorem applies to any type of population distribution in which a mean and variance exist, i.e., have finite values. Very few known distributions fail to meet this simple criterion.

In practice then, the normal curve is found in a variety of situations, both in the physical world and in the theoretical realm.

5–4. The z-Score Form of the Normal Curve. Any two sets of scores may both be normally distributed, but this correspondence between the two sets may be masked because of differences in their respective means and standard deviations. A standardized method for examining a normal distribution is to work with z-scores. If each score value in a normal distribution is transformed to its corresponding z-score equivalent, then the resulting relative frequency distribution will be a normal curve scaled in z-score units.

The characteristics of a set of z-scores have already been examined in Chapter 3. The mean of such a distribution is always 0 and the variance and σ are both equal to 1. Thus, any normal distribution of scores may be converted to a single standard distribution by converting the raw scores to z-scores.

Table C in the Appendix describes this standardized normal distribution of scores. Table C shows the proportion of z-scores which will be found with values between the mean and a z-score of a specified amount. The first column of numbers lists the units and tenths positions of the z-scores, and the top row of numbers lists the hundredths position of the z-scores. The tabled values for each such z-score are the proportions of cases (scores) between the mean of zero and the particular z-score in question.

Figure 5–2 illustrates the relationship between the z-score values and the corresponding proportions of scores found between the mean and that z-score value. Note especially that the normal curve is symmetrical and has .5 of the scores above the mean and .5 below the mean. Moreover, the symmetry of the curve implies that a positive z-score and a negative z-score of the same value need not both be given in Table C. It should be apparent that the proportion of scores encountered is the same when one goes a given distance in either direction from the mean.

In order to find the proportion of scores between the mean and a negative z-score value, simply consult Table C with the positive z-score of the same magnitude. For example, consider the z-scores shown in Fig. 5–2. Let us verify the proportions shown in the figure by use of Table C. The z-score value of $-.50$ is shown as a score below the mean, such that .1915 of the total scores in the distribution are between the mean and the z-score. In order to obtain this information from Table C, one must enter the table with a positive z-score, since no negative values are given. The value of $-.50$ is changed to its absolute value and .50 is found in the row and column headed z. The corresponding proportion entry is the desired answer. For a further example, consider the z-score of 1.00. When this value is found in Table C, the corresponding proportion is found to be .3413. This relationship is also indicated graphically in Fig. 5–2.

Table C may also be used in the reverse of this procedure. For instance, if we wish to find a particular z-score such that .4772 of the scores in a normal curve are between the mean and the given value of z, the proportion entries of Table C are consulted for a value as close as possible to .4772. In this case the value .4772 is found exactly. The corresponding z-score for this proportion is seen to be 2.00. Note that the tabled value of $z = 2.00$ is slightly ambiguous in that either a positive or a negative z-score will have the same proportion of cases between itself and the mean. The negative value -2.00 and the .4772 proportion of the scores are illustrated in the figure.

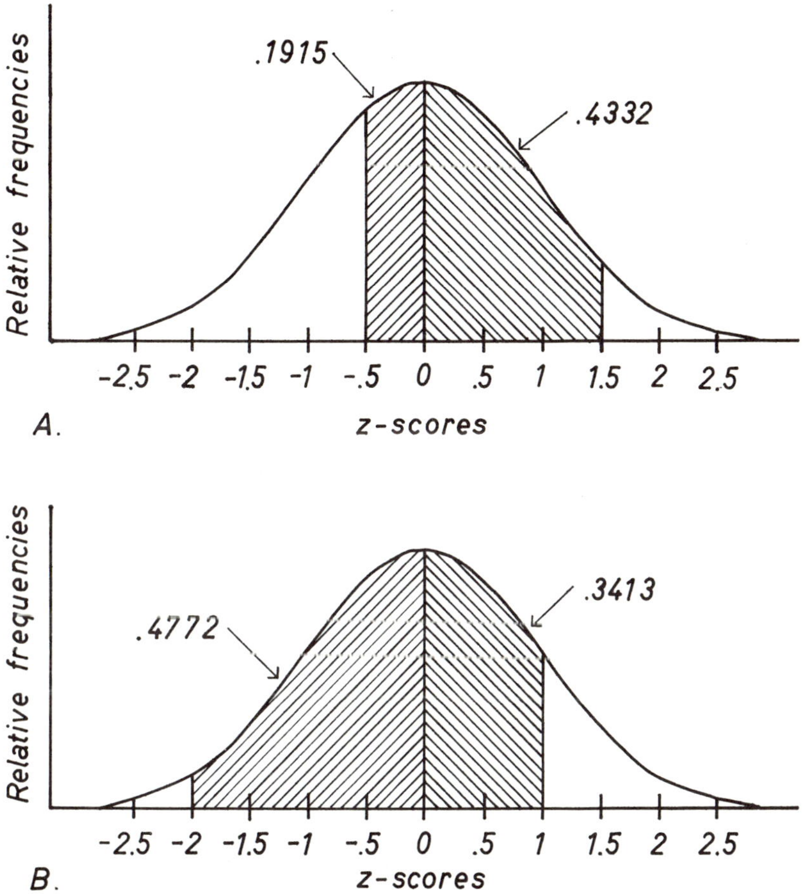

Fig. 5–2. Examples of the normal distribution in z-score form, showing proportions of cases between the mean of zero and selected z-score values.

Figure 5–3 may be used to illustrate some rather simple computations resulting from the use of Table C. The negative value of z shown in this figure illustrates the proportion of scores which are found in the tail end of a normal curve, starting with a particular z-score. The example shown considers a z-score of -1.25. Table C indicates that

.3944 of the scores in the curve are between the mean and this z-score value. Since .5000 of the scores are below the mean of the curve, only .1056 of the scores are left below the z-score of -1.25.

The positive side of Fig. 5–3 is used to indicate the method of finding the proportion of scores between two z-score values which are both on the same side of the mean. The shaded area of the figure shows the

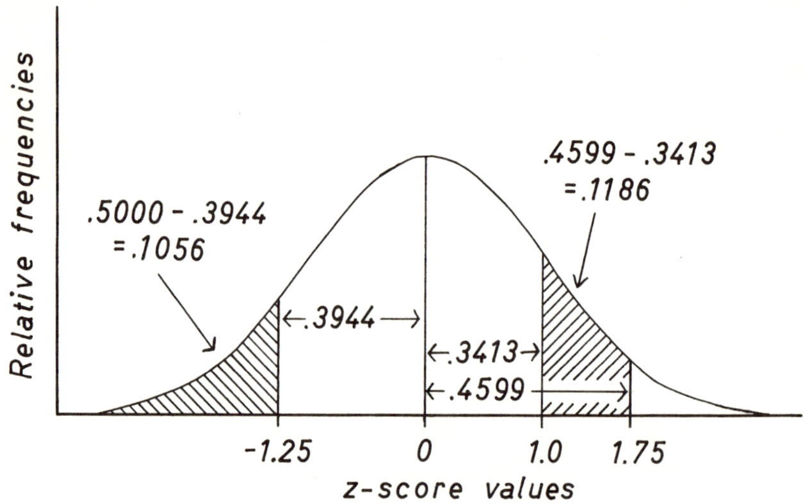

Fig. 5–3. A normal distribution in z-score form, illustrating the use of the normal curve table to find proportions of cases in selected parts of the distribution.

proportion of the scores which have z-score values between 1.00 and 1.75. This proportion is shown as .1186. The answer was obtained by noting that .4599 of the scores are between the mean and a z of $+1.75$ and that .3413 of the scores are between the mean and a z of $+1.00$. The latter scores are clearly contained in the .4599 considered first. Therefore, only .1186 $(.4599 - .3413)$ of the scores remain between the two z-scores.

The real limits of an interval. So far the fact that the normal curve is based on a continuous scale of numbers has not been greatly explored. At this point, consider the problems which arise if one attempts to treat the normal curve in the same way as a frequency distribution for a discrete variable. For example, Fig. 4–1 and Table 4–1 both describe the probability of obtaining a precise given value for the sum of four randomly selected digits. If one were to ask the probability of selecting some particular exact score at random from a true normal distribution, the answer would be vanishingly small! Since there are an infinite number of possible score values within any range of scores specified, the probability that a single randomly selected score will have some exact predetermined value is technically zero. *This technicality is readily overcome in practice because our measuring, recording, and calculating methods always convert scores to discrete scales which only approximate continuous scales.*

For instance, no matter how many decimal places of accuracy are used in a set of numbers, some numbers from a continuous scale would always be found between two adjacent scores used. These intermediate values are merely *rounded off* to the nearest score being used.

One might find a statement to the effect that a certain school contained 127 pupils of age 9 years. Clearly this cannot mean that 127 pupils are exactly 9 years old at the moment the statement is made. Rather, the ages of the students have been grouped into intervals, and any exact age within the confines or *limits* of one interval has been rounded off to 9 years. It is most likely, in this illustration, that anyone who is 9 years or over, but not yet 10 years of age, would be classed as a nine-year-old.

To return to the normal curve, consider the probability of selecting one score at random from a normal distribution and finding that the corresponding z-score is equal to $+1.50$. In this case, one must consider the value 1.50 as being an interval of exact scores, all of which equal 1.50 *when rounded off at the second decimal place.* Any value between 1.495000 . . . and 1.504999 . . . will be considered as satisfying the requirements. Although Table C does not give z-score listings to the degree of accuracy needed in this problem, the method of solution is straightforward. Just as in the case shown in Fig. 5–3, the proportions between the mean and each z-score (using the ends of the interval) are found. The difference between the two proportions may be taken as the answer.

In the example just made, a proportion has been used as equal to a probability. This conversion is justified so long as the one score being selected is chosen in a random fashion, because a random sampling technique implies equal opportunity of selection for each element in the population. The proportion of cases having a particular value or range of values is then seen to be equal to the probability that a single randomly selected score will have a value within that range.

When scores are grouped into intervals, the *apparent ends* or *limits* of the intervals are usually given. For example, 15 to 17, 18 to 20, and 21 to 23 inches, might be given as three score-intervals of interest in some statistical study. The apparent limits of the second of these intervals are 18 in. and 20 in. However, if similar exact values were used as end points for all three intervals, then certain score values would not be placed in any of the intervals. For instance the score 17.95 would be too high in value to be contained in the first interval given and too low to fall into the 18-to-20-in. interval. It is clear from this condition that the apparent limits of the intervals cannot be used when considering a continuous scale of numbers. The *real limits* of the interval 18 to 20 in. are 17.5 and 20.4999 . . . in., i.e., *halfway between the apparent limits of adjacent intervals.*

In using a continuous scale of values, such as the normal curve, one must take care to distinguish between the apparent and real limits of an interval of score values.

In the following examples and exercises in this chapter, the reader may assume that all scores are given as exact values. When an interval of scores is intended, rather than an exact value, special mention of the fact will be made.

Transformation problems with the normal curve. A variety of simple problems involving the normal distribution may be solved by knowledge of the z-score transformation formula, equation (3–4.1), and use of a normal curve table such as Table C.

All of these problems are essentially descriptive. They do not require statistical inferences of the type discussed in the three basic types of statistical questions. They are useful for descriptive purposes and form a basis for later inferential reasoning.

In each of the problems illustrated below, the z-score formula in its raw-score form is a starting point:

$$z_i = \frac{X_i - M}{s}$$

or (5–4.1)

$$z_i = \frac{X_i - \mu}{\sigma}$$

The problems will always involve one or more z-scores which may be found in Table C. There will usually be enough information given in the problem to fill in three of the four quantities in (5–4.1). Simple algebra will then yield the fourth quantity.

Only one additional point need be recalled. Each z-score corresponds to a proportion of the curve between the mean and itself, and each such proportion corresponds to a z-score.

It is advisable to sketch a normal curve and locate given information on the figure as an aid to reasoning.

Example 5–1. If the heights of men in a certain state are known to be normally distributed with a mean of 69 in., and $\sigma = 6$ in., what proportion of the men are taller than 60 in.?

A crude sketch of the normal curve can be made as a starting point. By filling in the mean of 69 in. and the score in question, 60 in., the required proportion of the curve desired is seen to be over .5. The exact amount over .5 would be given if the proportion of scores between the mean and a score of 60 in. were known.

The answer to this last question can be found in Table C if the score of 60 in. is converted to a z-score. Using (5–4.1),

$$z_i = \frac{60 - 69}{6} = -1.50$$

The corresponding proportion is found to be .4332. Adding the .5000 above the mean to the .4332 between the mean and our score of 60 in., the answer is found to be .9332.

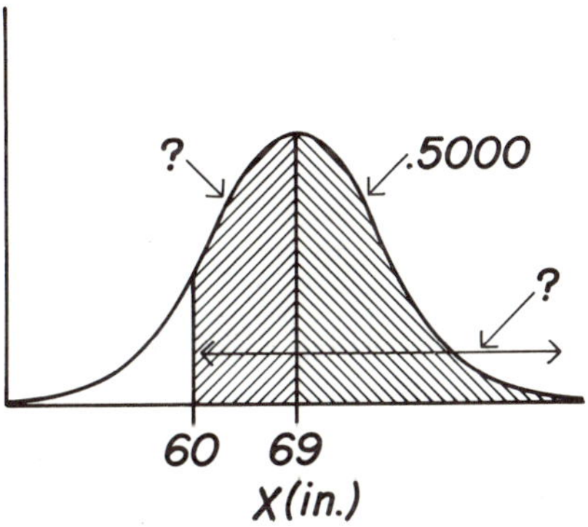

Fig. Ex. 5–1

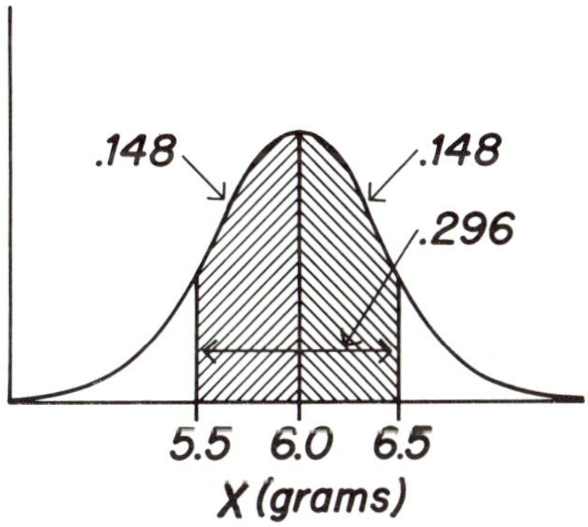

Fig. Ex. 5–2

Another variation of the same sort of problem is found in Example 5–2. Note that a proportion cannot be used in (5–4.1), only the related z-score value.

Example 5–2. The weights of lemon drops made in a candy factory are known to be normally distributed with a mean of 6 grams (g.) If .296 of the lemon drops weigh between 5.5 and 6.5 g., what is the standard deviation of the population?

Again a diagram of the distribution is helpful. The mean of 6 g. and the two limiting values of 5.5 and 6.5 g. are drawn on the x-axis. It may be seen that the .296 proportion of the scores between 5.5 and 6.5 is divided into two equal portions of .148 each on either side of the mean. The z-score which corresponds to this proportion is found to be .38. Equation (5–4.1) may now be used with either of the values 5.5 or 6.5. However, it must be remembered that the appropriate z-score for 5.5 g. would be $-.38$, since 5.5 is a score below the mean. Hence,

$$-.38 = \frac{5.5 - 6.0}{\sigma}$$
$$-.38\sigma = -0.5$$
$$\sigma = 1.32 \text{ g. (approx.)}$$

The next example requires the use of two z_i values, but otherwise is simply another variation of the first two examples.

Example 5–3. If a distribution of scores is known to be normal with $\mu = 128$ and $\sigma = 20$, what proportion of the scores will be found between the values 158 and 122.6?

The diagram shows that the two scores given are on opposite sides of the mean. The proportion between the two may then be found by finding the proportions between each score and the mean and adding the two proportions together. Transforming each raw score into its z-score equivalent allows the use of Table C:

$$z_1 = \frac{122.6 - 128}{20} = -.27$$

which yields a proportion of .1064, and

$$z_2 = \frac{158 - 128}{20} = 1.50$$

which yields a proportion of .4332. The total proportion of the distribution between 158 and 122.6 is then $.1064 + .4332 = .5396$.

In the final example of this sort, the deviation score form of the z-score equation, (3–4.1), is more appropriate than (5–4.1); but in either case, the use of a diagram of the normal curve is again suggested to provide structure to one's reasoning.

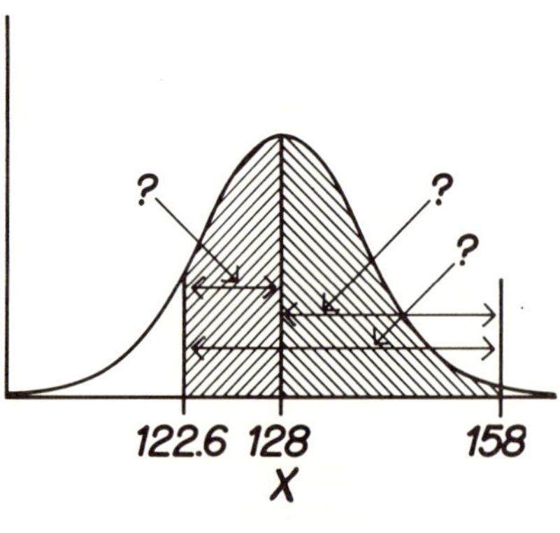

Fig. Ex. 5–3

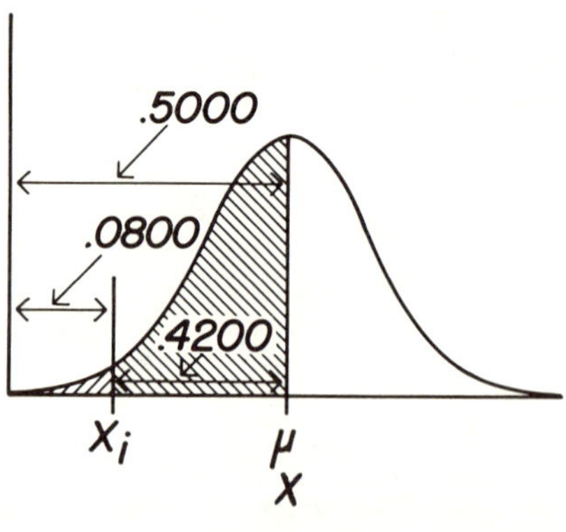

Fig. Ex. 5–4

Example 5–4. If a normal distribution has a standard deviation of 12 units and a score is given which exceeds 8 per cent of all other scores in the distribution, how many units above or below the mean is the given score?

The sketch made here indicates immediately that the score in question, whatever its value, must be below the mean of the set, otherwise it would exceed at least 50 per cent of the scores. Converting the 8-per cent figure to a proportion of .0800 and subtracting this from .5000 will give the proportion of the distribution between the score and the mean. The corresponding z-score is found to be about -1.41; the minus value is used since the score is known to be lower than the mean. In order to convert this z back into raw score terms, use (5–4.1), or (3–4.1), once again:

$$-1.41 = \frac{X_i - \mu}{12}$$

and

$$X_i - \mu = -16.92$$

or

$$-1.41 = \frac{x_i}{12}$$

$$x_i = -16.92$$

This indicates that the score is 16.92 units below the mean.

Note that each problem of the type just illustrated could be phrased in such a way that a probability statement is used. The translation of a proportion figure to a corresponding and equal probability value merely requires that random sampling be used.

In this sense, the problems just discussed may be considered as related to Question I. The assumed population is normal in shape with particular values for mean and σ given if needed. The question asked would be limited to random samples from the population in which the number of scores sampled is only one. The random sampling distribution of a single score is equal to the population distribution of scores, which should surprise no one.

For instance, using the data in Example 5–2, the probability of a randomly selected lemon drop having a weight less than 3.5 g. might be desired.

The appropriate $RSD(S)$ is the original population distribution, because only one score per random sample is used, i.e., $S = X_i$.

The use of the z-score transformation and the conversion of a proportion to a probability yields an answer of .0294 as the desired probability.

5–5. Unbiased Estimates. In order to proceed with normal-curve methods for answering basic statistical questions, the concept of an *unbiased estimate* must be discussed. Recall that a sample of scores may be used to guess at a population value of a statistic, S_*. Such an estimate is denoted as $\hat{S}_*$. The rule used to develop $\hat{S}_*$ need not be the same rule used in computing S_*; any method of obtaining the estimate of S_* is allowed. Naturally, some methods would be exceedingly poor, and others would lead to rather good estimates of S_*. We will not consider the general problem of finding the *best way of estimating* a given parameter, S_*, but we certainly would not wish to use a method of getting $\hat{S}_*$ which results in values which were consistently higher or consistently lower than S_*.

There are various ways of defining the word *good* in statistics. The following definition of an unbiased estimate implies that the use of the mean value of a distribution is somehow better than the use of the median or modal value. We shall find this assumption made in later material as well as in the present chapter. We will not discuss the topic in detail, but the reader may refer to Chapter 2 once again for review of the cases in which the various measures of central tendency are preferred to the others.

An unbiased estimate of S_* is any statistic ($\hat{S}_*$) for which the mean of the $RSD(\hat{S}_*)$ is equal in value to S_*. (5–5.1)

Definition (5–5.1) requires that the $RSD(\hat{S}_*)$ be examined for any method $\hat{S}_*$ of estimating the population value S_*. If the mean of the $RSD(\hat{S}_*)$, which is the average of all estimates, is equal to S_*, then $\hat{S}_*$ may be said to be an unbiased estimate of S_*. In general, a method of estimating $\hat{S}_*$ which is unbiased for one size sample will also be an unbiased method for another size sample. Although the above definition does not specify that this is the case, all of the unbiased estimates considered in this text will apply for any size sample unless otherwise noted.

The use of an unbiased estimate for answering the Question II problems assures that the average estimate made in repeated applications of the same type will be equal to the population value being estimated. It is certainly possible that the $RSD(\hat{S}_*)$ might be a skewed distribution, in which case the median estimate would not be equal to the mean estimate. The choice of the mean estimate's being equal to the population parameter, S_*, is not necessarily the best criterion, for all purposes, by which to define an unbiased estimate. However, it is the most common criterion used in statistical procedures in the behavioral sciences.

The method of developing and testing an unbiased estimate need not concern us in general, but two important examples of unbiased estimates will be considered.

5–6. Estimating a Population Mean. It is important to notice in the remaining theorems of this chapter just what types of assumptions about the population sets must be made in order that the theorems apply.

Theorem 5–2. If N scores are selected as a random sample from any population with a finite mean, μ, then the sample mean, M, will be an unbiased estimate of μ.

A formal proof of this theorem requires a few theorems which have been omitted from the present treatment of the subject. However, if one will accept the fact that the mean of an infinite number of scores can be described with the concept of a limit, as the notion of probability has been described, then Theorem 5–2 will appear reasonable.

For instance, consider the average of n random scores. As n approaches infinity, the average value will certainly approach the population mean. Since a $RSD(M)$, no matter what the sample size may be, contains an infinite number of samples, it must also contain an infinite number of scores. The mean of the $RSD(M)$ will therefore equal the population mean μ.

The result of Theorem 5–2 may be written in notation form as

$$\hat{\mu} = M \textit{ is an unbiased estimate of } \mu \tag{5–6.1}$$

The result of Theorem 5–2 should not be generalized to all other statistics in the sense that a sample value of a statistic is used as an unbiased estimate of the corresponding population statistic. As will be seen, sometimes the use of a sample value of a statistic as an estimate would give an average estimate over the long run which was not equal to the population value.

An explicit case in which the sample value of a statistic may be said to be *biased*, as an estimator, is the s^2, or s.

Notation. In later work it will be necessary to use special notation to indicate clearly what calculations are intended. A symbol will always indicate the type of statistic computed and a subscript, usually a symbol as well, will indicate what each number in the set used represents. For example, the notations

$$s_M,\ \sigma_M,\ M_s,\ \mu_M, \text{ and } M_{SS}$$

are simple combinations of symbols already encountered. The first of these, s_M, represents a standard deviation calculated on a set of numbers. The subscript, M, indicates that each number used is a mean value of other numbers. The computation of the s is done as if each mean were simply a raw score, the notation given is used to remind the reader that each score is actually a mean. This type of notation is used when dealing with random sampling distributions. The symbol σ_M would be used to indicate the standard deviation of the $RSD(M)$. This particular statistic has a special name because of its importance to later work. It is called the **standard error of the mean**, (σ_M). Keep in mind that it is still a standard deviation but has been computed on a special set of scores, each of which is a sample mean.

The next example, M_s, would be used to indicate an average based on a set of numbers, where each number is itself a standard deviation. For instance, if three samples of 10 scores each were found to have s's of 18, 21, and 15, respectively, then the value of M_s would be $\frac{54}{3}$ or 18.

The fourth example has already been encountered and represents the mean of a RSD of means, μ_M. The last example, M_{SS}, is read as the mean of a set of scores, each of which is a SS. When one of the above entities is used with reference to a random sampling distribution, it should be clear that a population value is implied since an infinite number of samples is involved in such distributions.

5–7. Estimating the Variance and Standard Deviation. The following derivations are given in order to illustrate that the sample s^2 is actually (on the average) an underestimate of a population variance, σ^2, and to develop an unbiased estimate of the σ^2 which may be used under appropriate conditions.

Let us start by defining the following terms which refer to a *single* sample of N scores, drawn from some population:

$$x_i' = X_i - \mu \qquad x_i = X_i - M \qquad E = M - \mu \tag{5–7.1}$$

$$SS' = \sum_{i=1}^{N} x_i'^2$$

$$SS = \sum_{i=1}^{N} x_i^2$$

These definitions permit the manipulations

$$SS' = \sum_{i=1}^{N} (X_i - \mu)^2 \tag{5–7.2}$$

$$= \sum_{i=1}^{N} (X_i - M + M - \mu)^2 = \sum_{i=1}^{N} (x_i + E)^2$$

$$= \sum_{i=1}^{N} x_i^2 + 2E \sum_{i=1}^{N} x_i + NE^2 \qquad (E \text{ is a constant})$$

$$= SS + NE^2 \qquad (\text{because } \Sigma\, x_i = 0) \tag{5–7.3}$$

Equation (5–7.3), if divided by N, will yield an average squared deviation from the population mean, μ, on the left side of the equal sign. Let us denote this as σ'^2, recalling that there would be one such quantity for each sample drawn. The right side of the equation, when divided by N, will yield the s^2 for this sample and an E^2 term for this sample. Hence (5–7.3) becomes

$$\sigma'^2 = s^2 + E^2 \tag{5–7.4}$$

Now consider that repeated random samples are drawn. Equation (5–7.4) values may be summed from each of K samples, and division by the number of samples will yield average sample values for the three quantities shown in the equation.

Since the quantity σ'^2 is based upon squared population deviation scores, we may consider that the population consists of squared deviation scores as well and apply Theorem 5–2. The average over many samples would approach the population average, which is by definition σ^2 (because the variance is the average squared deviation). Thus, as sampling continues,

$$\sigma^2 = M_{s^2} + M_{E^2} \tag{5–7.5}$$

Equation (5–7.5) may be modified still further by noting that the average E^2 term is actually equal to the $\sigma_M{}^2$, i.e., the *standard error of the mean* squared. The reasoning behind this relationship will be left as an exercise at the end of the chapter. The (5–7.1) definitions, the concept of a $RSD(M)$, and Theorem 5–2 will permit a reasonable proof to be accomplished. This would give

$$\sigma^2 = M_{s^2} + \sigma_M{}^2 \tag{5–7.6}$$

This equation clearly shows that the average sample variance is equal to the population variance only if the $\sigma_M{}^2$ is equal to zero. Since $\sigma_M{}^2$ is a variance, it cannot be negative and will only be 0 in the trivial case when σ^2 is also 0. This shows that the use of the sample value of the variance will result, on the average, in an underestimation of the population value.

In order to develop an unbiased estimate of σ^2, we may work with (5–7.6). We wish to modify the term $\sigma_M{}^2$ in order to obtain an expression which uses only data found in a single sample. There are two conditions under which a simple and useful solution is known. The first of these is when the samples are large enough to allow use of the central limit Theorem (5–1), which gives the $\sigma_M{}^2$ as equal to σ^2/N. The second case is more useful for statistical applications, because samples are rarely exceedingly large in practice. This condition is offered in the form of a theorem without proof.

Theorem 5–3. If a population of scores with mean μ and variance σ^2 has a normal distribution, then the $RSD(M)$ from this distribution, based on samples of size N, will be a normal distribution with

$$\mu_M = \mu \quad \text{and} \quad \sigma_M{}^2 = \sigma^2/N$$

The similarities and differences between Theorems 5–3 and 5–1 should be noted. The results of each theorem are the same: a normal distribution of sampling means with specified values for the mean of the means and the variance of the distribution. The difference is that Theorem 5–3 requires that the population of raw scores be normally distributed, whereas Theorem 5–1 makes no such assumptions. However, Theorem 5–3, while requiring a normal distribution of raw scores, makes no restriction on the sample size needed in order to apply the results.

When a population distribution can be considered normal, a number of statistical procedures become available, even when small-sized samples are used.

Let us apply the results of Theorems 5–3 and 5–1 to (5–7.6), using the fact that, under the conditions of the theorems, $\sigma_M{}^2 = \sigma^2/N$:

$$\sigma^2 = M_{s^2} + \sigma_M{}^2 = M_{s^2} + \frac{\sigma^2}{N}$$

and

$$M_{s^2} = \sigma^2 - \frac{\sigma^2}{N} = \sigma^2\left(1 - \frac{1}{N}\right) = \sigma^2\left(\frac{N}{N} - \frac{1}{N}\right)$$
$$= \sigma^2\left(\frac{N-1}{N}\right)$$

which gives

$$\sigma^2 = \left(\frac{N}{N-1}\right) M_{s^2} \qquad (5\text{–}7.7)$$

Equation (5–7.7) shows that the sample variance, when multiplied by $N/(N - 1)$, will serve as an unbiased estimate of the population variance. This is because the average of this estimator (with the constant applied) is equal to the population value as definition (5–5.1) requires.

The relationship shown in (5–7.7) actually holds for any size sample even if the population is not normal; however, this stronger statement will not be needed in later work.

The unbiased estimate of the population variance may be written in other forms, if a single sample variance and the correction factor are examined together. Hence

$$\hat{\sigma}^2 = \left(\frac{N}{N-1}\right)s^2 = \left(\frac{N}{N-1}\right)\frac{SS}{N} = \frac{SS}{N-1} \qquad (5\text{–}7.8)$$

and, approximately,

$$\hat{\sigma} = \sqrt{\frac{SS}{N-1}} \qquad (5\text{–}7.9)$$

Equation (5–7.8) is strictly true for random samples of any size drawn from any population with finite mean and variance. However, (5–7.9) is only approximately true and will only be employed when the population is normally distributed. Even then a small bias is present in the use of (5–7.9). The bias results from an elementary mathematical relationship. Given that (5–7.8) is true, one may see that $M_{\hat{\sigma}^2}$ is equal to σ^2, by the definition of an unbiased estimator. In order for (5–7.9) to be true as well, $M_{\hat{\sigma}}$ would have to equal σ. This would require

$$M_{\hat{\sigma}} = \sqrt{\sigma^2} = M_{\sqrt{\hat{\sigma}^2}} = \sqrt{M_{\hat{\sigma}^2}} \qquad (?)$$

but, letting $\hat{\sigma}_i^2 = Y_i$, it is clear that, generally

$$\sqrt{\text{mean } Y_i} \neq \text{mean } \sqrt{Y_i}$$

The appropriate corrections in Equation (5–7.9), when the population is normal, are so minor that they are usually ignored.

The fact that the average sample variance is less than the population variance, (5–7.6), arises from the characteristic of a sample variance being computed with a sample mean instead of the "true" population mean. It can be shown that the SS for any set of scores is less than the sum of squared deviations of the scores from any point other than this sample mean. In effect, this is what is shown in (5–7.1) through (5–7.6). One may say that the size of the squared deviation for any one score depends in part upon the other scores in the set chosen, because the sample mean depends upon the total of all N scores.

The concept of degrees of freedom may be recalled from Chapter 2. The degrees of freedom measure, denoted as df, for a sample variance or sum of squares, was defined as $N - 1$ in that chapter. It may be seen that (5–7.8) and (5–7.9) divide the SS by a quantity equal to the df for the set of scores used. This is an important relationship. In the next few chapters, we shall observe that the use of the concept of degrees of freedom will allow us to obtain unbiased estimates of the population variance, whereas the use of the sample size, N, would not.

It might be added in passing that some statistics textbooks define the standard deviation, computed from a sample of scores, according to (5–7.9). This technique creates a number of difficulties, including more complex raw-score equations; moreover, the value of a population standard deviation, as defined in this text, must then be computed in a different fashion. It may also be recalled that (5–7.9) does not even yield an exact unbiased estimate of σ. It is suggested that a sample "standard deviation" computed as shown in (5–7.9) is better thought of as an estimate of the population standard deviation than as a sample s.

5–8. Range Estimates of the Population Mean. The use of Theorem 5–2 will allow one to answer a Question II problem with a single value as an estimate of the population mean. The sample mean, M, is the value to use. However, it should be clear that the mean of one particular random sample is not very likely to be exactly equal to μ. Some estimate is desirable of the expected magnitude of error in our prediction. The $\hat{\mu}$ value may be called a *point estimate* of μ, and usually has a high probability of being inexact. One method of indicating the expected degree of error in our prediction indirectly is to make an estimate which is a range of values. Such an estimate is called a *range estimate*.

For instance, if a sample mean turns out to be equal to 47.986 ft., then the point estimate of μ would be that value. But the range estimate might be that μ is somewhere between 42.986 and 52.986 ft.

It is desirable that a range estimate contain the unbiased point estimate at the center of the range unless peculiarities of the related $RSD(M)$ are known in advance. Let us restrict attention to cases in which a range estimate contains the $\hat{\mu}$ at the midpoint of the interval (range).

In order to have some basis upon which to decide the width of the range estimate, we shall establish some degree of success which we wish to attain in making range estimates. As can be seen, if the chosen range is large, as opposed to small, we will be correct in our range estimate more often; i.e., the population mean will actually be more often within the range estimated. On the other hand, if we make the desired probability of successful prediction too high, then the range estimated for the μ would be so large as to be relatively useless. The latter case would be something like allowing the weight-guesser at a carnival to be as much as 50 lb. off in his guess; one would hardly need look at a person to succeed. In the range-estimate problem, a high probability of success can be assured if one selects a very large range of values, but such a range might be constructed without even drawing a sample.

A standard probability of success has been adopted by behavioral scientists for this type of problem. There is no magic rationale behind the value chosen, it is only a convention that has developed through the years. Let us remain in step with this tradition and chose the desired probability of success as no lower than .95. We wish to develop a method of estimating the population mean with a range of values such that we will be right 95 per cent of the time (on the average). For certain purposes one might wish to be right 99 per cent of the time, in which case a slightly larger range will result. Both of these two values will be found in use in applications to research in the behavioral sciences.

Range estimates with σ known. The method of selecting a range estimate that will contain the μ with a probability of .95 requires that certain assumptions be made about the problem:

1. Assume that the population is normally distributed or (for a good approximation) that the sample size is large enough to apply the central limit theorem.
2. Assume that the population value of the standard deviation is known.

Under these two conditions, the $RSD(M)$ will be a normal distribution with

$$\sigma_M = \frac{\sigma}{\sqrt{N}} \qquad \text{and} \qquad \mu_M = \mu \tag{5–8.1}$$

In which case the desired range estimate may be written as

$$\hat{\mu} - 1.96\,\sigma_M \leq \mu \leq \hat{\mu} + 1.96\,\sigma_M \tag{5–8.2}$$

or in a slightly different form, the estimate of

$$\mu = \hat{\mu} \pm 1.96\,\sigma_M = M \pm 1.96\,\sigma_M \tag{5–8.3}$$

Before showing that (5–8.2) or (5–8.3) will satisfy our demands on the range estimate, the number 1.96 should be identified. This number is actually a z-score on a normal curve such that .4750 of the cases are between the mean and the z-score. Since we use this value both as a positive and negative number, the proportion of scores on any normal curve between -1.96 and $+1.96$ z-score units will be .95.

For an illustration of the reasoning behind (5–8.2) and (5–8.3), examine Fig. 5–4. This figure represents the $RSD(M)$ which would result when the conditions for the above equations are satisfied. Notice that the μ_M is equal to μ, the σ_M is equal to $\sigma/\sqrt{N}$ and that the distribution is normal. The x-axis has been scaled in z-score (or standard deviation) units.

The two points of interest in Fig. 5–4 are those labelled -1.96 and 1.96. These represent mean values (in raw score units) which are 1.96 times the size of the σ_M away from the mean of the means.

If any single sample of scores is found to have a mean value between 1.96 and -1.96 z-score units on Fig. 5–4, then the addition and subtraction of 1.96 units to the sample mean will clearly form a range including μ_M; but $\mu_M = \mu$, so that the range would include μ as well.

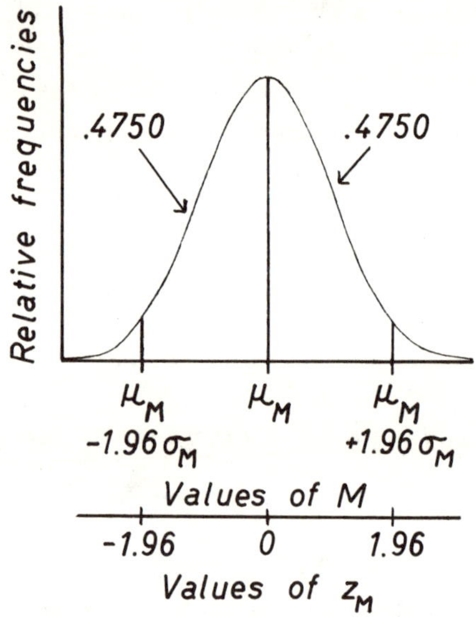

Fig. 5–4. The random sampling distribution of means in terms ot standard error of the mean units and raw score units. Of all means, .95 have values between —1.96 and 1.96 z_M units, where $z_M = (M - \mu_M)/\sigma_M$.

It follows that the probability of success in estimating the μ, using (5–8.2) and (5–8.3), will be equal to the probability of selecting a sample of scores having a mean between the two specified z-score values. Since the distribution of means is normal and sampling is random, the 95 per cent of the scores between -1.96 and $+1.96$ z-score units is converted to the desired probability value of .95.

As a more concrete illustration, examine Fig. 5–5. This figure is a $RSD(M)$ which would result from a normal population of I.Q. scores (as measured by an intelligence test) in which the mean equals 100 and the σ equals 15 points, assuming that the sample size used is 25.

The mean of the means is equal to 100 (μ) and the σ_M, which is the standard deviation of the $RSD(M)$, is equal to 3.00. This value for σ_M was found by use of (5–8.1):

$$\sigma_M = \frac{\sigma}{\sqrt{N}} = \frac{15}{\sqrt{25}} = 3.00$$

Now assume that a sample of 25 scores from this population has a mean of 96 points. Will the use of (5–8.2) or (5–8.3) give a range estimate which contains the population mean? The application of

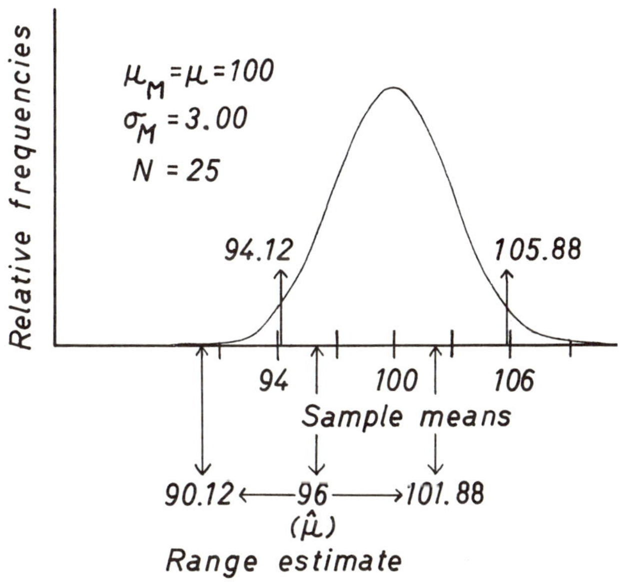

Fig. 5–5. The random sampling distribution of means illustrating the case of estimating the population mean from a single sample mean of 96, assuming knowledge of the standard error of the mean.

(5–8.2) gives

$$96 - 1.96(3.00) \leq \mu \leq 96 + 1.96(3.00)$$

or

$$90.12 \leq \mu \leq 101.88$$

which is a true statement.

The value of (1.96)(3.00) is 5.88. It will be seen that any mean that might have been selected would have produced the same sort of true range estimate just so long as that mean was not further from the μ_M than 5.88 points. The interval 94.12 to 105.88, the limits of which are shown on Fig. 5–5, is such that any mean in the interval will lead to a correct prediction of μ when our range estimate is applied. The probability of selecting such a mean value is equal to the proportion of the means within the interval, which is easily found to be .95.

The use of σ_M in estimating μ is what has led to the name standard error of the mean for σ_M.

When equations (5–8.2) and (5–8.3) are used to estimate a population mean, the resulting range estimate is also called a **confidence interval** for estimating μ. This is because each range estimate developed in the above fashion has a certain probability of success, and therefore, we have a certain degree of confidence in the estimate.

The term *confidence interval* is more common than range estimate and also has the connotation of the probability statement assigned to such estimation. We will use the term *confidence interval* to refer to any range estimate for which a probability of success has been calculated. The estimate given in (5–8.2) and (5–8.3) would be called a *95 per cent confidence interval* for estimating μ.

Confidence intervals with different degrees of confidence related to them may be constructed from (5–8.2) and (5–8.3) by replacing the z-score value of 1.96 by some other appropriate value. The value 2.58 may be used to obtain the *99 per cent confidence interval.*

Range estimates with σ not known. The foregoing method for computing a confidence interval for the mean of the population required that the value of σ be known so that σ_M could be calculated. However, suppose σ is not known. Is there any way of obtaining a confidence interval for the true mean of the population from just the data in a single random sample? The answer is yes; *but only when the shape of the population distribution is known.*

It might appear that the σ^2 might be estimated from the sample whenever (5–7.8) and (5–7.9) apply. This is true; however, we must also estimate μ from the same sample.

It may happen that values of s^2 and M in single samples are related to one another when sampled from populations of certain types. For example, samples with large means might tend to have large standard deviations. If this were to be the case, a rather precise description of this tendency would be needed in order to produce a confidence interval with a known probability of success.

The only type of population distribution which does not lead to the above relationship between sample M and s^2 values is the normal distribution.

When the population is normally distributed, Theorem 5–3 and

(5–7.9) apply, *and* the above problem does not occur. In this case, the estimation of σ leads directly to a good estimation of σ_M and, with additional labor, to the desired confidence interval for estimating μ:

$$\hat{\sigma}_M = \frac{\hat{\sigma}}{\sqrt{N}} = \frac{\sqrt{SS/(N-1)}}{\sqrt{N}} = \sqrt{\frac{SS}{N(N-1)}} \qquad (5\text{–}8.4)$$

$$\hat{\sigma}_M = \frac{s}{\sqrt{N-1}} \qquad (5\text{–}8.5)$$

It may be seen that the standard error of the mean may be estimated by *estimating* the value of σ and dividing by $\sqrt{N}$ (5–8.4) or by dividing the sample standard deviation by $\sqrt{N-1}$ (5–8.5).

However, this estimate, $\hat{\sigma}_M$, may not be used in exactly the same way as σ_M. The earlier confidence interval when σ_M was known was based upon the formation of a z-score value of M on a normal curve. The deviation of any random sampling mean from μ_M could have been expressed in z-score form as

$$z_M = \frac{M - \mu_M}{\sigma_M} = \frac{M - \mu}{\sigma_M} \qquad (5\text{–}8.6)$$

This is illustrated in Fig. 5–4, and one may see that these z_M values are normally distributed, since the M's are normally distributed.

The present case employs $\hat{\sigma}_M$ (which varies from sample to sample) rather than σ_M, and the analogous equation may be written as

$$t = \frac{M - \mu}{\hat{\sigma}_M} \qquad (5\text{–}8.7)$$

The t-distribution. The random sampling distribution of the values of t is not normal for small-sized samples, even though the population distribution is normal.

Figure 5–6 gives a comparison between the distribution of z as defined in (5–8.6) and the distribution of t as defined in (5–8.7). Both of the distributions shown are appropriate for the case in which the population is normal and the sample size is 6; df is then 5. As noted above, the z-distribution is normal but the t-distribution is not, unless the sample size is rather large.

The statistic t, defined in (5–8.7), is known as "Student's t," because its random sampling distribution was first used and published by W. S. Gosset under the psuedonym "Student."

The $RSD(t)$ is slightly different for each sample size (and related df) which might be considered. Because of this, a separate table of t, similar to Table C for the normal curve would have to be given for each sample size studied. Rather than doing this, a table may be given which gives only partial information about each such $RSD(t)$. Table D in the Appendix is such a table.

Table D gives a number of values of t for a number of different degrees of freedom. One of these, referred to as the .05 value, gives the t which will contain .95 of the $RSD(t)$ between its positive and negative values. For example, for samples of size 6, df is equal to 5 and the

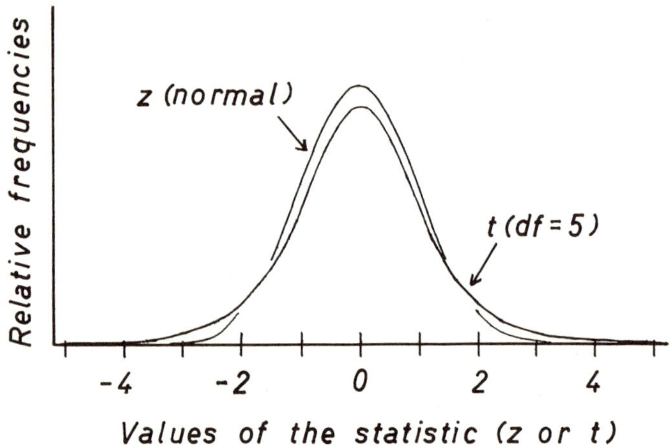

Fig. 5–6. A comparison of the distribution of t (for $df = 5$) with the standard form (z-score) of the normal distribution.

corresponding .05 value of t is read to be 2.571. This means that .95 of the random sampling values of t have values between -2.571 and $+2.571$. Conversely, only .05 of the t's fall outside this range, hence the term .05 value of t. A second tabled value of t, the .01 value, gives the same sort of information, but the corresponding range includes .99 of the t's and excludes .01 of them.

The reader should notice the correspondence between these values of t (2.571 and 4.032) and the normal curve z-scores of 1.96 and 2.58 which have been used to give similar ranges on the normal distribution.

It is now possible to form the confidence interval for the μ when σ is not known.

Theorem 5–4. If a population of scores is normally distributed, the 95 per cent confidence interval for estimating μ from a random sample of N scores is

$$\mu = \hat{\mu} \pm t_{.05}\,\hat{\sigma}_M$$

where $t_{.05}$ is the .05 value of t with $N - 1$ df, and

$$\hat{\sigma}_M = \frac{s}{\sqrt{N - 1}}$$

The 99 per cent confidence interval can be estimated by simply substituting the .01 value of t in the theorem.

The reason that the values of t in Theorem 5–4 are called the .05 and .01 values rather than the .95 and .99 values is that later work will emphasize the proportion of cases in the tails of $RSD(t)$ rather than the proportion closer to the mean.

Example 5–5. If the reaction times for different people are known to be normally distributed with $\sigma = .60$ sec., and a random sample of 100 such scores is found to have a mean of 1.30 sec., what is the 95 per cent confidence estimate of μ?

Since the population is normal and the σ is known, we may use (5–8.2). This gives

$$1.30 - 1.96 \frac{.60}{\sqrt{100}} \leq \mu \leq 1.30 + 1.96 \frac{.60}{\sqrt{100}}$$

or

$$1.1824 \text{ sec.} \leq \mu \leq 1.4176 \text{ sec.}$$

It should be noted that the sample size in Example 5–5 is large enough to allow the same computations to be reasonably valid even if the population were not normally distributed. In Example 5–6, the population must be normally distributed no matter how large the sample size might be.

Example 5–6. If reaction times of persons are known to be normally distributed and the following are such scores from 10 randomly selected people, what is the 99 per cent confidence interval for μ?

1.10 sec
1.35
1.20
1.15
1.28
1.19
0.99
1.31
1.18
1.16

The calculation of the mean and s for this sample may proceed as in earlier work.

$$\Sigma X_i = 11.91 \text{ sec.} \qquad \text{and} \qquad M = 1.191$$

Since the mean is not a convenient value, the raw score formulas for the SS and s are easiest to use:

$$SS = \Sigma X_i^2 - \frac{T^2}{N} = 14.2837 - 14.1848 = .0989$$

$$s = \sqrt{\frac{SS}{N}} = \sqrt{.00989} = .099448$$

The standard error of the mean may now be estimated by (5–8.5) or Theorem 5–4:

$$\hat{\sigma}_M = \frac{s}{\sqrt{N - 1}} = \frac{.099448}{3} = .0331$$

Since the population is known to be normal, the desired interval may be obtained by use of Theorem 5–4 with the .01 value of t used instead of the .05 value. The desired t for 9 degrees of freedom is found to be 3.250 from Table D, hence the 99 per cent confidence interval for μ is

$$1.191 - .1076 \leq \mu \leq 1.191 + .1076$$

or approximately

$$1.08 \text{ sec.} \leq \mu \leq 1.30 \text{ sec.}$$

Examination of (5–8.5) will show the important fact that the size of the confidence interval for estimating μ depends upon N, the sample size, for any given population. The larger the sample size, the smaller the estimated range for μ.

5–9. Testing Hypotheses about a Population Mean. The problems so far have illustrated answers to Questions I and II of our general types. With the information in this chapter, we may also consider problems of the type of Question III.

In such cases, certain assumptions are made about the population. Here let us assume the distribution of scores to be normal. If we wish to test the hypothesis that μ is equal to some particular value, we will further assume that the value of μ is in fact equal to the one hypothesized. A single random sample value of M will now be examined. The probability of obtaining that sample M (or one more unexpected than M) is now calculated with the assumptions made. If the answer turns out to be a low probability, we shall proceed to reject the assumed value for μ.

Let us first consider the possible outcomes of such an experiment when all of our hypotheses are actually true. The particular statistic to be computed from our one sample will be t, as defined in (5–8.7), with the assumed value of μ used. The $RSD(t)$, as illustrated in Fig. 5–6, is symmetrical with a mean of 0. The most likely values of t are relatively close to zero. Such a situation is expected when the value of μ is as specified. Let us refer to the hypothesis that μ is a particular value as H_0. In many cases, such an hypothesis will be called a **null hypothesis.** This may be verbally stated as the assumption that there is no difference between the population parameter and the value under test.

Now consider what might occur if H_0 is not true. This condition consists of many possible values for μ, all of which are different from those specified by H_0. Let us designate the class of such possibilities H_1. H_1 may be said to be true if H_0 is false and H_0 is true if H_1 is false.

Examine one possible value for μ which is different than the assumed value. This means that H_0 is false. The $RSD(t)$ which would now result may be considered by means of a linear transformation. Let us use the notation μ_0 for the hypothesized (but incorrect) value of μ, and the notation μ for the true population mean. The formation of the sample value (t') for each sample is

$$t' = \frac{M - \mu_0}{\hat{\sigma}_M} \tag{5–9.1}$$

But μ_0 and μ are fixed values so

$$\mu_0 = \mu - c$$

and

$$t' = \frac{M - \mu + c}{\hat{\sigma}_M} = t + \frac{c}{\hat{\sigma}_M} \tag{5–9.2}$$

Although $c/\hat{\sigma}_M$ will vary for different samples, if c is positive, μ_0 is smaller than μ and each random sample value of t is replaced by a somewhat larger value t'.

This analysis does not give the precise $RSD(t')$ but shows that such a distribution will have values higher than t when μ_0 is smaller than μ, and values of t' smaller than t when μ_0 is larger than μ.

We may now state that the sample value of t will tend to differ from zero when H_1 is true and H_0 is false.

The region of rejection. Before selecting a set of sample scores and testing some H_0, we will decide which sample values of the statistic (t, in this case) constitute reasonable evidence for the rejection of H_0. A set of sample values which will cause the rejection of H_0 is called the *region of rejection.* All other sample values of the statistic will not cause rejection of H_0. Sometimes the set of values not in the region of rejection is called the region of acceptance; however, as will be seen shortly, this terminology is misleading and will not be used in this text.

By convention, *we define the region of rejection as being those sample values of a statistic which comprise the most unexpected 5 per cent of the values, assuming H_0 to be true.* An appropriate statistical test is one which increases the probability of rejection when H_1 is true.

In the present case, the most unexpected values of t are the high and low (negative) ones. We will choose the highest positive values of t which occur .025 of the time and the lowest negative values which occur .025 of the time. This is the region of rejection and is illustrated in Fig. 5–7. It may be seen that the total region of rejection, formed from the two separate parts, contains the 5 per cent of the t's which are most different from zero (the expected average value). The table of t (Table D) will give the desired value for a particular sample size.

The use of this rejection region is referred to as the *5 per cent level of confidence* for testing an hypothesis.

If the sample value of t from a given random sample of scores falls into the region of rejection, we will reject the hypothesis that H_0 is true, which means that H_1 is accepted.

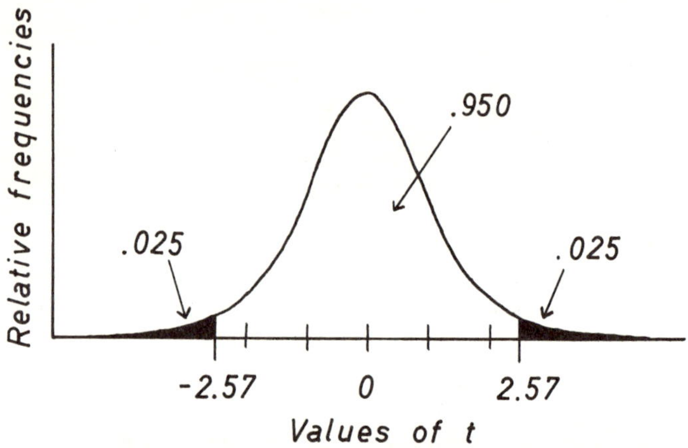

Fig. 5–7. The random sampling distribution of t, for 5 degrees of freedom, showing the region of rejection equivalent to the 5% level of confidence for a statistical test.

If the sample value of t does not fall in the region of rejection, we do not reject H_0. However, one must be careful not to claim that H_0 has been shown to be true. We have a case with insufficient evidence to reject H_0, but H_1 might still be true and yield a sample t well within the acceptable range. When μ is close to the hypothesized value, but still different, the sample values of t will not be far different from when H_0 is actually true. Examination of (5–9.2) for small c will show this clearly.

X	f	fX	fX^2
18	2	36	648
19	2	38	722
20	7	140	2,800
21	9	189	3,969
22	12	264	5,808
23	9	207	4,761
24	10	240	5,760
25	8	200	5,000
26	4	104	2,704
27	0	0	0
28	2	56	1,568
	65	1,474	33,740

Example 5–7. Assume that the following set of 65 scores has been randomly drawn from a normal population of scores. Test the hypothesis that μ is equal to 24.0.

The sample scores are represented in a frequency distribution in the first two columns. The df for this sample is $N - 1 = 64$. Consulting Table D, we find the closest entry as $df = 60$. The corresponding t at the .05 level is seen to be 2.00. Any sample t equal to 2.00 or greater (or equal to -2.00 or less) will be in the region of rejection.

The formation of the t for this sample requires the specification of μ_0, the hypothesized population mean, and the calculation of M and $\hat{\sigma}_M$.

The use of ΣfX and Σf yield the sample mean:

$$M = \frac{\Sigma fX}{N} = \frac{1{,}474}{65} = 22.677$$

The raw score formula for the variance may now be used:

$$s^2 = \frac{\Sigma X^2}{N} - M^2 = \frac{\Sigma fX^2}{N} - M^2 = \frac{33{,}740}{65} - (22.677)^2$$
$$= 519.077 - 514.246 = 4.831$$

which yields

$$s = 2.198$$

Application of (5–8.5) will give the estimated standard error of the mean

$$\hat{\sigma}_M = \frac{s}{\sqrt{N-1}} = \frac{2.198}{8} = .275$$

The formation of the sample value of t, as in (5–9.1), gives

$$t = \frac{M - \mu_0}{\hat{\sigma}_M} = \frac{22.677 - 24.000}{.275} = \frac{-1.323}{.275}$$
$$t = -4.81$$

This sample value of t is well into the region of rejection at the negative end of the expected distribution of t.

We may therefore reject the hypothesis, H_0, that the population mean is equal to 24.00. In fact a t of the magnitude obtained would occur less than one time in 100,000 samples if our assumptions were true.

The acceptance of H_1 is not very precise. Concluding that the population mean is not 24.00 is similar to stating that the average income in a community is not $5,500. One rather wonders what the population mean might be!

The 95 per cent confidence estimate of μ is easily accomplished now that we have $\hat{\sigma}_M$, $t_{.05}$, and M. From Theorem 5–4,

$$\mu = \hat{\mu} \pm t_{.05}\hat{\sigma}_M = 22.677 \pm (2.00)(.275)$$

or

$$22.127 \leq \mu \leq 23.227$$

The test of an hypothesis often is closely related to the estimation of a confidence interval as shown in this example.

A similar procedure is used to test H_0 when σ is known rather than estimated. In this case, the test statistic is a z-score on the expected *RSD* of sample means. The region of rejection (at the .05 level of confidence) is defined as in the case of t, but the *critical value* of 1.96 for the sample value of z does not change for different size samples. Hence if

$$z_M = \frac{M - \mu_0}{\sigma_M} \tag{5-9.3}$$

has a value of 1.96 or greater (in absolute value), H_0 may be rejected.

Errors in hypothesis testing. The procedures just illustrated for testing an hypothesis about the population lead to a decision about H_0. This decision may be correct or incorrect. Table 5–1 illustrates four possible logical outcomes of the hypothesis testing procedure. Notice that the truth of the situation corresponds to the actual population value. This is either as hypothesized, or it is not. The two decisions possible are to reject H_0 or not to reject H_0. It may be seen that two of these four combinations are given positive evaluations, which simply means that a correct decision has been made.

TABLE 5–1

Possible Results of Two Decisions Made in Hypothesis Testing

		Decisions	
		Reject H_0	Fail to reject H_0
Truth in Population	H_0	− − Type I error	+ Little information. No error
	H_1	+ + Correct rejection	− Type II error

However, Table 5–1 indicates two types of errors may result from our decision process. The first of these, called a type I error, results from rejecting H_0 when it is actually true. The probability of making this type of error, when H_0 is true, is equal to the level of confidence related to the size of the region of rejection. In the case of the .05 (5 per cent) level of confidence, we would expect to make a type I error 5 per cent of the time that H_0 is actually true. In other words, a t value falling in the region of rejection (commonly called a **significant** value) does not guarantee that H_0 is false. We run a 5-per cent risk of making a type I error.

This analysis should not be interpreted to the effect that 5 per cent of all rejected hypotheses are mistakes. The .05 value only applies to cases where H_0 is actually true. It is suggested that most research applications test H_0 hypotheses which are actually false. The purpose of investigation in most cases, is to reject H_0. One merely demonstrates a difference by assuming no difference (H_0) and showing that this hypothesis should be rejected. This may seem to be a rather backward method of proving something, but man uses devious methods when they work.

The second type of error, called a type II error, arises when H_1 is true but our decision is not to reject H_0. If we are careful to refrain from the decision that H_0 is true and to simply withold judgment, then we may class this error as one of omission. It is not a serious error, in most cases, because we have not really decided that either H_0 or H_1 is actually true. Later work may demonstrate the truth. We might state that H_0 is innocent until proven guilty, beyond reasonable doubt.

5–10. Violations of Assumptions. The validity of the procedures illustrated in this chapter rests upon the truth of certain assumptions. Our procedures for estimating μ and testing H_0 often assume that the population is normally distributed. Moreover, the assumption of random sampling from the population is always made. If either of these is not true, our conclusions may be in error; at least the logic becomes suspect.

For instance, in the hypothesis-testing case, the finding of a significant value of t leads to the rejection of H_0. How can we be sure that this single assumption is wrong? Only if all other assumptions are correct and all computations correct, can we validly point to a false H_0 as the reason for an unreasonable sample value for our test statistic. Even then we run a certain risk of error. Fortunately, one can control this risk by setting a more stringent level of confidence when needed.

EXERCISES

Assume that all populations are normally distributed unless otherwise stated.

1. If $\sigma = 14$ and $\mu = 85$, what proportion of the population is found with scores between 70 and 78?
2. If $\mu = 60$ and $\sigma = 5$, what is the probability of selecting one score at random which is within 2 points of the mean?
3. If .20 of the scores are above a score of 28 and $\sigma = 6.2$, what is μ?
4. Given $\mu = 127$ and $\sigma = 16$, what is the proportion of scores between 118 and 139?
5. Find the 99 per cent confidence interval for estimating μ if a random sample of 50 scores has $M = 75$ and $s = 14$.

6. A population has $\mu = 50$ and $\sigma = 10$; if this population is linearly transformed with $a = 2$ and $b = -20$, what proportion of the transformed scores will be below a value which exceeds 90 per cent of the original scores?
7. Given $\sigma = 10$ and a random sample of 100 scores has a mean, M, of 26, can you reject the hypothesis that $\mu = 28$?
8. If a population has $\sigma = 15$ and $\mu = 100$, what is the probability that a single random sample with 25 scores will have a mean below 90?
9. A manufacturer knows that the number of employees absent from work on an average day is approximately normally distributed with $\mu = 28.5$ and $\sigma = 5$. If he finds that a six-day period during the world series shows a per day average of 33.20, should he conclude that these are not typical days?
10. If a random sample of 37 cases has a $SS = 29.6$, what is the unbiased point estimate of σ^2?
11. The following 10 scores are a random sample from a normal population. What is the 95 per cent confidence interval for μ?

    4, 8, 8, 9, 10, 12, 14, 14, 17, 18

12. Using the data from exercise 11, test the hypothesis that $\mu = 15.5$. Test the hypothesis that $\mu = 5.5$.
13. If a normal distribution has $\mu = 4.5$ and $\sigma = 2.872$, what proportion of sample means (with $N = 4$) would have values of 2.375 or less?

    Table 4–2 in the last chapter shows a $RSD(M)$ from a population with the above values but a non-normal distribution. Compare the proportions of sample means from Table 4–2 less than 2.375 and the expected proportion from a normal population as just computed. The real class limit should be considered when entering Table 4–2.
14. If E is defined as $M - \mu$, where M is a sample mean based on N scores, show that the average of E^2 for K samples is $\sigma_M{}^2$, as K becomes infinite.

6

Analysis of Variance: One Variable

6–1. The Experiment. The behavioral scientist usually thinks in terms of stimuli which impinge upon an organism and responses which that organism produces. The basic question asked in research, ignoring theory for the moment, is whether or not different stimulus conditions cause or are related to different responses by the organism.

Independent and dependent variables. In order to establish a relationship between stimulus conditions and the behavior (responses) of the organisms, at least two different stimulus conditions are used in an experiment. Responses of the subjects are then measured and examined for existing differences between the stimulus conditions.

The conditions which are varied in an experiment are called *independent variables.* This term suggests the usual requirement that any subject could be given any of the different stimulus conditions used.

For instance, if an investigator were interested in room temperature as a stimulus condition, he might decide to use three different temperature conditions. With appropriate equipment, he could assign any subject used in the experiment to any one of the three temperatures employed. Usually such assignment would be made at random (using a table of random numbers is a typical technique). The assignment of subjects to temperature *treatments* is independent of any characteristic of the subjects.

A single type of stimulus variation, such as temperature, is called the independent variable. The fact that a number of different temperatures are used may be described by using the term *levels* of the independent variable. The experimental conditions described above would then be indicated by saying that the experiment used one independent variable (temperature) with three levels (the different temperatures).

The type of independent variable just discussed is also called a *manipulated* independent variable, in the sense that the stimulus conditions are under the control of the investigator. Another type of independent variable is the *normative* independent variable. In this second type, subjects are *classified* on the basis of some characteristic of themselves or their normal environment or past history. A classic example is the gender of the subject. If differences between men and women in some type of behavior are being studied, it is clear that men are assigned to the men's group or condition on the basis of their being men, not on a purely random basis. The fact that a man is a man is not under the control of the investigator. The use of a normative variable in an experiment does not allow a cause and effect relationship to be established between the independent variable and the behavior being measured. However, the appropriate use of a manipulated independent variable allows one to assert that differences obtained under different levels of the variable have been caused by these different stimulus conditions, if the explanation of chance can be rejected. The rejection of the hypothesis that obtained differences were due to chance is one purpose for using statistical tests.

The behavior measured by the investigator is called the *dependent variable*. That is, measurements (scores) obtained from each subject depend upon his behavior in the situation. The measurement of behavior yields two basic types of results: a *score* which may vary from person to person; or a classification of the behavior into a category or class of behavior. For the moment, we will consider only the first of these sorts of dependent variables, and will term such scores **variates**.

In order to meaningfully use the analysis of variance techniques to be discussed later, the dependent variable must be measured in such a way that the scores (variates) constitute an equal-interval or ratio scale.

If the variates meet the above criterion, the analysis of variance tests will allow us to answer Questions II and III of our general types.

In summary, we may define an experiment as: *manipulating one or more independent variables, measuring one or more (usually one) dependent variables, while controlling all other factors.*

The methods of controlling all other factors are beyond the scope of this book, but it should be mentioned that exact equality between experimental conditions on all other factors not under study is *not* required. For instance, when subjects are randomly assigned to three temperature conditions, it may well be that one group of subjects has more ability to perform the experimental task or test than do subjects in some other group. The fact that subjects are randomly assigned to conditions permits the statistical tests, in a sense, to automatically allow for a certain amount of difference between conditions on the variate measured. When observed differences become large enough, the inference will be made that chance differences caused by random assignment or limitations of the measuring instrument are not a sufficient explanation for the results of the experiment.

Statistical versus experimental decisions. A c'ear distinction must be made between the results of a statistical test and the meaning of such results. For example, if one were to test a null hypothesis, H_0, and find that rejection of this hypothesis is in order, the decision that the null hypothesis is false is a *statistical decision.* To supply an explanation of why the hypothesis is false is an *experimental decision.* In the earlier illustration of an experiment using three temperature conditions, we may have required the subjects in each group (condition) to work 100 algebra problems. The dependent variable would probably be the number of correct problems for each person. Let us say that subjects working under the highest temperature condition obtained a mean number of solutions which was quite a bit less than the means of the other two groups. The analysis of variance technique might lead to the rejection of the hypothesis that scores in the three groups were drawn from populations of scores with equal means. This rejection of the null hypothesis of no difference between population means is a statistical decision.

Care must be taken in trying to give a reason why the null hypothesis is not true. The interpretation of the statistical decision always depends upon how well the experiment has been conducted. For instance, we might like to conclude that high temperature conditions reduce the efficiency of solving algebra problems of the type used. However, suppose that the equipment used to raise the temperature somehow lowered the room lighting. In such a case, the lower illumination may be the cause of the results, or the temperature may be the cause, or perhaps both factors had an influence on the results.

The statistical decision may allow the rejection of some hypothesis about the populations of scores, but the interpretation of the results (an experimental decision) requires careful examination of the way in which the experiment was conducted. The experimental decision will not be considered in a systematic fashion in this text, but the student should keep in mind that a statistical test "proves" something about populations of numbers of a certain type and does not directly answer the question of what may have caused obtained differences in the dependent variable measured.

The use of the statistical tests is meant to rule out the possibility, as conclusively as possible, that obtained differences between conditions of an experiment were due to chance factors. Only after this conclusion has been made, can we assume that some factor or factors caused differences in behavior and begin to identify the causes.

The effects of an independent variable. When a variate is used to measure some aspect of behavior, changes in the level of an independent variable may cause changes in the dependent measures in one or both of two major ways. Compared to the scores obtained from some standard level of the independent variable, the scores obtained under a different condition may be higher or lower with respect to the measures of central tendency and may be larger or smaller in their dispersion about the mean. Let us denote a person's expected score under a standard condition as X_A and his expected score under some different experimental condition as X_B. We may then relate these two scores by a linear transformation:

$$X_B = aX_A + b \tag{6–1.1}$$

When (6–1.1) applies for each person in a population of subjects, we find that the mean and variance are both changed when the independent variable is changed in value, i.e., when scores are taken under different levels of the independent variable. A statement of the constants a and b combined with (6–1.1) would then suffice as a description of the functional relationship between the independent and dependent variables. Although such a description would be very useful, we will place a further restriction on the effects of the independent variable. We will assume that (6–1.1) applies to our data and that the constant a is equal to unity.

6–2. The Additive Model. We have now a basis for the analysis-of-variance method of hypothesis testing in a number of research designs. The effects of changing the level of an independent variable upon the dependent variable are assumed to correspond to the formula

$$X_B = X_A + b + e \tag{6–2.1}$$

where X_B, X_A, and b are defined as in (6–1.1), and e is defined as a positive or negative number reflecting errors inherent in the measurement of the dependent variable. It must also be assumed that the size of e is not related to the independent variable.

Under these conditions, we may say that the effect of changes in the level of the independent variable is to cause changes in the mean of the dependent measure equal to the constant b. Note that the independent variable does not cause changes in the s or s^2, as shown in Chapter 3. Since the values of e are not related to the independent variable, they will produce no average change in the dependent variable.

It should be noted that (6–2.1) may not always apply to the effects of an independent variable on a dependent variable. In this case, the analysis of variance tests to be presented may still be used, but the interpretation of the results, even at the statistical decision level, will be quite difficult unless *group means* on the variate are the only matter of concern.

6–3. The F-Ratio. The test statistic used in the analysis of variance technique is called an *F-ratio* or simply F. The F-ratio will be defined here as

$$F = \frac{\hat{\sigma}_1^2}{\hat{\sigma}_2^2} \tag{6–3.1}$$

where $\hat{\sigma}_1^2$ and $\hat{\sigma}_2^2$ are *independent* unbiased estimates of population variance. The term independent, in this context, means that the value of one variance estimate does not affect the value of the other variance estimate. Another way of saying the same thing is that one variance estimate might be changed without changing the other variance estimate.

As an example, let us assume that two groups of scores are selected at random from a population of scores. Equation (5–7.8) indicates that an unbiased estimate of σ^2 may be obtained from each set of scores by dividing the respective SS by $(N - 1)$. The test statistic F is formed by dividing the first estimate by the second. The evaluation of the value of F requires the formulation of a hypothesis and knowledge of the random sampling distribution of F.

The $RSD(F)$. The random sampling distribution of F rests upon the assumption that each variance estimate is based upon populations with equal means and variances. This is a sort of null hypothesis. If the value of F is too high (or too low in some cases) we will reject part or all of this null hypothesis. For the moment, let us consider the $RSD(F)$ which results from repeated sampling values of F.

In the same way that the $RSD(t)$ depends upon the degrees of freedom in an estimate of the standard error of the mean, the distribution of F is different for different sample sizes used in estimating the population variance. In the case of the F-ratio, two variance estimates are employed, each of which has a certain number of degrees of freedom

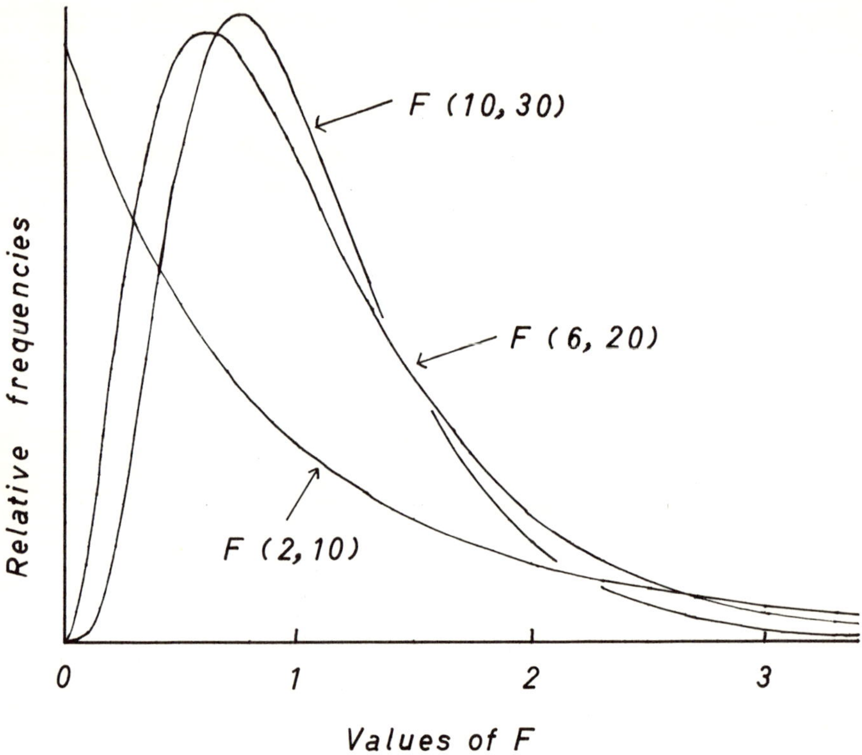

Fig. 6–1. The random sampling distributions of F for three different combinations of degrees of freedom. The curves are drawn to slightly different scale on the *y-axis*.

associated with it. Figure 6–1 shows the shapes of three of the many different possible distributions of F. Since there are so many possible distributions of F, one for each possible combination of degrees of freedom for variance estimates one and two, only limited information about any one distribution may be given in a table because of space limitations.

Table E in the Appendix gives values of F corresponding to the highest 5 per cent and 1 per cent of the $RSD(F)$ for each of many combinations of degrees of freedom. Care must be used in entering Table E. Two separate degrees of freedom values are associated with each F-ratio, one for the numerator estimate of σ^2 and one for the denominator estimate of σ^2. First locate the column of Table E which is headed by the correct numerator *df*, then, within that column, find the row entries which correspond to the denominator *df*. The two values found in this way give the 5 per cent and 1 per cent (or .05 and .01) values on this particular $RSD(F)$. The larger of the two values is always the .01 value of F. This may be understood if one notices that the two variance estimates are unbiased estimates of the same population value and, therefore, will tend to be about equal. The resulting ratio will usually be around one; higher and higher ratios becoming less and less likely.

Let us illustrate the use of Table E with hypothetical data. Assume that one of two randomly selected sets with 15 scores has a SS of 56. If the second set of 21 scores has a SS of 30, we will compute F by

$$F = \frac{\hat{\sigma}_1^2}{\hat{\sigma}_2^2} = \frac{\frac{56}{14}}{\frac{30}{20}} = \frac{4.00}{1.50} = 2.67$$

In order to find the approximate probability of an F value's being this high or higher, we must consult Table E. Using the *df* of 14 for the numerator of the F-ratio and a *df* of 20 for the denominator, Table E gives two values of F: 2.23 and 3.13. The first of these tabled values indicates that only .05 of the random sampling values of F are 2.23 or higher (the other 95 per cent are lower). The second value of 3.13 means that only .01 of the distribution is 3.13 or higher. Our obtained value of F exceeds the tabled value at the .05 level. If all assumptions are true, the probability of obtaining an F as high or higher than our obtained value is less than .05. The conclusion to be drawn from this result would depend upon the hypothesis being tested.

As a statement of the result of this test, let us include the degrees of freedom as well as the resulting value of F. The following summary of our test might be given: $F(14,20) = 2.67$, $P \leq .05$. This notation system indicates the statistic, F, the associated degrees of freedom, and the approximate probability of a value of F as high or higher than obtained, assuming the truth of our assumptions.

6–4. The Single-Variable Case. We will now use the F-ratio to test the hypothesis that a single independent variable has no effect upon the mean of a dependent variable. We typically hope to reject this hypothesis and to further argue for a causal relationship between the independent variable and the dependent measures. But first, we must demonstrate that the differences between the mean scores in the different conditions are too large to be attributed to chance. In other words, we wish to show that the hypothesis of no differences between the population means of the experimental conditions is suspect and should be rejected.

Assumptions. A number of assumptions are made when applying the analysis of variance technique to be given.

1. We assume that the additive model (6–2.1) holds. In particular, this means that the different sets of population scores related to each experimental treatment have equal variances.
2. We assume that each group of subjects is randomly selected from its population.
3. We assume that the populations of scores are normally distributed.
4. We assume (and test) that all populations have equal means.

Although this may seem to be a restrictive list of assumptions, certain violations are not at all serious. These will be discussed later in the chapter.

In order to satisfy assumption 2, it is required in this type of analysis that each subject serve in only one of the experimental conditions, i.e., levels of the independent variable, *and* that only a single score, X_i, be used for each subject. This single score may be a total of other scores in some cases, but in any case each and every score in the experiment must represent the performance of a different subject.

The design. The type of experiment to be considered may be described in terms of the number of levels of the independent variable and the number of subjects (and hence the number of scores) within each condition (level). Let us denote the number of levels of the independent variable as K, and the number of scores within each *group* or level as n. Later we shall consider a modification in which each condition has a different number of subjects. It may be seen that the total number of scores in the experiment, N, will be equal to nK.

The between-groups variance estimate. We may conceptualize the selection of scores for the experiment in two ways. The entire set of N scores may be considered as being randomly selected from a single normally distributed population, or each subset (group) of scores may be considered as a random sample from its own normal population, in which case we assume that all population means are equal and all population variances are equal. These two methods of thinking about the experiment are equivalent to one another for our purposes; however, let us speak for the moment as if a single population were involved.

Each of the K different conditions will yield a mean, M_k, by dividing the total of the scores in each group by n. Since random sampling has been assumed, we may consider that the set of *means* is a random selection of scores from a population we have called the $RSD(M)$. Theorem 5–3 indicates that the $RSD(M)$ will be a normal distribution, and that (using n instead of N)

$$\sigma_M{}^2 = \frac{\sigma^2}{n} \tag{6–4.1}$$

or

$$\sigma^2 = n\sigma_M{}^2 \tag{6–4.2}$$

We may now obtain an estimate of σ^2 by estimating the variance of the $RSD(M)$ distribution, i.e., by estimating $\sigma_M{}^2$:

$$\hat{\sigma}^2 = n\hat{\sigma}_M{}^2 \tag{6–4.3}$$

Since $\hat{\sigma}_M{}^2$ is the variance of a normal distribution of means, we may use our set of K means and (5–7.8) to obtain the estimate

$$\hat{\sigma}_M{}^2 = \frac{SS_M}{K - 1} \tag{6–4.4}$$

where SS_M is the sum of squared deviations of the K means from the mean of these means. Let us denote the mean of the means as the *grand mean*, GM, of the experiment. The GM may be found by adding the K means together and dividing by K or by dividing the total of all N scores (call it GT) by N.

The value of the SS_M may be found in the same way that any SS is computed on a set of scores:

$$SS_M = (M_1 - GM)^2 + (M_2 - GM)^2 + \cdots + (M_K - GM)^2 \tag{6–4.5}$$

Let us now define a quantity called the *between-groups sum of squares*, SS_{BG}, which will be useful in later computational methods:

$$SS_{BG} = nSS_M \tag{6–4.6}$$

We may now combine (6–4.3), (6–4.4) and (6–4.6) in order to express the estimate of the population variance based on the group means.

$$\hat{\sigma}^2 = \frac{nSS_M}{K - 1} = \frac{SS_{BG}}{K - 1} \tag{6–4.7}$$

The common term for a variance estimate in the analysis of variance method is a **mean square** estimate, or simply MS. Since our estimate in (6–4.7) is based on the variation of means between the groups, we will call the estimate a *between-groups mean square*, MS_{BG}. With this change in notation, (6–4.7) becomes

$$MS_{BG} = \frac{SS_{BG}}{K - 1} \tag{6–4.8}$$

The computation of this variance estimate is not difficult. The foregoing equations are given to indicate that we are indeed obtaining an estimate of the population variance.

The within-groups variance estimate. The distribution of the scores within any group in the experiment has not been examined or used in any way in the calculation of the MS_{BG}. No matter what the value of a particular mean may be, the variance of the scores around that mean may be small or large. In order that there be no tendency for the size of a mean to be related to the size of the variance within that group of scores, we use assumption 3, that the population(s) is(are) normally distributed. It has been stated in Chapter 5 that such a population will yield samples in which the mean and variance are not related. This aspect of the normal distribution permits us to obtain a second estimate of the population variance which is independent of the first estimate (MS_{BG}). This is required when using the $RSD(F)$ as given in Table E.

Notice that a single group of scores obtained under one level of the independent variable may be used to obtain an estimate of the population variance by use of (5–7.8). One merely needs to compute the SS for that set of scores and divide by $(n - 1)$, n being the number of scores in the group. However, in the present design there are K different groups, each of which will probably have a slightly different SS. Which of the various possible estimates is to be used? The answer is to use all of the K possible estimates of σ^2.

In the same way that the mean of large sets of scores is usually closer to the population mean than the mean of a smaller set, the average of a number of variance estimates will usually be closer to the population value than a single variance estimate. Let us proceed to obtain an average value of these estimates. The average variance estimate based upon the estimates from within each group of scores is called the *within-group* mean square estimate, MS_{WG}:

$$MS_{WG} = \frac{\hat{\sigma}_1^2 + \hat{\sigma}_2^2 + \cdots + \hat{\sigma}_K^2}{K} \tag{6–4.9}$$

Expressing each variance estimate in terms of (5–7.8), we obtain

$$MS_{WG} = \frac{1}{K}\left(\frac{SS_1}{n-1} + \frac{SS_2}{n-1} + \cdots + \frac{SS_K}{n-1}\right)$$

$$MS_{WG} = \frac{(SS_1 + SS_2 + \cdots + SS_K)}{K(n-1)} \qquad (6\text{–}4.10)$$

Let us now define the *sum of squares within groups*, SS_{WG}, which is the sum of the SS values from all K conditions:

$$SS_{WG} = SS_1 + SS_2 + \cdots + SS_K$$

Finally, MS_{WG} may be expressed as

$$MS_{WG} = \frac{SS_{WG}}{K(n-1)} = \frac{SS_{WG}}{N-K} \qquad (6\text{–}4.11)$$

Use of the F-ratio. The formation of the F-ratio in this type of analysis of variance implies a test of a null hypothesis. Explicitly, we make assumptions 1, 2, and 3. We further assume, for purposes of testing, that all scores have been drawn from a population with a single mean, or that all groups of scores have been drawn from separate populations with equal means. This assumption constitutes the null hypothesis, H_0, for this experimental design. Before formulating the F-ratio, let us consider what might be expected if H_0 is false. If the different groups of scores were drawn from populations with different means, it can be shown that the means will tend to be further from one another than if H_0 were true. The SS_{BG} will tend to be larger than expected. The conclusion may be stated: *When the null hypothesis is false, the MS_{BG} will tend to be larger than the MS_{WG}.*

In this sort of single variable design, there is no reason to expect that the between-groups mean square will be smaller than the within-groups mean square, unless the experiment has been poorly conducted or the random sampling assumption violated. A result of this type could be obtained, but the statistical explanation would have to be that the result was due to chance. On the other hand, if the MS_{BG} is larger than the MS_{WG}, we have reason to suspect that H_0 is false and that the groups of scores were drawn from populations with different means. The F-ratio in this design will always be formed by dividing the MS_{BG} by the MS_{WG}, and the highest 5 per cent of the $RSD(F)$ will be chosen as the region of rejection:

$$F = \frac{MS_{BG}}{MS_{WG}} \qquad (6\text{–}4.12)$$

6–5. Raw Score Computations. The actual analysis of an experiment of the single-variable type is somewhat easier than the above equations might lead one to expect. As in the case of earlier measures of dispersion, raw-score formulas are available which reduce the labor needed to perform an analysis of variance test.

We first wish to calculate three types of sum of squares. The first of these is a *total sum of squares* for the entire experiment, SS_{TOT}. This value will be useful for checking purposes. The total SS is merely the SS based on the entire set of N scores, ignoring the division into experimental groups. Denoting the sum of all N scores as GT (the grand total), we may use (3–2.1) to obtain

$$SS_{TOT} = \sum_{i=1}^{N} X_i^2 - \frac{(GT)^2}{N} \qquad (6\text{–}5.1)$$

Let T_k refer to the sum of scores for group k. Manipulation of (6–4.5) and (6–4.6) will yield the raw-score formula for the SS_{BG} term, which turns out to be

$$SS_{BG} = \frac{\sum_{k=1}^{K} T_k^2}{n} - \frac{(GT)^2}{N} = \frac{(T_1^2 + T_2^2 + \cdots + T_K^2)}{n} - \frac{(GT)^2}{N} \qquad (6\text{–}5.2)$$

The SS_{WG} is the sum of the SS's from each group of scores, as classified on the basis of the levels of the independent variable. We need to calculate each of these individual sum of squares and then add them together. Let us call the SS within group k the SS_{WGk}. For any group we then have

$$SS_{WGk} = \sum_{i=1}^{n} X_i^2 - \frac{T_k^2}{n} \qquad (6\text{–}5.3)$$

Equation (6–5.3) is simply a SS based on the n scores in group k. After (6–5.3) has been used on each of the K sets of scores, we find SS_{WG} by addition:

$$SS_{WG} = SS_{WG1} + SS_{WG2} + \cdots + SS_{WGK} \qquad (6\text{–}5.4)$$

An important relationship between these three sums of squares terms is that

$$SS_{TOT} = SS_{BG} + SS_{WG} \qquad (6\text{–}5.5)$$

Although (6–5.5) will not be derived here, the proof is straightforward but somewhat long. We may use (6–5.5) as a check on computations. If SS_{BG} plus SS_{WG} does not equal the SS_{TOT}, allowing for a small error in rounding numbers, then some error in computation has taken place.

It may be seen that the use of (6–5.5) would allow one to compute the SS_{TOT} and the SS_{BG} and then obtain the SS_{WG} by subtraction. This technique is not recommended, because errors in computation are easy to make and will go undiscovered.

An additional gross check on errors is available. Since all of our SS terms are based upon squared deviation scores, no SS can be a negative number. If a negative SS is obtained, it is a certain signal that an error has been made.

Degrees of freedom. In order to convert the SS measures to corresponding MS values, let us examine (6–4.8) and (6–4.11). It will be seen that the SS_{BG} is based upon K numbers, each a mean. The dispersion of these K numbers is calculated by placing one restriction on the set; i.e., that the mean of the K numbers is equal to the obtained mean. Hence we have $(K - 1)$ degrees of freedom associated with the SS_{BG}. The MS_{BG} may then be expressed as

$$MS_{BG} = \frac{SS_{BG}}{df_{BG}} \tag{6–5.6}$$

where df_{BG} refers to the degrees of freedom associated with the SS_{BG}.

To obtain the mean square estimate from within the groups, equation (6–4.11) indicates that the SS_{WG} must be divided by $[K(n - 1)]$ or $(N - K)$, which is the same number. The SS_{WG} is based on K parts, each of which is a SS from a set of n scores. Each SS has $(n - 1)$ degrees of freedom. We obtain the df_{WG} by adding the df's for each group. The result is clearly $K(n - 1)$, or the total N minus one restriction for each of the K groups, $N - K$. We may then express the MS_{WG} as

$$MS_{WG} = \frac{SS_{WG}}{df_{WG}} \tag{6–5.7}$$

The resulting F-ratio found by dividing MS_{BG} by MS_{WG} is evaluated by using Table E with the numerator df equal to df_{BG} and the denominator df equal to df_{WG}. An F which equals or exceeds the .05 value in the table is in the region of rejection; therefore, the hypothesis of equal population means would be rejected. The degrees of freedom associated with SS_{TOT} is equal to $N - 1$. But since we have no need for a mean square based on the SS_{TOT}, one usually does not obtain a MS_{TOT}.

In the same way that the SS_{BG} and SS_{WG} add to the SS_{TOT}, the df_{BG} plus the df_{WG} will always equal the df_{TOT}. This allows a check on one's calculation of df's.

Table 6–1 gives a format for presenting a summary of an analysis-of-variance test as well as giving the basic computational formulas employed.

TABLE 6–1

Summary of the Analysis of Variance for the Single-Variable Case

Source	df	SS	MS	F
Between Groups	$K - 1$	$\frac{1}{n}\sum_{k=1}^{K} T_k^2 - \frac{(GT)^2}{N}$	$\frac{SS_{BG}}{df_{BG}}$	$\frac{MS_{BG}}{MS_{WG}}$
Within Groups	$N - K$	$SS_{WG1} + SS_{WG2} + \cdots + SS_{WGK}$	$\frac{SS_{WG}}{df_{WG}}$	–
Total	$N - 1$	$\sum_{i=1}^{N} X_i^2 - \frac{(GT)^2}{N}$	–	–

Unequal n's. Before presenting examples of the analysis-of-variance test, let us consider the possibility that an unequal number of subjects might be used in the different groups of an experiment. If such an experiment is to be subjected to the analysis-of-variance procedure, the SS_{TOT} may be calculated as in (6–5.1). The SS_{WG} may be computed as in (6–5.3) and (6–5.4), making sure to use the value of n which is appropriate for the group being considered. Only the SS_{BG} need be found in a slightly different manner. The correct calculations are given by

$$SS_{BG} = \frac{T_1^2}{n_1} + \frac{T_2^2}{n_2} + \cdots + \frac{T_K^2}{n_K} - \frac{(GT)^2}{N} \qquad (6\text{–}5.8)$$

where the n_k's refer to the number of scores in each group.

All other calculations are done as in the case of equal n's, as shown in Table 6–1.

Example 6–1. Let us illustrate the application of the analysis-of-variance technique by using a three-condition experiment. Assume, as in earlier illustrations, that an investigator wishes to study the influence of room temperature on efficiency in working algebra problems. Suppose he selects the three temperature conditions of 70, 80, and 90 degrees Fahrenheit. He might then select 30 subjects at random from some available population of persons and randomly assign 10 to each of three different groups, corresponding to the three temperature conditions. The dependent variable will be the number of problems correctly done by each subject, out of 100 possible. The following are hypothetical data which might result.

Group 1 (70°)	Group 2 (80°)	Group 3 (90°)	
70	59	62	
50	67	60	
61	67	57	
65	78	65	
70	68	54	
75	71	56	
76	65	58	
70	64	48	
69	70	46	
68	63	55	
$T_1 = 674$ $M_1 = 67.4$	$T_2 = 672$ $M_2 = 67.2$	$T_3 = 561$ $M_3 = 56.1$	$GT = 1{,}907$

It may be seen that the means for groups 1 and 2 are very close in value but the mean number of correct solutions in group 3 is well below that of the other two conditions. We must now determine whether or not the existing differences between the three means are too large to be accounted for by random assignment of subjects to conditions, and errors inherent in the task.

First, using (6–5.1), we calculate the SS_{TOT}:

$$SS_{TOT} = \sum_{i=1}^{N} X_i^2 - \frac{(GT)^2}{N} = 70^2 + 50^2 + \cdots + 55^2 - \frac{(1{,}907)^2}{30}$$
$$= 123{,}109.00 - 121{,}221.63$$
$$= 1{,}887.37$$

The SS_{BG} may be found by using (6–5.2):

$$SS_{BG} = \frac{\sum_{k=1}^{K} T_k^2}{n} - \frac{(GT)^2}{N} = \frac{674^2 + 672^2 + 561^2}{10} - \frac{(1{,}907)^2}{30}$$
$$= 122{,}058.10 - 121{,}221.63$$
$$= 836.47$$

In order to obtain the SS_{WG}, we must find the SS within each of the three groups. Using (6–5.3),

$$SS_{WG1} = 70^2 + 50^2 + \cdots + 68^2 - \frac{674^2}{10} = 45{,}932.0 - 45{,}427.6$$
$$= 504.4$$

Similarly, for groups 2 and 3,

$$SS_{WG2} = 59^2 + 67^2 + \cdots + 63^2 - \frac{672^2}{10} = 239.6$$

$$SS_{WG3} = 62^2 + 60^2 + \cdots + 55^2 - \frac{561^2}{10} = 306.9$$

Use of (6–5.4) gives

$$SS_{WG} = 504.4 + 239.6 + 306.9 = 1{,}050.9$$

Adding the SS_{BG} and SS_{WG} together we obtain 1,887.37, which is equal to SS_{TOT}.

Here, K is 3 and N is 30, so that the df_{BG} is 2 and the df_{WG} is 27. Dividing the sum-of-squares values by corresponding degrees of freedom, we obtain the two mean squares, i.e., the two estimates of the population variance:

$$MS_{BG} = \frac{836.47}{2} = 418.23$$

and

$$MS_{WG} = \frac{1{,}050.9}{27} = 38.92$$

The F-ratio, using (6–4.12) is

$$F\ (2,27) = \frac{418.23}{38.92} = 10.75$$

Entering Table E with 2 *df* for the numerator mean square and 27 *df* for the denominator mean square, we find the two values 3.35 and 5.49, which correspond to the .05 and .01 values of F. Clearly our sample value of 10.75 is significant, and the hypothesis of equal population means is rejected. The obtained F exceeds the .01 value given in the table. We may therefore state that the probability of such a large F-ratio is less than .01, if H_0 were true. A summary of our results may be given in tabular form as follows:

Source	*df*	*SS*	*MS*	*F*
Between Groups	2	836.47	418.23	10.75*
Within Groups	27	1,050.90	38.92	–
Total	29	1,887.37	–	–

$^*P < .01$.

A common practice in reporting the results of an analysis of variance is to identify the sources of variation in a more explicit fashion than simply "between groups" and "within groups." Since we hope that variation in Example 6–1 for the between-groups term is caused by variations in the temperature variable, we might use the term "Temperature" instead of "between groups." In the case of the within-group source of variation, the present design employs different subjects within each condition. The variation of scores within any one group may then be logically assigned to differences between subjects. The within-groups source of variation is referred to as "subjects within groups." We will often use this means of identifying sources of variation in later examples.

Example 6–2. The following data represent the results of a hypothetical experiment in which the investigator wished to find if the intake of certain drugs would affect the rate with which rats would press a bar for a food reward in a standard instrumental learning situation. He employed 4 different drugs, call them W, X, Y, and Z, and a control condition in which the rats were given an inert substance. This design may be considered a single-variable design with 5 conditions. The experimenter randomly assigned 6 different animals to each treatment condition and measured their average number of responses per minute over a 30-min. period. The results were:

Group 1 Control	Group 2 Drug W	Group 3 Drug X	Group 4 Drug Y	Group 5 Drug Z	
7.6	8.9	4.1	3.0	10.7	
7.0	8.1	5.5	4.3	9.6	
6.8	7.5	4.8	4.9	9.8	
8.1	7.8	7.0	5.6	8.7	
7.8	9.2	6.1	5.1	7.1	
6.7	7.8	5.2	4.5	9.0	
$T_1 = 44.0$ $M_1 = 7.33$	$T_2 = 49.3$ $M_2 = 8.22$	$T_3 = 32.7$ $M_3 = 5.45$	$T_4 = 27.4$ $M_4 = 4.57$	$T_5 = 54.9$ $M_5 = 9.15$	$GT = 208.3$

The analysis of variance of these data follows the same pattern as in the previous example. There is ample opportunity for arithmetic mistakes in such procedures. The student is cautioned that mistakes in obtaining the totals of the groups and grand total of scores are perceptually difficult to find after the more laborious procedures have been started. Check all totals carefully before starting the analysis.

The total sum of squares requires squaring and adding of all raw scores. It is often convenient to find the sum of squared scores separately for each group, since these subtotals are needed when the SS_{WG} is found. For the present problem, with $N = 30$,

$$SS_{TOT} = 1{,}553.99 - \frac{208.3^2}{30} = 1{,}553.99 - 1{,}446.30$$
$$= 107.69$$

The between-groups SS, using (6–5.2) is

$$SS_{BG} = \frac{44.0^2 + 49.3^2 + 32.7^2 + 27.4^2 + 54.9^2}{6} - 1{,}446.30$$
$$= 87.13$$

The calculation of the within-groups SS's is

$$SS_{WG1} = 324.34 - \frac{44.0^2}{6} = 1.67$$

$$SS_{WG2} = 407.39 - \frac{49.3^2}{6} = 2.31$$

$$SS_{WG3} = 183.35 - \frac{32.7^2}{6} = 5.13$$

$$SS_{WG4} = 129.12 - \frac{27.4^2}{6} = 3.99$$

$$SS_{WG5} = 509.79 - \frac{54.9^2}{6} = 7.45$$

$$SS_{WG} = 1.67 + 2.31 + 5.13 + 3.99 + 7.45 = 20.55$$

The sum of the between- and within-groups terms is 107.68, which is equal to the SS_{TOT}, allowing for the small error from rounding numbers.

Since we have 5 groups in the experiment, the df_{BG} is $5 - 1$, or 4. The df_{WG} is based on $6 - 1$, or 5, df for each of 5 groups, which gives a total of 25. This may also be found from $N - K$. The total df is 29 since N is 30. The conversion of SS's to MS's and the formation of the F-ratio again follow Table 6–1. The summary of the analysis might be given as:

Source	*df*	*SS*	*MS*	*F*
Drugs	4	87.13	21.78	26.50*
Subjects within Groups	25	20.55	.822	–
Total	29	107.69	–	–

$^*P < .01$.

The evaluation of the obtained value of F has been accomplished by finding the tabled values of F corresponding to 4 and 25 degrees of freedom for this experiment. They are found to be 2.76 and 4.18. Our value of 26.50 is obviously far bigger than any reasonable allowance for chance would suggest.

In the final example of this type of design, let us illustrate with the smallest number of groups used in an experiment, just two. Let us also use an unequal number of subjects in the two conditions in order to give an application of (6–5.8).

Example 6–3. An experimenter studying problem solving behavior feels that the instructions to subjects used in previous studies were not very clear.

He rewords certain portions of the instructions but wishes to test their effectiveness before making a change for a large experiment. He randomly assigns subjects to the two conditions of a standard type of problem, one group being given the old set of instructions and the other group the new instructions. The end of the semester arrives before he has completed as many subjects as desired, and student subjects become unavailable until two weeks later. Since he already has run 7 subjects with the old instructions and 9 with the new instructions, he decides to test for differences at this point. The following scores are times to solve the problem for each of the 16 people, measured in minutes and hundredths of a minute. Do the two groups have significantly different means?

Old Instructions	New Instructions	
4.85	5.20	
3.78	4.33	
6.40	6.42	
5.73	3.56	
5.04	4.87	
7.21	5.42	
5.69	5.85	
–	5.55	
–	4.77	
$T_1 = 38.70$	$T_2 = 45.97$	$GT = 84.67$
$n_1 = 7$	$n_2 = 9$	$N = 16$
$M_1 = 5.53$	$M_2 = 5.11$	

The SS_{TOT} is found as in the equal n case.

$$SS_{TOT} = 4.85^2 + 3.78^2 + \cdots + 4.77^2 - \frac{(84.67)^2}{16}$$

$$= 461.9157 - 448.0631$$
$$= 13.8526$$

The sum of squares between the groups must now be found by (6–5.8), because we do not have an equal number of subjects in each condition:

$$SS_{BG} = \frac{38.70^2}{7} + \frac{45.97^2}{9} - \frac{84.67^2}{16} = 213.9557 + 234.8045 - 448.0631$$
$$= .6971$$

The calculations for SS_{WG} are given next:

$$SS_{WG1} = 4.85^2 + 3.78^2 + \cdots + 5.69^2 - \frac{38.70^2}{7} = 7.4099$$

$$SS_{WG2} = 5.20^2 + 4.33^2 + \cdots + 4.77^2 - \frac{45.97^2}{9} = 5.7456$$

$$SS_{WG} = 7.4099 + 5.7456 = 13.1555$$

The sum of the last two major terms does equal the SS_{TOT}, as it must. The df_{BG} is $K - 1$, or 1, and the df_{TOT} is $N - 1$, or 15. Because the df_{BG} and df_{WG} must equal the df_{TOT}, we can obtain the degrees of freedom associated with the within-groups variability by subtraction, the answer being 14. This figure may also be obtained by taking $N - K$, or we may add the *df*'s from within each group together, i.e., $(n_1 - 1) + (n_2 - 1) = 14$.

The remaining computations are the same as in the equal *n* case:

Source	*df*	*SS*	*MS*	*F*
Instructions	1	.6971	.6971	.74
Subjects within Groups	14	13.1555	.9397	–
Total	15	13.8526	–	–

When an *F*-ratio in this type of design is less than 1.00, it is never sufficient to reject the null hypothesis. However, for practice in using the table of *RSD*(*F*), find the values for this experiment. With 1 and 14 *df*'s, we find 4.60 and 8.86 as the .05 and .01 values of *F*. The experimenter would be on very shaky ground were he to conclude that the two sets of instructions had differential effects on mean time to solution.

A final note on the above problem is that the number of decimal places carried along in the *SS* computations was more than needed for the final two-place accuracy in the *F*-ratio. Experience with such analyses will allow one to decide how many digits are required. In more complex designs to be considered later, the decision is not as simple as one might think.

6–6. Procedures Following Analysis of Variance. Interpretations. Once the null hypothesis has been rejected, on the basis of a large *F*, we must determine which conclusions are legitimate. The rejection of H_0 implies that H_1 is considered to be true. However, when more than two conditions are used in this type of design, the decision that all of the population means are not equal does not mean that they are all different. In Example 6–1 one may see that the means of groups 1 and 2 are quite close. It may be that they were based upon populations with equal means and that only the population from which group-3 scores were selected is different from the others. The *F*-ratio is designed to test a rather global hypothesis.

There are simple techniques available that may be used following the analysis of variance which are designed to discover just which population means may be considered as different from the other population means. We will not study this refinement, but the interested reader will find reference to these procedures in many advanced texts in statistics.

Once again, the statistical decision, the rejection of H_0, does not permit one to automatically assign the cause of such rejection to the independent variable under study. A close examination of experimental procedures is required for this experimental decision to be made.

Estimation of a population mean. Following an analysis of variance, one may wish to obtain a confidence interval for estimating the population mean for one or more of the experimental conditions. If we may assume that the hypothesis of equal population variances is true, the MS_{WG} may be used to develop an estimate of the standard error of the mean which is better than such an estimate based on any single group of scores within the experiment. Since the effect of the independent variable is assumed to be of the form shown in (6–2.1), the MS_{WG} is an unbiased estimate of σ^2, even when the F is significant. Use of (5–8.4) will yield the following estimate for the standard error of the mean for a group of means based on n scores:

$$\hat{\sigma}_M = \sqrt{\frac{MS_{WG}}{n}} \qquad (6\text{–}6.1)$$

Equation (6–6.1) may be used in Theorem 5–4 to obtain the confidence interval for estimating any μ.

6–7. Violations of Assumptions. There are a number of assumptions which might not be strictly true in the use of analysis of variance. The random assignment of subjects to conditions is one which absolutely must not be violated. If it were, there is no way of knowing whether the conclusions reached are close to the truth or not.

However, the assumptions of equal population variance and normal distributions of scores may be violated to some extent, under certain restrictions, and the F-ratio will still be distributed very nearly as given in Table E. The F-test is very sensitive to violations of the null hypothesis about population means, but relatively insensitive to violations of the equal variance and normal distribution assumptions, *if an equal number of subjects is used in each experimental condition.*

If the reader will note that the dependent variable in Example 6–1 was a discrete, as opposed to a continuous, variable, it is clear that the population of scores could not possibly have been normally distributed. The fact that the F-test is not sensitive to this violation permitted the analysis shown in the example. Moreover, the conclusion to reject the null hypothesis remains a sound one.

As a rule of thumb, use an equal number of subjects in each experimental condition. Unless the scores in some groups are badly skewed in one direction and the scores in some other group or groups are badly skewed in the *other* direction, one may have confidence that the .05 value of F as found in Table E is close to being correct.

Prior to the knowledge of these characteristics of the $RSD(F)$, rather involved tests were recommended for checking on the assumption of equal population variances before proceeding with the analysis of variance. One author, after considering the characteristics of the F-ratio, concluded that to ". . . make the preliminary test on variances is rather like putting to sea in a rowing boat to find out whether conditions are sufficiently calm for an ocean liner to leave port!"*

EXERCISES

1. For each of the following sets of group totals, find the SS_{BG}, df_{BG}, and df_{WG}:
   (a) $T_1 = 285$, $T_2 = 311$, $T_3 = 294$; $n = 10$.
   (b) $T_1 = 12.75$, $T_2 = 16.80$; $n = 24$.
   (c) $T_1 = 97.4$, $T_2 = 101.0$, $T_3 = 85.6$, $T_4 = 67.5$, $T_5 = 95.4$; $n = 8$.
   (d) $T_1 = 45.7$, $n_1 = 10$; $T_2 = 68.5$, $n_2 = 8$; $T_3 = 49.0$, $n_3 = 11$.
2. Find the F-ratio, and assess its significance, for each part of Exercise 1, given the SS_{WG} terms:

   for (a), 250.5; for (b), 5.013; for (c), 1245.; for (d), 318.0

3. Perform the analysis-of-variance test for differences between means on the following data.

Group 1	Group 2	Group 3
16.0	17.6	14.4
17.5	19.4	14.6
18.1	20.5	16.3
18.4	21.0	16.5
19.5	22.3	18.2

4. Test for differences between the means of the following two independent groups of scores.

Group 1	Group 2
145	180
158	188
180	191
185	199
187	206
192	214
198	–
202	–
210	–

* G. E. P. Box, "Non-normality and tests on variances," *Biometrika*, 1953, *40*, 318–335.

5. If $n_1 = n_2$, show that $SS_{BG} = (T_1 - T_2)^2/N$.

Applications: Use information from all previous chapters where relevant.

6. An experiment is conducted to assess the effects of three methods of presenting visual materials in a lecture course in modern art. A class is randomly divided into three groups. The same instructor gives a single lecture to each group, but Group 1 views the pictures on closed-circuit television screens, Group 2 on a large motion-picture screen, and Group 3 subjects are given individual reproductions of the pictures. Following the lecture, a test is given on the material with the following results, given as the number of correct answers per person.

Group 1	Group 2	Group 3
28	34	30
32	36	31
33	36	31
34	39	33
35	40	35
37	40	36
38	41	38
39	43	38
40	44	41
41	48	43
42	51	44
45	51	47

(a) Test for differential effects of the methods on test results.

(b) If we assume that all scores were taken from populations with equal variances, what is the unbiased point estimate of σ^2?

(c) Find the 95 per cent confidence intervals for estimating the population means for each method; use the single best $\hat{\sigma}$ as found from $\hat{\sigma}^2$ in part *b*.

7. An experimenter has completed half of a two group study on group cohesiveness. He decides to perform an analysis of variance on his dependent variable just to find out how the study is progressing. With 10 subjects in each group (a single score each), he has $T_1 = 140$, $T_2 = 180$, and $SS_{WG} = 576$. What is his F-ratio?

If the second half of the experiment gives the same results as the first half, what would be his final F-ratio?

7

Analysis of Variance: Two Variables

7–1. The Two-Variable Case. One of the facts of life to be faced by the behavioral scientist is that the behavior of organisms takes place under complex conditions, except (perhaps) in laboratory situations. The knowledge that one single independent variable is related to a type of behavior in a causative way is usually insufficient to predict or control behavior very well. Most of the time, a number of stimulus factors (independent variables) have simultaneous effects upon a dependent variable under study.

Even a moderate degree of accuracy in predicting behavior under complex conditions requires, at least, knowledge of the effects of variations in each of a number of independent variables. For example, one may know that a certain type of psychotherapy causes faster improvement in manic depressive patients than other forms of psychotherapy. However, the prediction that such patients will show rapid improvement when given the appropriate treatment in a certain hospital may not prove correct. The failure of such a treatment in a particular case may result from the actions of other independent variables operating at the same time. In the present example, one might suggest that the environmental conditions in the hospital ward had a detrimental or offsetting effect upon the patients.

Although it is impossible to study all independent variables which may have effects upon a given behavior, we may assume that certain variables have greater effects on the behavior than others and attempt to gain knowledge of the more important ones. One possible research plan which results from such thinking is to conduct a number of experiments in which a single independent variable is studied in each one. However, it is often more efficient to study two or more independent variables in the same experiment. In this chapter, we shall consider the analysis of variance which results from an experiment using two independent variables at the same time.

The proper use of two independent variables in the same experiment, which we will call the *two-variable case,* allows one to assess the effects, if any, of each of these variables on a given dependent measure. Moreover, it is also possible to gain information about the *combined* effects of both variables from such an experiment. This latter information cannot be obtained from two single-variable experiments.

The combined effect of particular levels of two independent variables may simply be one of adding the individual effects of these two variables together, or something very special may be obtained. When such a combined effect of two variables is not a simple addition of two separate effects, the resulting relationship is called an **interaction** of the two independent variables.

An example of an *interaction* effect might be made in the area of medicine. The physician must be aware that a number of drugs have beneficial effects for the patient when given singly, but when given in certain combinations, may result in death for the patient. Or, one may know that each of two types of psychotherapeutic treatments has a small but beneficial effect on certain patients. We might discover that when both of these treatments are given to the patients, the improvement which results is far greater than the sum of the two treatments when given separately. This too would be termed an *interaction.* Clearly, when interactions between independent variables are discovered, we may predict resulting behavior better by using such information, than by using information about each variable separately.

The design. In the two-variable design, two or more levels of each of two independent variables are used. In general, let us refer to one of these as *variable A* and to the other as *variable B.* Each of these variables will have two or more different levels which are used in the experiment. The exact size of such an experiment may be designated by giving the number of levels used for each variable. For instance, the smallest possible example of a two-variable experiment would be called a 2×2 design, meaning that two levels of each variable are used. For a larger experiment, we might consider a 5×3 design. In this case, five levels of variable A are combined with three levels of variable B in the same experiment.

Such experiments are called **factorial designs**. The term orthogonal design is also used to describe such experiments in some texts, but we will use the former consistently. In order to qualify for the term *factorial design,* a two-variable experiment must contain *all possible combinations* of the levels of the two independent variables. In a 2 × 2 design, this would mean that four different groups or treatments would be used. Subjects in one condition of the experiment would be given level number one of variable A and level number one of variable B. Subjects in another condition would be given level one of variable A and level number two of variable B; the third condition would involve level two of the first variable and level one of the second variable. The final condition would involve the second level of each variable.

To better illustrate these various combinations of the independent variables, let us consider a 3 × 4 design in which variable A consists of three different amounts of the drug chlorpromazine (a tranquilizer), and variable B is four different reaction-time tests that vary in the number of stimuli (lights) which might appear and the number of responses (different keys) which must be made to the stimuli. The dependent variable would be the time taken to react appropriately. Table 7–1 shows a simple method of portraying the twelve different experimental conditions which would result from combining each level of variable A with each level of variable B.

TABLE 7–1

An Example of a 3 × 4 Factorial Design

		Variable A Dosages of Chlorpromazine		
		Zero dose ($A1$)	Half dose ($A2$)	Full dose ($A3$)
Variable B	1 ($B1$)	Group 1	Group 2	Group 3
Number of	2 ($B2$)	Group 4	Group 5	Group 6
stimuli and	3 ($B3$)	Group 7	Group 8	Group 9
responses	4 ($B4$)	Group 10	Group 11	Group 12

The subjects in group 7, for example, would be given an inert substance, zero amount of chlorpromazine, and would be tested with a three-light-key-stimulus-and-response task. The subjects in group 9 would receive a full dose of chlorpromazine and be tested with the same three-light-key task. As a final particular case, the subjects in group 2 would receive a half-dose of chlorpromazine and would be tested on the simple reaction-time task of a single light and single key.

One additional important requirement of the factorial design is that each treatment (group) must contain *different subjects.* No single subject would serve in more than one of the combinations of the two-independent-variable conditions. As in the single-variable case, only one score from each subject is used in the analysis of variance of the two-variable case.

7–2. The Main Effects. The analysis of variance of the two-variable experiment may be considered to be an extension of the single-variable analysis. However, rather than simply considering the between-groups variation as a single entity, we will analyze the between-group variability by dividing it into three component parts. One of these parts will reflect the average effect of variation in variable A, the second part will reflect the average effect of variable B, and the third part will reflect the existence of any interaction between the two variables.

The effect of variable A, ignoring the other variable, is called the *main effect of variable A*. In order to assess a main effect, we will examine the means of the dependent variable from the different levels of the independent variable under consideration. Referring again to Table 7–1, let us see how the main effect of variable A might be examined.

Since we wish to find the means from the three different levels of variable A, all scores from the zero-dose condition are added together and divided by the number of such scores. Denoting the first level of variable A as $A1$, we must include all of the scores in groups 1, 4, 7, and 10, as shown in Table 7–1. All of these scores are used in getting the mean for the zero-dose condition, because all of these scores were collected under the zero-dose level of the variable. In a similar fashion, the means for the second and third conditions of variable A are based on all of the scores in the $A2$ and $A3$ columns of Table 7–1, respectively. As will be shown shortly, differences between these three means will reflect the average effects of variable A, independent of the effects of the other variable! Likewise, the main effect of variable B may be examined by finding the four means based on scores in the four separate rows of Table 7–1.

Let us consider the possibility that variable B has a real effect upon the dependent variable; i.e., that the scores for the different levels of B are in fact from populations with different means. Assuming this state of affairs to be true, we wish to show that expected differences in the means of variable A conditions will remain unchanged, under certain conditions.

This independence of the effects of variables A and B may be demonstrated if we require that an *equal number of scores* (subjects) be used for each of the experimental groups. Let us denote the population means of variable A conditions as $\mu_{A1}, \mu_{A2}, \ldots, \mu_{AC}$ *when variable B is held at the B1 level.* Use of (6–2.1) and our knowledge of linear transformations will permit the expression of the expected means for A (μ_{Ac}') from the whole experiment as related to the main effects of B:

$$\begin{aligned} \mu_{A1}' &= \mu_{A1} + b_{B2} + b_{B3} + \cdots + b_{BR} \\ \mu_{A2}' &= \mu_{A2} + b_{B2} + b_{B3} + \cdots + b_{BR} \\ &\;\;\vdots \\ \mu_{AC}' &= \mu_{AC} + b_{B2} + b_{B3} + \cdots + b_{BR} \end{aligned} \tag{7–2.1}$$

where the b_{B2}, b_{B3}, etc. are the mean score changes caused by changing from condition $B1$ to the other levels of variable B. It may readily be seen that the difference between any two μ' values for variable A is the same as the corresponding difference between the respective μ values. This shows that the main effects of variable B, whatever their magnitude, have no effect upon the differences between variable A means. The same type of reasoning will show that variable A effects have no effect upon differences between variable B means.

One should also notice that the assumption of the linear model, (6–2.1), allows the conclusion that the variance within any single group of scores remains unchanged no matter what constants are added to the scores by virtue of the main effects of either independent variable.

7–3. Interaction Effects. The model for changing scores when changing levels of the independent variables in the two-variable case must be modified in two ways from (6–2.1). We will allow the addition of a constant for *each* main effect and the addition of a constant, $b_{A\times B}$, which reflects a unique contribution due to the particular combination of the levels of A and B. The model might then be written as

$$X' = X + b_A + b_B + b_{A\times B} + e \tag{7–3.1}$$

where X is a score from some standard condition.

Again, errors in measurement and from other factors, e's, are assumed to be unrelated to all experimental conditions.

Working with (7–3.1), let us graph some possible relationships between the dependent and independent variables. Figure 7–1 illustrates a situation in which variable A has an effect on the dependent

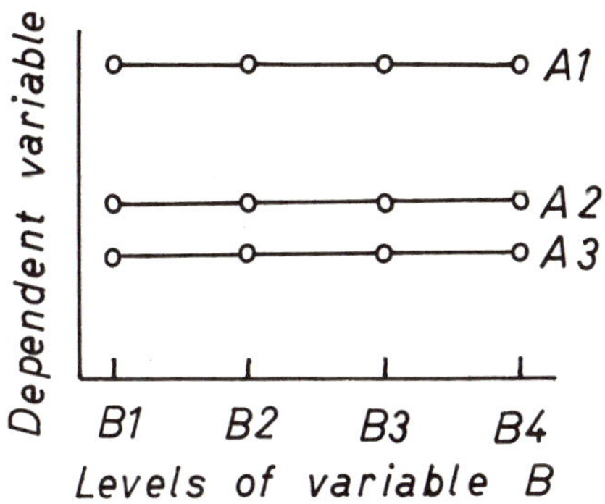

Fig. 7–1. A case in which variable A has an effect upon the dependent variable, but in which variable B and the interaction have zero effects.

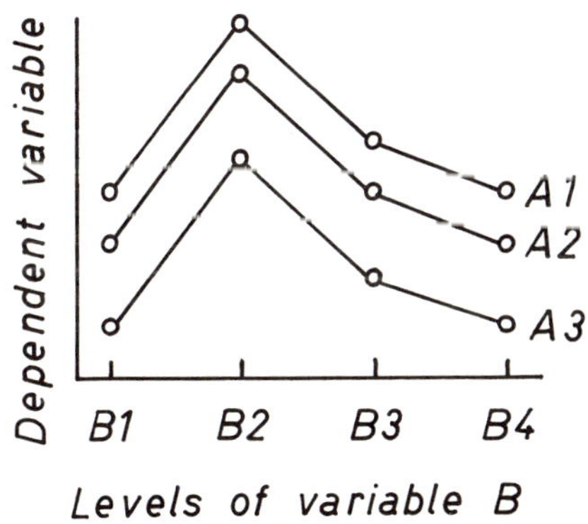

Fig. 7–2. A case in which both variables A and B have effects, but in which there is no interaction.

variable, but the effects of variable B and the interaction are both zero. Note that the levels of one independent variable are designated on the x-axis and each level of the other independent variable is drawn separately from the others. The fact that any two curves are equal distances apart at each point reflects the absence of interaction effects.

Figure 7–2 illustrates a case in which both independent variables have significant effects but in which the interaction is still zero.

Notice that any two curves are still equal distances apart at each point.

Figure 7–3 shows a strong interaction effect. Here the average effect of variable A does not apply to each level of variable B. Whenever an interaction exists, the effects of one independent variable depend

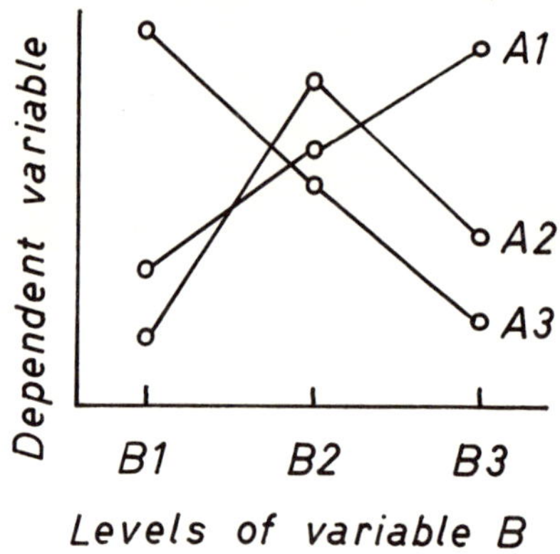

Fig. 7–3. A case illustrating a strong interaction effect.

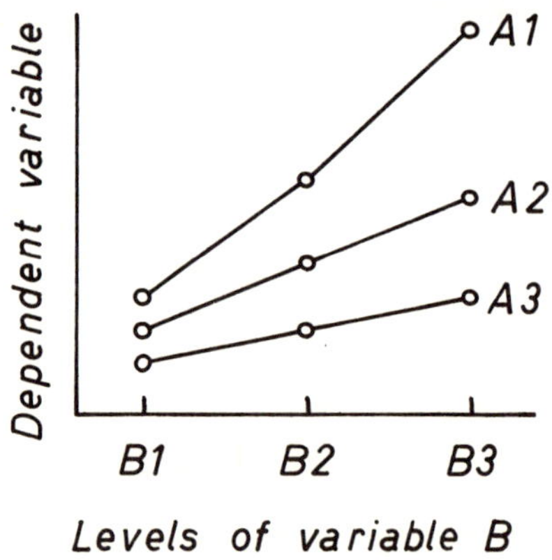

Fig. 7–4. An illustration of a weak interaction between variables A and B, both of which also have some effect upon the dependent variable.

upon the level of the other independent variable. Figure 7–4 shows another type of interaction which, compared to Fig. 7–3, is a weak interaction. Verbal interpretations of interaction effects will be discussed in the examples later in the chapter.

7–4. Computations for the Two-Variable Case. **Assumptions.** The same general assumptions must be made here as in the single-variable analysis of variance. The model (7–3.1) is also assumed to be valid. We will treat only the type of experiment in which each condition has an equal number, n, of scores. This restriction not only aids computations but helps to overcome problems arising from the violation of the assumption of equal population variances. The discussion in Section 6–7 is again appropriate to such violations.

Preliminary steps. We will denote the number of levels of variable A as C (the number of columns in Table 7–1), and the number of levels of variable B as R (the number of rows in Table 7–1). The total number of groups, or treatment combinations, is K. Here, K is equal to RC. The total number of scores in the experiment, N, is then equal to Kn, or RCn, where n is the number of scores in each group.

We start the analysis as if a single-variable design had been used!

$$SS_{TOT} = \sum_{i=1}^{N} X_i^2 - \frac{(GT)^2}{N}$$

$$SS_{BG} = \frac{T_1^2 + T_2^2 + \cdots + T_K^2}{n} - \frac{(GT)^2}{N}$$

$$SS_{WG} = SS_{WG1} + SS_{WG2} + \cdots + SS_{WGK}$$

These computations are done in exactly the same way as shown in the last chapter by considering the K conditions to be K groups in a single-variable design.

Analysis of between-groups variation. Once the SS_{BG} has been correctly computed, we may proceed to the computations of three different sums of squares which reflect the two main effects and the interaction. These will be called the SS_A, related to variable A, the SS_B, related to variable B, and the $SS_{A\times B}$, which will reflect the degree of interaction between the two independent variables. These three sums of squares are actually three independent parts of the SS_{BG}. In the same way that the between- and within-groups sums of squares add up to the total sum of squares, we find that

$$SS_{BG} = SS_A + SS_B + SS_{A\times B} \qquad (7\text{–}4.1)$$

The totals from each level of variable A, referred to as T_{A1}, T_{A2}, . . . , T_{AC}, are used to obtain the SS_A:

$$SS_A = \frac{T_{A1}^2 + T_{A2}^2 + \cdots + T_{AC}^2}{nR} - \frac{(GT)^2}{N} \qquad (7\text{–}4.2)$$

The divisor in (7–4.2), nR, is the number of scores in each of the totals in the numerator of the term.

Now, using the totals of the levels of variable B, $T_{B1}, T_{B2}, \ldots, T_{BR}$, we may obtain the sum of squares for the other main effect by

$$SS_B = \frac{T_{B1}^2 + T_{B2}^2 + \cdots + T_{BR}^2}{nC} - \frac{(GT)^2}{N} \qquad (7\text{–}4.3)$$

Again the denominator of the sum of squared totals is equal to the number of scores in each of the totals.

The SS_A and SS_B are parts of what we originally calculated as SS_{BG}. The remaining amount of the SS_{BG} is assigned to the interaction effects present in the experiment. The general method of obtaining the $SS_{A \times B}$ is to use (7–4.1). Hence,

$$SS_{A \times B} = SS_{BG} - SS_A - SS_B \qquad (7\text{–}4.4)$$

Degrees of freedom. The degrees of freedom associated with SS_{TOT}, SS_{WG}, and SS_{BG} are found as in the single-variable analysis. However, since we have broken down the SS_{BG} term into its three component parts, we also need to consider the degrees of freedom associated with these separate factors.

The SS_A is based upon variation in the means (totals having been used for computational ease) from the different levels of variable A. Since there are C such mean values, in general terms, the unbiased estimate of the population variance which is to be found from the SS_A is found by dividing the SS_A by the number of means minus one, $C - 1$. Similar reasoning for variable B allows

$$df_A = C - 1 \quad \text{and} \quad df_B = R - 1 \qquad (7\text{–}4.5)$$

The degrees of freedom associated with the between-groups source of variability have been reduced after the degrees of freedom in (7–4.5) are assigned to variables A and B. The remainder of the $K - 1$ degrees of freedom for between-groups variation is related to the $SS_{A \times B}$. We obtain

$$\begin{aligned} df_{A \times B} &= df_{BG} - df_A - df_B \\ &= (K - 1) - (C - 1) - (R - 1) \\ &= (RC - 1) - (C - 1) - (R - 1) \\ &= RC - C - R + 1 \\ df_{A \times B} &= (C - 1)(R - 1) = (df_A)(df_B) \end{aligned} \qquad (7\text{–}4.6)$$

We find then that the SS_{BG} has been broken into three additive parts and that the df_{BG} has also been broken down into three additive and corresponding degrees-of-freedom terms.

The F-ratios. The analysis of variance in this two-variable experiment results in four major sources of variability: the two main effects, the interaction, and the within-groups source. The degrees of freedom related to each of these terms is used to convert the corresponding sums of squares to mean squares, i.e., estimates of population variances.

TABLE 7–2

Summary of the Analysis of Variance for a Two-Variable Factorial Design

Source	df	SS	MS	F
Variable A	$C - 1$	$\left(\sum_{c=1}^{C} T_{A_c}^2\right)/nR - \frac{(GT)^2}{N}$	$\frac{SS_A}{df_A}$	$\frac{MS_A}{MS_{WG}}$
Variable B	$R - 1$	$\left(\sum_{r=1}^{R} T_{B_r}^2\right)/nC - \frac{(GT)^2}{N}$	$\frac{SS_B}{df_B}$	$\frac{MS_B}{MS_{WG}}$
$A \times B$	$(C - 1)(R - 1)$	$SS_{BG} - SS_A - SS_B$	$\frac{SS_{A\times B}}{df_{A\times B}}$	$\frac{MS_{A\times B}}{MS_{WG}}$
Within Groups	$N - K$	$SS_{WG_1} + SS_{WG_2} + \cdots + SS_{WG_K}$	$\frac{SS_{WG}}{df_{WG}}$	–
Total	$N - 1$	$\sum_{i=1}^{N} X_i^2 - \frac{(GT)^2}{N}$	–	–

Three separate F-ratios will be formed and evaluated for significance in the two-variable design. The appropriate summary table for this design, Table 7–2, indicates the formation of these three F-ratios and shows the general method of the entire analysis. Explicitly, the F-ratios are

$$F_A = \frac{MS_A}{MS_{WG}}, \qquad F_B = \frac{MS_B}{MS_{WG}}, \qquad \text{and} \qquad F_{A\times B} = \frac{MS_{A\times B}}{MS_{WG}} \tag{7–4.7}$$

In order to illustrate the exact nature of the null hypothesis under test in each of these F-tests, consider a 3 × 3 factorial design. We may first think in terms of 9 populations of scores corresponding to the 9 combinations of the levels of A and B. Denoting the mean of the $A1$-$B1$ population as μ, Table 7–3 shows what the means of the

TABLE 7–3

Relationships Between Population Means in a 3 × 3 Factorial Design

		Variable A		
		$A1$	$A2$	$A3$
	$B1$	μ	$\mu + b_{A2} + b_{A2 \times B1}$	$\mu + b_{A3} + b_{A3 \times B1}$
Variable B	$B2$	$\mu + b_{B2} + b_{A1 \times B2}$	$\mu + b_{A2} + b_{B2} + b_{A2 \times B2}$	$\mu + b_{A3} + b_{B2} + b_{A3 \times B2}$
	$B3$	$\mu + b_{B3} + b_{A1 \times B3}$	$\mu + b_{A2} + b_{B3} + b_{A2 \times B3}$	$\mu + b_{A3} + b_{B3} + b_{A3 \times B3}$

other populations should be in terms of the constants of the additive model (7–3.1).

The null hypothesis related to the main effect of variable A may be perceived if each *column* is considered as a single population rather than as three popluations. We must now consider that the interaction constants in the model sum to zero within any row or within any column. If they did not, we could simply increase or decrease the constants related to the main effects until this requirement was satisfied. When this manipulation is done, one finds that the means from each column or level of A are equal to each other if the main effects of A are zero. In other words, if the constants b_{A2} and b_{A3} are both zero, all three populations related to the conditions of variable A will have equal means. The existence of this situation is tested by the F_A in this analysis of variance. In an analogous fashion, the F_B is based on the hypothesis that the row means are equal; i.e., that the constants b_{B2} and b_{B3} are both zero.

The F-test for the interaction of variables A and B automatically allows for the existence of main effects of A and B and specifically tests the hypothesis that *all* of the interaction constants are zero. The interaction-test hypothesis may also be likened to a test of equal differences between the means of corresponding row (or column) populations. Table 7–4 illustrates this concept in a 4 × 2 factorial design. In this table realistic means have been used which suggest that both main effects are significant. The four differences shown in the last row of the table reflect the degree of interaction present. The null hypothesis for the interaction is that the populations in the eight conditions are such that the true differences between row entries are equal. The statistical question, then, is whether or not the differences shown in Table 7–4 are close enough to one another to permit the explanation of chance to be considered reasonable. The $F_{A \times B}$ provides an answer

TABLE 7–4

Hypothetical Sample Means in a 4 × 2 Factorial Design Showing the Interaction Effects by Difference Scores

		Variable *A*			
		*A*1	*A*2	*A*3	*A*4
Variable *B*	*B*1	31.2	14.5	18.5	27.6
	*B*2	21.5	3.8	10.4	20.3
Differences (Row *B*1 − *B*2)		9.7	10.7	8.1	7.3

to this question because the two mean square estimates used in the F-ratio are independent estimates of variances presumed to be equal.

As earlier, each F-ratio is evaluated by consulting Table E, using the degrees of freedom associated with both numerator and denominator mean squares. It is important to note that any of these three F-ratios may be significant without changing the probability level of either of the other F-ratios. It is possible that none, any one, any two, or all three of the null hypotheses related to these tests may be rejected. Interpretations of these possible outcomes will be suggested in the examples which follow.

Example 7–1. Suppose that the experiment outlined in Table 7–1 were conducted using 4 different subjects in each of the 12 treatments of the design. The following are hypothetical data, measured in seconds, showing the reaction time for each of the 48 subjects. The totals of the scores are given for each of the 12 groups as well as the totals and corresponding means for each of the rows and columns. Note that the *GT* is 40.00 sec. Let us perform an analysis of variance appropriate to this design.

First, calculate the value of $(GT)^2/N$, which is used a number of times:

Number of stimuli and responses	Dosages of chlorpromazine			
	Zero dose ($A1$)	Half dose ($A2$)	Full dose ($A3$)	
1 ($B1$)	.27 .41 .35 .37	.45 .38 .51 .43	.47 .49 .61 .55	
	$T_1 = 1.40$	$T_2 = 1.77$	$T_3 = 2.12$	$T_{B1} = 5.29$ $M_{B1} = .441$
2 ($B2$)	.55 .61 .58 .82	.62 .71 .68 .49	.81 .75 .66 .71	
	$T_4 = 2.56$	$T_5 = 2.50$	$T_6 = 2.93$	$T_{B2} = 7.99$ $M_{B2} = .666$
3 ($B3$)	.87 .95 .72 .63	1.07 .87 .98 1.15	1.25 1.04 .97 1.15	
	$T_7 = 3.17$	$T_8 = 4.07$	$T_9 = 4.41$	$T_{B3} = 11.65$ $M_{B3} = .971$
4 ($B4$)	1.32 1.05 1.16 1.20	1.37 1.45 1.00 1.27	1.11 1.48 1.47 1.29	
	$T_{10} = 4.73$	$T_{11} = 5.09$	$T_{12} = 5.25$	$T_{B4} = 15.07$ $M_{B4} = 1.256$
	$T_{A1} = 11.86$ $M_{A1} = .741$	$T_{A2} = 13.43$ $M_{A2} = .839$	$T_{A3} = 14.71$ $M_{A3} = .919$	$GT = 40.00$

$$\frac{GT^2}{N} = \frac{(40.)^2}{48} = 33.3333$$

The SS_{TOT} is found as usual by adding all of the squared raw scores together and subtracting the above value. Hence,

$$SS_{TOT} = 38.7138 - 33.3333 = 5.3805$$

The SS_{WG} is found by obtaining a sum of squares within each of the 12 groups and adding these together. These 12 values are shown, respectively, in the following calculations:

$$\begin{aligned} SS_{WG} &= .0104 + .0087 + .0120 \\ &\quad + .0450 + .0285 + .0121 \\ &\quad + .0625 + .0435 + .0455 \\ &\quad + .0373 + .1153 + .0729 \\ &= .4937 \end{aligned}$$

The SS_{BG} is calculated from the 12 group totals as

$$\begin{aligned} SS_{BG} &= \frac{(1.40)^2 + (1.77)^2 + \cdots + (5.25)^2}{4} - 33.3333 \\ &= 38.2203 - 33.3333 = 4.8870 \end{aligned}$$

The sum of the SS_{WG} and SS_{BG} is 5.3807, which checks with the SS_{TOT} value within the margin allowed for errors of rounding.

We may now use (7–4.2) and (7–4.3) to find the sums of squares associated with the main effects of the two independent variables:

$$\begin{aligned} SS_A &= \frac{(11.86)^2 + (13.43)^2 + (14.71)^2}{(4)(4)} - 33.3333 \\ &= 33.5880 - 33.3333 = .2547 \\ SS_B &= \frac{(5.29)^2 + (7.99)^2 + (11.65)^2 + (15.07)^2}{(4)(3)} - 33.3333 \\ &= 37.8876 - 33.3333 = 4.5543 \end{aligned}$$

Use of (7–4.4), the SS_{BG}, and the main effects permit

$$SS_{A \times B} = 4.8870 - .2547 - 4.5543 = .0780$$

The completion of the analysis and summary of results follow the model shown in Table 7–2:

Source	*df*	*SS*	*MS*	*F*
Dose (*D*)	2	.2547	.1274	9.30*
Task (*T*)	3	4.5543	1.5181	110.81*
$D \times T$	6	.0780	.0130	.95
Within Groups	36	.4937	.0137	–
Total	47	5.3805	–	–

* $P < .01$.

The F-ratios for both of the main effects are much higher than the tabled values of F associated with the 1 per cent level of confidence for rejecting the null hypothesis. The first of these is based on 2 and 36 *df* while the second is based on 3 and 36 *df*. We may rather safely conclude that both of the null hypotheses associated with the main effects of the independent variables are false. The F-ratio associated with the interaction effects is clearly about what one should expect if the related null hypothesis for the interaction were true. We have no basis for concluding that interaction effects are present.

The lack of a significant interaction allows us to state the relationship between independent and dependent variables in a simple fashion. If we may assume that the experiment has been conducted correctly, then we may make the experimental decision that the *degree of difference* between the means for the three drug dosages represents a true effect of these dosages which is applicable to all levels of the task variable studied. We may also conclude that the average reaction times are affected by the type of task. The lack of an interaction suggests that the *changes* in mean reaction time associated with changes in the task are about the same for any of the drug levels.

If desired, one might use the row and column means shown in this example and the model shown in Table 7–3 and, assuming that all interaction effects are zero, calculate the best point estimates of the population means of each of the 12 conditions of the experiment. However, this procedure is beyond the hypothesis-testing function of the analysis of variance and will not be illustrated.

In the next case, Example 7–2, we will consider an experiment in which neither main effect is significant but in which a strong interaction between the two variables is present. This example illustrates the importance of the interaction effect which is assessed in the two-variable factorial design.

Example 7–2. An investigator interested in concept identification has been using geometric designs as stimuli in his problems. One characteristic of the designs which has not been too obvious to previous subjects is the distinction between parallelograms with right angles (rectangles) and those with slightly different angles. He decides that past experience with this distinction might be an important variable.

Twenty-four fifth-grade students are used as subjects in a 2×2 factorial design. One independent variable is the degree of training in simple aspects of geometry. Half of the students receive one hour of special instruction in simple geometry with emphasis on parallelograms and the special case of rectangles. The other students are given an hour of training in English grammar.

Each training group is further subdivided in terms of the concept task the subjects are required to solve. The stimulus dimension of form (using rectangles and non-rectangular parallelograms) is one of the characteristics used. For some subjects, the form of the stimuli is important to the solution of the problem; i.e., form is relevant to solution, and in the other conditions, form is irrelevant and the subjects may solve the problem without noting form differences from trial to trial.

Each student is given repeated trials on the task until he reaches a solution to the problem. The dependent variable is the number of trials needed by a subject to attain the solution. The design and the scores, with 6 subjects in each group, are given below.

	Problem		
Prior training in geometry	Form relevant (*A*1)	Form irrelevant (*A*2)	
None (*B*1)	18	12	
	25	8	
	14	17	
	21	11	
	21	20	
	26	12	
	$T_1 = 125$	$T_2 = 80$	$T_{B1} = 205$
One hour (*B*2)	9	23	
	16	18	
	13	18	
	8	14	
	10	23	
	6	20	
	$T_3 = 62$	$T_4 = 116$	$T_{B2} = 178$
	$T_{A1} = 187$	$T_{A2} = 196$	$GT = 383$

The analysis of variance for these data follows:

$$SS_{TOT} = 6{,}873 - 6{,}112.04 = 760.96$$

$$SS_{WG} = 98.83 + 95.33 + 65.33 + 59.33 = 318.82$$

$$SS_{BG} = \frac{(125)^2 + (80)^2 + (62)^2 + (116)^2}{6} - 6{,}112.04$$

$$= 6{,}554.17 - 6{,}112.04 = 442.13$$

Before proceeding, we check that $SS_{BG} + SS_{WG}$ is equal to SS_{TOT}, which it is.

$$SS_A = \frac{(187)^2 + (196)^2}{(6)(2)} - 6{,}112.04$$

$$= 3.38$$

$$SS_B = \frac{(205)^2 + (178)^2}{(6)(2)} - 6{,}112.04$$

$$= 30.38$$

$$SS_{A\times B} = 442.13 - 3.38 - 30.38$$

$$= 408.37$$

The summary of the analysis may be given as

Source	df	SS	MS	F
Problem (P)	1	3.38	3.38	.21
Training (T)	1	30.38	30.38	1.91
$P \times T$	1	408.37	408.37	25.62*
Within Groups	20	318.82	15.94	–
Total	23	760.96	–	–

$*P < .01$.

It is apparent from comparing the F-ratios for the main effects with the values tabled for 1 and 20 degrees of freedom that neither main effect is large enough to permit rejection of the null hypothesis of equal population means. Note that the F for the interaction is far above the tabled value of 8.10 needed for rejection at the 1 per cent level of confidence. We conclude that these two independent variables interact in their effects upon trials to solution.

In general, the problems do not appear to be of unequal difficulty. Also, the average effects of the two types of training do not seem to be different. However, the significance of the interaction term requires modification of these two general statements. We may say that the effect of prior training on this type of task depends upon (or is different for) the type of problem to be solved. An alternative verbal description of the interaction would be that the difficulty of the two types of problems depends upon (or is different for) the type of prior training given the subjects.

Example 7–3 illustrates an experiment in which one independent variable is significant and interacts in a weak way with the other independent variable. The particular dependent variable used in the example is such that negative values for the raw scores are found in the data. The use of a linear transformation, the addition of a constant, is made in order to permit ease in the analysis of variance. It should be carefully noted that this type of transformation has no effect upon any of the analysis-of-variance results. Since all scores in the experiment are increased by the same constant, all means are increased by the same constant. All of the measures of dispersion, including those based upon differences between group means, will remain unchanged when a constant of addition is used on the entire set of scores.

Example 7–3. When a stationary light is viewed in an otherwise dark room, the movements of one's eyes produce an illusion of movement of the light. This phenomemon is called the autokinetic effect. It is well known that the report of a person as to the extent of movement he perceives may be influenced by the reports of other people.

An investigator studying the autokinetic effect feels that the influence of other people's judgments on a subject might depend upon the size of the stimulus light used in the experiment. In order to test this idea, he designs a 3 × 2 factorial experiment in which both the size of the stimulus light and the amount of movement reported by other people (who pretend to be subjects) are varied. Specifically, two sizes of light, of .2 in. and 1.0 in. diameters, are used. He instructs his helpers to give estimates of movement of the light which average 4 in., 12 in., and 20 in. for three levels of this variable.

The subjects, 5 in each of the resulting 6 groups, first give an estimate of movement seen before the other "subjects" report. Then, following the rigged reports, which the subject hears, the subject gives a second judgment of movement. The dependent variable used by the experimenter is the difference between each subject's first judgment and his second judgment, always subtracting the first report from the second. The following data, then, are changes in each subject's judgments, measured in inches.

Diameter of light	Average report of others		
	4 in. ($A1$)	12 in. ($A2$)	20 in. ($A3$)
0.2 in. ($B1$)	5 −5 −6 −2 −1	8 3 1 5 5	8 13 11 9 10
1.0 in. ($B2$)	3 −2 2 2 4	5 −1 4 3 7	5 3 1 7 5

The analysis of variance may be done directly on these data, taking care with the negative scores when finding totals. However, the probability of making an error is usually increased when using negative numbers in such an analysis. We may, for convenience only, transform the entire set of 30 scores to non-negative scores merely by adding a constant to each score. Since the smallest score found in this experiment is −6, let us add a constant of 6 to each score. The analysis of the resulting set of scores will give the same answers as the analysis of the untransformed scores. The values of the means based on the transformed scores would, of course, need to be corrected in order that they be equal to the untransformed measures. The resulting scores for analysis are given in the following array.

Variable B	Variable A (A1)	(A2)	(A3)	
(B1)	11 1 0 4 5	14 9 7 11 11	14 19 17 15 16	
	$T_1 = 21$	$T_2 = 52$	$T_3 = 81$	$T_{B1} = 154$
(B2)	9 4 8 8 10	11 5 10 9 13	11 9 7 13 11	
	$T_4 = 39$	$T_5 = 48$	$T_6 = 51$	$T_{B2} = 138$
	$T_{A1} = 60$	$T_{A2} = 100$	$T_{A3} = 132$	$GT = 292$

The value of $(GT)^2/N$, which is a correction used to obtain squared deviations from squared raw scores, is equal to 2,842.13, in this example.

As usual, the total and between-groups and within-groups sources are worked with first:

$$
\begin{aligned}
SS_{TOT} &= 3420 - 2842.13 = 577.87 \\
SS_{WG} &= SS_{WG1} + SS_{WG2} + SS_{WG3} + SS_{WG4} + SS_{WG5} + SS_{WG6} \\
&= 74.8 + 27.2 + 14.8 + 20.8 + 35.2 + 20.8 \\
&= 193.6 \\
SS_{BG} &= \frac{(21)^2 + (52)^2 + \cdots + (51)^2}{5} - 2,842.13 \\
&= 384.27
\end{aligned}
$$

Since the between and within sums of squares total the SS_{TOT}, we now break down the between-groups sum of squares into its component parts.

$$
\begin{aligned}
SS_A &= \frac{(60)^2 + (100)^2 + (132)^2}{(5)(2)} - 2,842.13 \\
&= 260.27 \\
SS_B &= \frac{(154)^2 + (138)^2}{(5)(3)} - 2,842.13 \\
&= 8.54 \\
SS_{A \times B} &= 384.27 - 260.27 - 8.54 = 115.46
\end{aligned}
$$

After converting the sum of squares for each source to a mean square estimate, the F-ratios may be calculated in the usual manner. The summary of results is:

Source	*df*	*SS*	*MS*	*F*
Reports (R)	2	260.27	130.14	16.13*
Lights (L)	1	8.54	8.54	1.06
$R \times L$	2	115.46	57.73	7.15*
Within	24	193.60	8.07	–
Total	29	577.87	–	–

* $P < .01$.

It may be seen that the effects of reports of others are significantly different from zero in this experiment. Generally, one may observe that the scores get larger for the groups which heard larger reports of apparent movement from the helpers in the experiment. This indicates that the subjects' second scores were progressively larger than their first scores, as one moves from condition $A1$ to $A3$. The average dependent measures for the two diameters of light show no great tendency to be different. Certainly the observed difference in these two means cannot be judged to be reliable, using the F of 1.06 as our index.

Note the significance of the interaction term in this analysis. This indicates that the effect of the "Reports" variable is not the same for both levels of the "Lights" variable. Inspection of the group totals shows this lack of uniform effect. For the smaller light conditions, changes in the magnitude of reports by the false "subjects" produced greater shifts in the dependent variable than was the case for the larger light conditions. Note that the average effect of "Reports" was in the same direction for each level of the "Lights" variable, but that the degree of change was not the same. This can be termed as a weak interaction as opposed to the type shown in Example 7–2.

7–5. Generalizations of Results. We have already noted the distinction between the statistical decision regarding a null hypothesis and the experimental decision dealing with the interpretation of what may have caused differences judged to be significant. However, it should be mentioned here that generalization of either statistical or experimental decisions must be made with care. The human tends to summarize experience (in this case the results of an experiment) by use of language which is much more abstract than the situations to which it applies. This tendency to describe the knowledge gained in an experiment in rather broad terms is what is meant (in this context) by generalization. It should be clear to the reader that the results obtained in a particular experiment should be repeatable if the exact conditions were used again (ignoring the probabilities of the two types of errors in our statistical tests). However, it is actually not possible to reproduce any situation exactly! Of more importance is the case in which results from one set of circumstances are used to predict the results which will be obtained in different conditions involving the same types of independent and dependent variables. Consider, for instance, the experiment described in Example 7–3. A condition in which the "Reports" of the other persons were greater than the highest level used in the example might be predicted to produce a still greater mean in the dependent measure than found here. This is not justified logically. Strictly speaking, our results should be used to predict behavior in conditions where the same levels of the independent variable are used. Even then, variation in constant conditions of the experimental situation, such as room temperature, personality of the experimenter, etc., may produce changes in the results.

About the best the scientist can do in applying results to new situations is to make a judgment as to whether or not he perceives any aspects of the new situation which might change the results and, if not, make a prediction that the same type of results will occur in the new situation. Such procedures are, at present, more of an art of the scientist than a branch of the logic of science. There are, of course, some guiding principles available in these decisions, but they fall beyond the purpose of this introductory text.

7–6. The Nature of the Error Term. The denominator of the F-ratio, which we will call the *error term*, is an unbiased estimate of the variance of scores in the populations. In the first two types of designs considered, this error term reflects the variance between scores from different subjects in the experiment. The fact that such population scores are different illustrates that certain differences between subjects occur even though all subjects are defined as being in a common population. A multitude of individual differences, none of which are under study in a given experiment, affect the scores measured in an experiment, even when considering scores obtained under the same experimental conditions.

Equation (5–8.1) shows that the standard error of a sample mean depends upon the magnitude of the population variance (or σ). When the scores in a population are highly dispersed, random samples of a given size may vary widely in their means. Conversely, a relatively low dispersion of population scores produces small variability of the sample means. In a very real sense, the estimate of the population variance, the error term, indicates how large differences between group means should be if only chance variation is involved. It may readily be seen that differences between the means of the groups of an experiment may yield a rejection of the null hypothesis in one case (where the error term is small), but the same set of means may not be significant when the error term is large.

As will be discussed in a later chapter, the sensitivity of an experiment to effects of the independent variable of a given amount depends upon the variance of the scores within each population sampled in the experiment.

The error term reflects the degree of individual differences in the dependent measure, but it also reflects errors of measurement inherent to our instruments or other measuring devices. Even though two subjects "ought" to be equal in their scores, some difference may be obtained because of errors in the measurement instrument or procedures. One reason that accurate measurements are useful is to increase the sensitivity of an experiment.

Sample size and the physical sciences. The need for inferential statistics comes directly from the fact that variability in results of sampling from the same population is as large as commonly found. One may examine (5–8.1) from the viewpoint of sample size rather than population dispersion. Clearly, as the sample size grows very large, any degree of population dispersion is overcome in the sense that the variability of sample means becomes very small.

Perhaps one reason that the physical sciences have been termed "exact" by the layman (and scientists as well) is that the sample sizes dealt with have been, historically, very large. Consider the predictions made and tested in a laboratory when working with Boyle's Law relating pressure and volume of gasses. Even a small container has millions (conservatively) of molecules, which may be considered the elements of the sample set. The pressure, for example, is a type of average effect per moment of this large sample. The "exactness" of the results stems from the large size sample and not from the inherent similarity of the "scores" of the elements. For a further illustration, consider the half-life of a radioactive substance. This half-life, when measured in a small amount of the substance, may be validated by appropriate measurements almost exactly. But ask the physicist to predict the half-life of a sample of only 100 atoms of the same substance. The answer will most likely be given in terms already familiar to the reader: a range of times with a probability value corresponding to the range.

There are probably many laws of behavior which are "exact," if one could test them with 10,000,000 subjects.

The statistical point related to this discussion is that very small effects of independent variables upon dependent measures may be found if the sample sizes are quite large. Such effects will be missed with small samples. This is an additional reason for withholding one's opinion about the state of nature when a null hypothesis cannot be rejected. There may exist some relatively small effects which are not statistically apparent with the sample sizes used.

EXERCISES

1. Find the sum of squares and *df* for each variable and the interaction term:

   (a) Cell totals based on $n = 6$ are:

	$A1$	$A2$	$A3$	$A4$
$B1$	94	107	145	82
$B2$	67	98	101	45
$B3$	63	87	95	55

   (b) Cell totals based on $n = 10$ are:

	$A1$	$A2$	$A3$
$B1$	12.7	18.3	21.2
$B2$	18.5	17.7	13.4
$B3$	15.7	16.4	16.4

   (c) Cell totals based on $n = 15$ are:

	$A1$	$A2$
$B1$	412	333
$B2$	353	397

2. If the SS_{WG} is as given below for each part of Exercise 1, complete each analysis of variance and assess the significance of each F.

   (a) $SS_{WG} = 6{,}512$
   (b) $SS_{WG} = 62.53$
   (c) $SS_{WG} = 247$

3. Use the two-variable type of analysis of variance to test for significant effects in the following data.

	$A1$					$A2$				
$B1$	14	18	19	20	20	18	19	20	20	22
$B2$	16	18	19	19	21	17	18	21	21	24
$B3$	14	15	17	18	19	16	18	19	20	22

4. Using the data from Exercise 1, part (c), verify that the $SS_{A\times B}$ in a 2×2 factorial design may be found directly by using two totals:

$$T_1 = T_{A1,B1} + T_{A2,B2} \qquad T_2 = T_{A1,B2} + T_{A2,B1}$$

Applications: Use information from all previous chapters where relevant.

5. An investigator wishes to show that a teaching machine and program will improve the rate of learning to spell English words. He uses 20 randomly selected fourth-grade children, divided into two groups of 10. The first group (called the experimental group) studies a set of 50 words on a teaching machine with a special program designed to make students produce the correct response to each word through standard techniques. The control subjects are simply given the words on a piece of paper and allowed to study them for a time equal to the average time used in the experimental group.

   Each student is tested 24 hours later. The number of words spelled correctly is given below for each student. Test the hypothesis of no mean difference between conditions.

   Experimental: 34, 38, 39, 40, 41, 38, 40, 41, 42, 44
   Control: 32, 34, 35, 37, 38, 37, 38, 40, 41, 42

6. The investigator in Exercise 5 considered that there might be a difference in spelling ability between boys and girls. He had the foresight to use 5 boys and 5 girls in each group shown in Exercise 5. The first 5 scores in each group are from boys while the second 5 are from girls. Reanalyze the data in light of this new information.

8

Analysis of Repeated Measurements

8–1. The Repeated-Measures Design. The previous chapters on analysis of variance tests have dealt exclusively with designs in which a single score from each subject is used. As indicated earlier, such a single score may be a summary of the behavior of a subject in an experimental situation. In effect, such a summary ignores the manner in which the composite score may have been obtained. For instance, in a verbal learning experiment each subject may be given 15 trials. On each trial, the subject may make some variable number of correct responses. The single summary score given a subject might be the total number of correct responses made over the entire 15 trials. Such variates, one for each subject, could be used for an analysis of variance of either of the types already examined.

However, instead of simply adding the results of each trial together for a single subject, suppose that we examine the trial scores separately. For instance, we might obtain the average number of correct responses on trial number 1 for all subjects working under the same experimental conditions. We might find that large differences exist between such means calculated for different experimental conditions. A similar comparison might be made for each of the 15 trials.

It is quite possible that differences exist between experimental groups on the first trial or two but all later trial means may be essentially equal. Under this set of circumstances, the total correct responses of each subject, when used as the dependent variable, might yield results which were not significant, even though true differences existed on the first few trials.

Another interesting possibility which may occur in cases where a series of measurements are taken on each subject is that one group of subjects may start with lower means than another group on early trials but increase so that it has higher means on the later trials. The over-all means for each group would tend to be equal, but the manner of obtaining the means would be different when the separate trials were considered. Figure 8–1 illustrates the above situation for a two group experiment with 10 trials per group. The data are artificial.

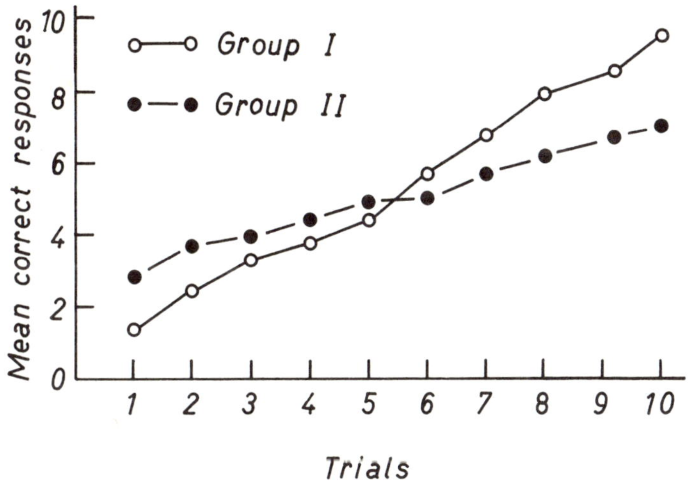

Fig. 8–1. An illustration of a groups-by-trials interaction. Note the different slopes for the two groups.

Notation. We will consider the general design involving repeated measures on the same subject to have the same number of measurements for each subject in the experiment. Let us refer to this number of measures per subject as L. Again, let us use the letter K to refer to the number of experimental groups, or levels of a single independent variable. Each subject may be randomly assigned by the experimenter to only one of these K conditions, but each subject will receive a series of L measures, usually taken in a precise order which is the same for all subjects. Restricting attention to cases in which an equal number of subjects, n, is used in each of the K treatments, we find that N, the total number of scores, is equal to nLK.

Prior to giving the appropriate analysis-of-variance procedures for this type of design, let us examine the comparisons which may be of primary interest.

Sources of interest. As in the single-variable case, one important aspect of the repeated-measures design as described above will be the average scores from the K different groups, one for each of the levels of the independent variable. We shall refer to this factor as between-group differences.

A second comparison of some importance is the one involving the average scores, over all of the nK subjects, from each of the L separate measurements. We shall refer to such comparisons as the *trials* effects. One should note that the term *trials* should not be interpreted too strictly. Any ordered sequence of scores from a single subject may be termed trials even though the scores are differentiated on some other basis.

Of particular interest is the interaction of the trials effects and groups. This is illustrated in Fig. 8–1 as the lack of parallel trial curves for the two groups diagramed. This interaction may be examined by finding trial means separately for each of the K independent groups. We will have KL such means which can easily be arranged in the familiar row-and-column pattern used in the last chapter. One may think of the rows as corresponding to the K levels of one independent variable and the columns as corresponding to the L levels of another "independent" variable, trials. The interaction of groups by trials is then analogous to the interaction of the two independent variables as described in the last chapter. However, there is one important difference in the present interaction term: a new type of error term must be used in forming the F-ratio related to a significance test.

8–2. Correlated and Uncorrelated Means. Consider a simple two-condition experimental design, in order that the need for a different type of error term may be illustrated in the repeated-measures experiment. Table 8–1 gives two sets of hypothetical scores. Assume that each score given in the table is the number of words spelled correctly on a 30-item spelling test and that the two conditions are two different lists of words to be compared in difficulty.

If each of the 16 scores in Table 8–1 represents a different subject, the design is a single-classification type appropriate for the analysis-of-variance technique illustrated in Chapter 6. Note that the mean from Group I is 15.75 and the mean for Group II is 16.50. The difference between these means will be reflected in the SS_{BG}. Now suppose that we add two more scores to each group. We will do this in two different ways. The first way is to assume that 4 new subjects are used, two on each list. The set of scores in Table 8–1 called "additional scores-different subjects" might be the results. As pictured, we have obtained two very good spellers in Group I and two average spellers in Group II. The new group means are shown as 18.4 and 16.7, for the two conditions, respectively. Note that our "random" selection of two good spellers in Group I has rather greatly changed the difference between the group means.

Now suppose that the 4 new scores added to the basic data in Table 8–1 come from only two new subjects, each being tested once on each list of words. The subject making the perfect score in Group I under the heading "same subjects" continues to the second condition, *and because of his superior spelling ability*, also contributes a high score to the other condition. The same holds true for the second subject.

TABLE 8–1

An Example of Differences in Means for Independent Groups Compared to Correlated Groups

	Group I	Group II
	16	18
	18	15
	14	17
	17	19
	17	15
	15	18
	13	16
	16	14
	$M_1 = 15.75$	$M_2 = 16.50$
	$T_1 = 126$	$T_2 = 132$
	ADDITIONAL SCORES	
	Group I	Group II
	30	17
Different subjects	28	18
	$M_1' = 18.4$	$M_2' = 16.7$
	30	29
Same subjects	28	30
	$M_1' = 18.4$	$M_2' = 19.1$

Now the new means are 18.4 and 19.1. The *difference* between the means of the two conditions remains about the same as in the original case. This is because the selection of a person with deviant ability (compared to the average subject) has been overcome to some extent by virtue of the fact that the person has been higher than average on both conditions.

In the case of independent subjects in each group, the variation of the between-group means, assuming the null hypothesis, depends upon the between-subjects' variability. However, when each subject serves in each condition, the variation of group means (assuming H_0) depends upon the degree to which single subjects vary in their performance from one condition to the next. If one considers any two scores in an independent-group design, we say that the two scores are unrelated or **uncorrelated**, because the value of one of these scores gives no information as to what the value of the other score might be. However, when the same subject is used in both conditions of a simple experiment, knowledge of one of his scores (noting whether he is above or below the mean of the condition) will often give information about whether his other score is above or below the mean of the second condition. The two scores are then related to one another or **correlated**.

The means from any independent-group experiment, one subject serving under only one condition, are called *uncorrelated means*. The means from conditions such that each subject has given a score to each condition are called *correlated means*.

Correlated means in the repeated-measures design. Using the above guides, we may now consider the repeated-measures design once again and find which sources of variation arise from correlated means and which arise from uncorrelated means.

The analysis of differences between the means of the K conditions will be considered to give an uncorrelated source of variation since a single subject's performance can affect only the mean of one group.

However, the L different means for the trials variable are clearly correlated, since each subject in the experiment gives one score to each trial mean.

The analysis of the groups by trials interaction effect is not so simple. But one may observe that the interaction effect represents a lack of equal differences between any two groups for each of the L trials, as illustrated in Fig. 8–1. It may then be claimed that the interaction depends upon how much each group changes from trial to trial, rather than upon how different the groups are in their over-all means. We conclude that the groups-by-trial interaction term is based upon correlated scores or means.

In general, any interaction effect, in which one or more of the variables contributing to the interaction is correlated, is correlated too.

The significance of identifying whether a source of variation is based on correlated or uncorrelated means is that the random-sampling variation of correlated means is different than for uncorrelated means. For the types of illustrations and examples used in this chapter, the variation of uncorrelated means will be greater than the variation of correlated means, assuming the appropriate null hypotheses.

8–3. The Correlated Error Term. In order to determine the significance of some measured effects, one needs to know how much variation between means is likely to result from random sampling. This estimate is made in the F-ratio by using the appropriate mean-square estimate (the error term) as the denominator of the ratio. In the case of testing for significant trial effects, for example, we have seen that the estimate of a population variance which reflects differences between subjects will not be appropriate as an error term. This estimate would typically be too large compared to the variance of the correlated means. In order to test the null hypotheses related to the correlated means in the repeated-measures design, we must develop an error term which reflects the degree of relationship between a subject's scores from trial to trial.

Let us look at a simple model for the score from a single subject (i) in group k on trial l:

$$X = M_k + C_l + A_i + e_j \qquad (8\text{–}3.1)$$

where M_k is the grand mean of all scores in the group, C_l is a deviation resulting from the particular trial considered, A_i is a deviation caused by that subject's average ability on the experimental task, and e_j is the error arising from the subject's behavior and its measurement. The variance of the e_j terms from each score in the group will yield the error term desired, i.e., the correlated error term. The computational methods for obtaining this error term will be considered in due course, but let us examine these e_j scores in a particular case in order to understand the type of variability being measured.

Table 8–2 illustrates a hypothetical set of data from a single group of subjects in a repeated measures design. Only 3 trials and 5 subjects are used in this illustration. Note that means (on a per-score basis) are given for each trial, each subject, and for the entire group (the grand mean). In addition, the deviations of each subject's mean from the grand mean, and each trial mean from the grand mean, are given.

TABLE 8–2

Hypothetical Data for Subjects in a Repeated-Measures Design

Subjects (i)	Trials (l) 1	2	3	Subject Means	A_i
1	16	18	23	19	5
2	7	7	10	8	−6
3	13	18	20	17	3
4	10	11	15	12	−2
5	12	15	15	14	0
Trial Means	11.6	13.8	16.6	Grand Mean = 14.0	
C_l	−2.4	−.2	2.6		

In order to obtain the e_j's from the raw scores, (8–3.1) is solved for e_j:

$$e_j = X - M_k - C_l - A_i \tag{8–3.2}$$

If (8–3.2) is applied to each score in Table 8–2, we obtain an e_j related to each raw score. The results of this procedure are shown in Table 8–3. To illustrate, the e_j for the fourth subject on his third trial is obtained by the calculations

$$e_j = 15. - 14. - 2.6 - (-2.0) = 0.4$$

By squaring and adding the e_j entries in Table 8–3, we obtain a sum of squares (note that the mean e_j is zero) within this group which

TABLE 8–3

Errors in Each Score from Table 8–2

Subjects	Trials 1	2	3
1	−0.6	−0.8	1.4
2	1.4	−0.8	−0.6
3	−1.6	1.2	0.4
4	0.4	−0.8	0.4
5	0.4	1.2	−1.6

may be used to get the estimate of population variance for the correlated terms of this design. We will refer to this sum of squares by calling the source of variation it measures the *subjects-by-trials interaction within groups.*

In the same way that a group-by-trials interaction may be shown to represent a lack of equal distance between groups on the different trials, the subjects-by-trials interaction within any one group is equivalent to the degree of departure from equal differences between subjects on the different trials.

Note that all subjects show an improvement over trials in Table 8–2. In no case did a subject get a lower score than on an earlier trial. This consistency of improving from trial to trial results in a mean square estimate related to the correlated error variance which is substantially less than the mean square based on differences between subjects. Consequently, the improvement in trial means shown in Table 8–2 is highly significant when the appropriate F-test is made with the correlated error term. If the same data were considered to be from an independent-group design (trials being the variable), the differences between trial means would be judged not significant.

As in our earlier work with the analysis of variance procedures, an uncorrelated error term will be developed in the repeated-measures analysis. This uncorrelated source of variation is related to differences between subjects within the groups and will be used to test the significance of the one main effect in this design which is uncorrelated, i.e., the between-groups effect associated with the K treatments.

8–4. The Analysis of the Repeated-Measures Design. The analysis of variance technique appropriate to the repeated-measures design involves five sources of variation in addition to the total variation. Three of these are of primary interest and the other two serve as the error terms in the F-ratios. Table 8–4 illustrates the general form of the experimental design.

The between-groups source. We have K different groups of subjects working under different levels of the independent variable. The comparison of the K resulting means will be termed the between-groups source.

As in all previous raw-score formulas for sums of squares, the set of totals corresponding to the means to be tested are used. Each total is squared and divided by the number of scores entering into the total; after adding these terms together, the correction factor is subtracted. This general method yields the following computational formula for

TABLE 8–4

General Arrangement of Data and Notation for a Repeated-Measures Design

Groups	Subjects	Trials 1	2	...	L	Totals
	1	X	X	...	X	T_{S1}
	2	X	X	...	X	T_{S2}
1	·	·	·	...	·	·
	·	·	·	...	·	·
	n	X	X	...	X	T_{Sn}
	Cell totals	$T_{K1,L1}$	$T_{K1,L2}$	...	$T_{K1,LL}$	T_{I}
	1	X	X	...	X	T_{S1}
	2	X	X	...	X	T_{S2}
2	·	·	·	...	·	·
	·	·	·	...	·	·
	n	X	X	...	X	T_{Sn}
	Cell totals	$T_{K2,L1}$	$T_{K2,L2}$	...	$T_{K2,LL}$	T_{II}
·				·		·
·				·		·
·				·		·
	1	X	X	...	X	T_{S1}
	2	X	X	...	X	T_{S2}
K	·	·	·	...	·	·
	·	·	·	...	·	·
	n	X	X	...	X	T_{Sn}
	Cell totals	$T_{KK,L1}$	$T_{KK,L2}$	...	$T_{KK,LL}$	T_K
Trial totals		T_1	T_2	...	T_L	GT

the sum of squares between groups:

$$SS_{BG} = \frac{T_{\text{I}}^2 + T_{\text{II}}^2 + \cdots + T_K^2}{nL} - \frac{(GT)^2}{N} \tag{8-4.1}$$

where the T_k entries represent the totals from the K different groups, nL is equal to number of scores in a single group, and GT and N are defined as usual.

The mean square based on the between-groups source of variation is found as in previous work. There are K means being compared, so the df_{BG} is equal to $K - 1$. This gives

$$MS_{BG} = \frac{SS_{BG}}{K - 1} \tag{8-4.2}$$

The trials source. The average score on each of the L trials is found by adding together all nK scores from a given trial and dividing this total by nK. The raw-score computation of the associated sum of squares is

$$SS_T = \frac{T_1^2 + T_2^2 + \cdots + T_L^2}{nK} - \frac{(GT)^2}{N} \tag{8-4.3}$$

where the T_l's refer to the trial totals.

The mean square related to (8–4.3) is obtained by noting that variation between the L trial means has $L - 1$ degrees of freedom:

$$MS_T = \frac{SS_T}{L - 1} \tag{8-4.4}$$

The groups-by-trials interaction. Let us refer to any set of scores in a single group and from the same trial as a **cell** of this design. The resulting KL cells will contain n scores each. As in the last chapter, the interaction sum of squares is obtained by subtraction. What we have called *cells* here is quite analogous to what was called groups in the previous chapter. We obtain the $SS_{G\times T}$ (the sum of squares for groups by trials interaction) by first obtaining a sum of squares between the cells.

These cell totals are given as $T_{K1,L1}$ through $T_{KK,LL}$ in Table 8–4. Let us number these KL cell totals as $T_1, T_2, \ldots, T_M$. The between-cells sum of squares may then be given as

$$SS_{BC} = \frac{T_1^2 + T_2^2 + \cdots + T_M^2}{n} - \frac{(GT)^2}{N} \tag{8–4.5}$$

We now have the SS related to the means from all combinations of groups and trials. The interaction of these two variables is obtained by subtracting the main-effects sums of squares from the between-cells sum of squares:

$$SS_{G\times T} = SS_{BC} - SS_{BG} - SS_T \tag{8–4.6}$$

As in the interaction term from the two-variable case, the $df_{G\times T}$ may be found by multiplying the df's of each main effect interacting. This gives $df_{G\times T} = (K - 1)(L - 1)$. The $MS_{G\times T}$ may now be found as

$$MS_{G\times T} = \frac{SS_{G\times T}}{df_{G\times T}} = \frac{SS_{G\times T}}{(K - 1)(L - 1)} \tag{8–4.7}$$

The uncorrelated-error term. The degree of variability between subjects treated alike serves as the basis for estimating the uncorrelated population variance. We obtain this information in much the same way as in the previous independent-group designs, i.e., by considering variability between subjects within the groups. We begin by finding the total of the L trial scores for each subject in the experiment. Since there are n subjects in each of the K groups, we obtain nK subject totals, which may be denoted as T_{Si}. Let us calculate a sum of squares between subjects in a single group, k, using the n subject totals from this group.

$$SS_{Ss/\text{gr}.k} = \frac{T_{S1}^2 + T_{S2}^2 + \cdots + T_{Sn}^2}{L} - \frac{(T_k)^2}{nL} \qquad (8\text{–}4.8)$$

Equation (8–4.8) is used on each of the K groups. The pooled sum of squares for subjects within groups is found by adding the K parts:

$$SS_{Ss} = SS_{Ss/\text{gr}.\,1} + SS_{Ss/\text{gr}.\,2} + \cdots + SS_{Ss/\text{gr}.\,K} \qquad (8\text{–}4.9)$$

An alternate computational method for calculating this SS_{Ss} is given in (8–4.10). This method is strictly equivalent to the one given above and will give the same answer:

$$SS_{Ss} = \frac{T_{S1}^2 + T_{S2}^2 + \cdots + T_{S(nK)}^2}{L} - \frac{(GT)^2}{N} - SS_{BG} \qquad (8\text{–}4.10)$$

In (8–4.10) the squared totals of all subjects in the experiment are added together, whereas use of (8–4.8) requires separate treatment of the subject totals from each group. In either case, the sum of the squared subject totals is divided by L since each subject total is based on L raw scores.

The degrees of freedom associated with this uncorrelated sum of squares is equal to $K(n - 1)$, since there are $n - 1$ df's related to the n subject totals in each group. The mean-square estimate which results from the SS_{Ss} is

$$MS_{Ss} = \frac{SS_{Ss}}{df_{Ss}} = \frac{SS_{Ss}}{K(n-1)} \qquad (8\text{–}4.11)$$

The correlated-error term. The most convenient method of obtaining the correlated-error term in the repeated-measures design is to subtract the sums of squares for groups, trials, groups-by-trials interaction, and subjects within groups from the total sum of squares, which is computed as usual on the entire set of nKL scores. The source of variation so measured will be called the *subjects-by-trials interaction within groups*, which we may simply denote as $Ss \times T$.

This sum of squares may be given as

$$SS_{Ss \times T} = SS_{TOT} - SS_{BG} - SS_T - SS_{G \times T} - SS_{Ss} \quad (8\text{–}4.12)$$

Use of (8–4.12) will allow certain computational errors to go undetected. However, the more direct method of calculating the $SS_{Ss \times T}$ is often quite laborious. If the direct method, given next, is used, then the sum of the component sums of squares in this design should be equal to the SS_{TOT}, except for rounding errors.

In order to obtain the $SS_{Ss \times T}$ directly, each group of scores must be analyzed separately. Using just the nL scores from one group, obtain the total sum of squares for this set of scores. Next obtain the sum of squares for the trial totals by using just the trial totals from this one group. Then obtain the sum of squares based on the subject totals, using just the subjects from this group. Each of these calculations requires the use of a correction factor for this group, which is equal to the group total squared divided by nL. Finally the $SS_{Ss \times T}$ for this group is found by subtracting the subjects and trials terms from the total sum of squares.

The above analysis is repeated for each of the K groups. The $SS_{Ss \times T}$ is now obtained by adding the results from all groups.

Although this direct method of finding the $SS_{Ss \times T}$ is quite time consuming, the general approach allows the calculation of the related degrees of freedom to be easily expressed.

Within each group, the subjects by trials interaction may be seen to be a type of row-by-column interaction, in which each row is a subject and each column is a trial. From our analysis of the two-variable case, recall that such an interaction has $(R - 1)(C - 1)$ degrees of freedom. Here, this calculation of degrees of freedom would be given as $(n - 1)(L - 1)$, since the single group has n rows and L columns. Therefore, adding the $(n - 1)(L - 1)$ degrees of freedom from each group together, we obtain $K(n - 1)(L - 1)$ as the total number of degrees of freedom related to the correlated error term. The appropriate mean square estimate is now found as

$$MS_{Ss \times T} = \frac{SS_{Ss \times T}}{df_{Ss \times T}} = \frac{SS_{Ss \times T}}{K(n - 1)(L - 1)} \quad (8\text{–}4.13)$$

As in earlier work, the df_{TOT} is $N - 1$ and should equal the sum of df's from the separate sources of variation in the analysis. We have obtained three sources of major interest and two error terms: these are *groups, trials, groups by trials, subjects within groups,* and *subjects by trials within groups.* Let us add the corresponding df's for these terms in order to show that the sum is actually equal to $N - 1$. We obtain

$$\begin{aligned}(K - 1) + (L - 1) + (K - 1)(L - 1) + K(n - 1)& \\ + K(n - 1)(L - 1)& \\ = -1 + KL + Kn - K + K(nL - n - L + 1)& \\ = -1 + KnL = N - 1&\end{aligned}$$

It may also be shown that the addition of the SS terms from these parts will equal the total SS in the experiment.

The F-ratios. In order to complete the analysis, we must develop the appropriate F-ratios for testing the null hypotheses related to the groups, trials, and groups-by-trials effects. These F tests require the division of terms into correlated and uncorrelated sources of variation.

We have noted that the K group means are uncorrelated since each subject contributes scores to only one such mean. Hence the between-groups mean square is an estimate of an uncorrelated population variance. The uncorrelated-error term, subjects within groups, is used to form the F-ratio related to differences between the groups:

$$F_{BG} = \frac{MS_{BG}}{MS_{Ss}} \tag{8–4.14}$$

The other F-ratios of interest in this type of design involve terms which are all based on correlated means. The correlated error MS becomes the denominator of the F-ratios used to assess the significance of the trials and the groups-by-trials effects:

$$F_T = \frac{MS_T}{MS_{Ss \times T}} \tag{8–4.15}$$

and

$$F_{G \times T} = \frac{MS_{G \times T}}{MS_{Ss \times T}} \tag{8–4.16}$$

The summary table for the repeated-measures design, with some of the computational formulas, is given in Table 8–5. It may be noted in Table 8–5 that the two uncorrelated terms are given first, followed by the correlated terms. This method of presenting the results is widely used; however, the series of calculations may actually be done in other orders.

Example 8–1. An investigator is interested in testing the transfer effects resulting from practicing a difficult vs. an easier motor task. He chooses a rotary-pursuit task in which subjects must try to keep a stylus on a small target mounted in a turntable which rotates continuously. On the first day of practice, 12 subjects receive 30 trials of practice at the task, with each trial lasting 30 sec. Half of the subjects perform with the turntable rotating at

TABLE 8–5

Summary of the Repeated-Measures Analysis of Variance

Source	df	SS	MS	F
Between groups (G)	$K - 1$	$\sum^{K} (T_k^2/nL) - C$[a]	SS_{BG}/df_{BG}	MS_{BG}/MS_{Ss}
Subjects within groups (Ss)	$K(n - 1)$	$\sum^{K} \left(\sum^{n} T_{Si}^2/n - T_k^2/nL \right)$	SS_{Ss}/df_{Ss}	–
Trials (T)	$L - 1$	$\sum^{L} (T_l^2/nK) - C$	SS_T/df_T	$MS_T/MS_{Ss \times T}$
$G \times T$	$(K - 1)(L - 1)$	$SS_{BC} - SS_G - SS_T$	$SS_{G \times T}/df_{G \times T}$	$MS_{G \times T}/MS_{Ss \times T}$
$Ss \times T$	$K(n - 1)(L - 1)$	SS_{TOT} – above SS' s	$SS_{Ss \times T}/df_{Ss \times T}$	–
Totals	$N - 1$	$\sum^{N} X_i^2 - C$	–	–

[a] $C = \dfrac{(GT)^2}{N}$.

40 rpm and the other half perform with a rotation rate of 80 rpm, a much more difficult task.

The second day of the experiment is used for testing transfer. All subjects are given 5 trials in which the speed of the turntable is set at 60 rpm. Trials are again of 30 sec. duration, and the experimenter, with an electric circuit, records the time spent "on target" by each subject on each trial. The following are the data from the experiment, in which the scores are grouped on the basis of the preceding day's practice conditions.

Group	Subject	Trials 1	2	3	4	5	Subject totals
	1	5.8	8.8	9.3	11.5	10.7	46.1
I	2	9.7	11.7	10.6	12.5	13.1	57.6
(40 rpm)	3	6.0	8.2	9.6	10.1	11.8	45.7
	4	8.7	8.9	10.4	11.7	11.2	50.9
	5	7.3	9.7	11.6	12.4	12.9	53.9
	6	4.4	6.1	7.2	6.9	8.8	33.4
Group-I Totals		41.9	53.4	58.7	65.1	68.5	$T_{\text{I}} = 287.6$
	1	3.2	5.8	7.5	8.7	10.5	35.7
	2	4.0	5.9	7.8	7.9	9.3	34.9
II	3	5.9	7.6	9.1	10.0	11.3	43.9
(80 rpm)	4	1.4	3.6	5.8	7.0	9.6	27.4
	5	3.2	6.4	5.7	8.9	10.7	34.9
	6	4.8	6.1	8.1	9.6	11.8	40.4
Group-II Totals		22.5	35.4	44.0	52.1	63.2	$T_{\text{II}} = 217.2$
Trial Totals		64.4	88.8	102.7	117.2	131.7	$GT = 504.8$

We wish to use the repeated-measures analysis-of-variance method to test the null hypotheses regarding group means, trial means, and the groups-by-trials interaction.

The correction factor, C, is found to be $(504.8)^2/60$ which is equal to 4,247.05. Hence,

$$SS_{TOT} = \sum_{i=1}^{N} X_i^2 - C = 4{,}687.18 - 4{,}247.05 = 440.13$$

$$SS_{BG} = \frac{(287.6)^2 + (217.2)^2}{(6)(5)} - C = 4{,}329.65 - 4{,}247.05$$
$$= 82.60$$

We may now use (8–4.3) to obtain the sum of squares for the trials effect:

$$SS_T = \frac{(64.4)^2 + (88.8)^2 + \cdots + (131.7)^2}{(6)(2)} - C$$
$$= 4{,}471.74 - 4{,}247.05 = 224.69$$

In order to obtain the SS for the groups-by-trials interaction, we first obtain a sum of squares between the *cells.* Using (8–4.5) we find

$$SS_{BC} = \frac{(41.9)^2 + (53.4)^2 + \cdots + (63.2)^2}{6} - C = 317.48$$

The interaction sum of squares may now be obtained by subtracting the SS_{BG} and SS_T from the above value (8–4.6):

$$SS_{G \times T} = 317.48 - 82.60 - 224.69 = 10.19$$

Let us now use (8–4.8) and (8–4.9) to find the variability related to differences between subject means within each of the two groups:

$$SS_{Ss/\text{gr. I}} = \frac{(46.1)^2 + (57.6)^2 + \cdots + (33.4)^2}{5} - \frac{(287.6)^2}{(6)(5)}$$
$$= 71.48$$
$$SS_{Ss/\text{gr. II}} = \frac{(35.7)^2 + (34.9)^2 + \cdots + (40.4)^2}{5} - \frac{(217.2)^2}{(6)(5)}$$
$$= 31.60$$
$$SS_{Ss} = 71.48 + 31.60 = 103.08$$

The alternate method, (8–4.10), will give the same result, within errors of rounding, and is given merely for illustration.

$$SS_{Ss} = \frac{(46.1)^2 + (57.6)^2 + \cdots + (40.4)^2}{5} - 4247.05 - 82.60$$
$$= 103.09$$

Finally, the correlated-error sum of squares must be obtained. Usually the method shown in (8–4.12) is used:

$$SS_{Ss \times T} = 440.13 - 82.60 - 224.69 - 10.19 - 103.08$$
$$= 19.57$$

The direct method of finding the $SS_{Ss \times T}$ will be given as well. This method does provide somewhat of a check on calculations.

First, using only the data from Group I, find the SS within this group (let us omit the notation which indicates that only Group I is being considered):

$$SS_{TOT} = (5.8)^2 + (8.8)^2 + \cdots + (8.8)^2 - \frac{(287.6)^2}{30} = 155.99$$

Now obtain SS_{Ss} for this group. This value has already been calculated as 71.48.

The SS_T for Group I is found as follows:

$$SS_T = \frac{(41.9)^2 + (53.4)^2 + \cdots + (68.5)^2}{6} - \frac{(287.6)^2}{30} = 73.39$$

The desired value of $SS_{Ss \times T}$ from this group is then

$$SS_{Ss \times T} = 155.99 - 71.48 - 73.39 = 11.12$$

Using the identical routine on Group II data yields

$$SS_{TOT} = 201.53, \quad SS_{Ss} = 31.60, \quad \text{and} \quad SS_T = 161.48$$
$$SS_{Ss\times T} = 201.53 - 31.60 - 161.48 = 8.45$$

Finally, we obtain the pooled $SS_{Ss\times T}$ by adding these parts:

$$SS_{Ss\times T} = 11.12 + 8.45 = 19.57$$

which equals the result found by use of (8–4.12).

The various mean square estimates are now obtained by dividing the sums of squares by their respective degrees of freedom.

Since this experiment has only two groups, the df_{BG} equals one. The use of five trials gives the df_T of four. The interaction *df* is found by multiplying the *df*'s for the two main effects. This gives $df_{G\times T} = 1 \times 4 = 4$.

The df_{Ss} is a combination of the degrees of freedom for subject variation within each group. We find $(n - 1) + (n - 1)$ equals 10 for this experiment.

The $df_{Ss\times T}$ is also composed of two parts, the degrees of freedom of the subjects-by-trials interaction within each group. Since we have 6 subjects and 5 trials within each group, the total $df_{Ss\times T}$ may be expressed as $(6 - 1)(5 - 1) + (6 - 1)(5 - 1) = 40$.

The total degrees of freedom, as always, is $N - 1 = 59$.

Use of (8–4.14), (8–4.15), (8–4.16), and Table 8–5 will allow the reader to complete the steps of this analysis. The summary takes the following form.

Source	*df*	*SS*	*MS*	*F*
Between Groups	1	82.60	82.60	8.01*
Subjects	10	103.08	10.31	
Trials	4	224.69	56.17	114.87**
Groups × Trials	4	10.19	2.548	5.21**
Ss × Trials	40	19.57	.489	–
Totals	59	440.13	–	–

$*P < .05 \qquad **P < .01.$

In this type of analysis, care must be taken in computing the *F*-ratios and assessing them for significance. The MS_{BG} is based on uncorrelated means and must be divided by the MS_{Ss} which is the estimate of the variance of uncorrelated population scores.

The mean squares for both the trials and groups-by-trials interaction effects are based on correlated means and must be tested with the $MS_{Ss\times T}$.

When using the tabled values of the *RSD*(*F*), it is often the case that a different combination of numerator and denominator degrees of freedom must be sought for each *F*-ratio to be tested.

In the present example, the F_{BG} is based on 1 and 10 *df* and is significant at only the .05 level of confidence, while both the F_T and $F_{G\times T}$ are based on 4 and 40 *df*. The $F_{G\times T}$ is found to be significant beyond the .01 level of confidence, even though this *F* is smaller than the F_{BG}.

The interpretation of these results at the verbal level of description might be given. The mean for the 40-rpm condition is found to be significantly higher than that for the 80-rpm condition. The apparent increase in performance over trials is highly significant. The significance of the groups-by-trials interaction indicates that the observed tendency for Group II scores to approach the Group I scores late in practice is a statistically reliable (non-chance) aspect of the results.

Further interpretation at the experimental-decision level is irrelevant to our purposes.

In the next example, 8–2, differences between the groups of subjects are identified as differences between strains of laboratory animals. It should be pointed out that this is a normative independent variable. If subjects are randomly selected from the respective populations of animals, no statistical problem is presented by the use of such an independent variable, but one must keep in mind that the experimental interpretation of results is often difficult, if cause and effect relationships are desired.

Example 8–2. The following data represent emotionality scores obtained from rats in a standardized "open-field" test situation. The larger scores are presumed to be related to higher degrees of overt emotionality on the part of the subjects as compared to the lower scores.

The three groups of subjects are supposed to represent three different strains of rats. The investigator is interested in strain differences in the development of emotional responding in young animals. For this reason he tests each animal at 1, 2, 3, and 4 weeks of age. Only three subjects are used from each strain.

Group (Strain)	Subject	Weeks of age 1	2	3	4	Subject Totals
I	1	7	12	18	11	48
	2	5	8	10	7	30
	3	12	19	27	24	82
	Totals	24	39	55	42	$T_{\text{I}} = 160$
II	1	4	7	8	12	31
	2	9	11	13	14	47
	3	8	12	15	19	54
	Totals	21	30	36	45	$T_{\text{II}} = 132$
III	1	14	13	15	21	63
	2	17	20	18	24	79
	3	10	11	9	13	43
	Totals	41	44	42	58	$T_{\text{III}} = 185$
Trial Totals		86	113	133	145	$GT = 477$

Since the animals in each group were tested repeatedly and the resulting scores are variates, we may use the repeated measures analysis of variance on these data.

The GT of 477 and the N of 36 give the value of the correction factor as $C = 6{,}320.25$.

$$SS_{TOT} = 7{,}411 - C = 1{,}090.75$$

The two main effects are calculated before the interaction:

$$SS_{BG} = \frac{(160)^2 + (132)^2 + (185)^2}{(3)(4)} - \mathrm{C} = 6{,}437.42 - 6{,}320.25$$

$$SS_{BG} = 117.17$$

$$SS_T = \frac{(86)^2 + (113)^2 + (133)^2 + (145)^2}{(3)(3)} - \mathrm{C} = 6{,}542.11 - 6{,}320.25$$

$$SS_T = 221.86$$

Now the interaction term may be developed. We first obtain the between cells sum of squares:

$$SS_{BC} = \frac{(24)^2 + (39)^2 + \cdots + (58)^2}{3} - C = 444.08$$

and

$$SS_{G \times T} = 444.08 - 117.17 - 221.86 = 105.05$$

The subject variability is found separately for each group:

$$SS_{Ss/\text{gr.I}} = \frac{(48)^2 + (30)^2 + (82)^2}{4} - \frac{(160)^2}{12} = 348.67$$

$$SS_{Ss/\text{gr.II}} = \frac{(31)^2 + (47)^2 + (54)^2}{4} - \frac{(132)^2}{12} = 69.50$$

$$SS_{Ss/\text{gr.III}} = \frac{(63)^2 + (79)^2 + (43)^2}{4} - \frac{(185)^2}{12} = 162.67$$

These parts are added together to obtain

$$SS_{Ss} = 580.84$$

Assuming no errors in the previous work, we calculate the correlated error sum of squares by subtraction:

$$\begin{aligned} SS_{Ss \times T} &= 1{,}090.75 - 117.17 - 221.86 - 105.05 - 580.84 \\ &= 65.83 \end{aligned}$$

The results of the remaining calculations are given in the summary table, in which groups are designated as "strains" and trials as "weeks."

Source	*df*	*SS*	*MS*	*F*
Strains (*St*)	2	117.17	58.59	.61
Subjects	6	580.84	96.81	–
Weeks (*W*)	3	221.86	73.95	20.20*
$W \times St$	6	105.05	17.51	4.78*
$Ss \times W$	18	65.83	3.66	–
Total	35	1,090.75	–	–

* $P < .01$.

Although no average differences between the strains may be claimed, it is clear that overt emotionality scores increased significantly over the period of testing and that the course of such increase was not equal for the different strains. This latter conclusion is based on the significant interaction effect. If means for each strain were graphed as a function of weeks, one could see clearly that the curves are not equally distant apart at each time period. These departures have been tested by the $F_{W \times St}$.

In the final example of this chapter, 8–3, a very useful application of the repeated-measures type of analysis of variance is shown. In this example, the "repeated" measures on each subject do not form any type of sequence. Rather, the L scores for each subject arise from the classification of his responses into L different types. So long as a fairly large range of possible scores is involved, the violations of assumptions are not of sufficient magnitude to greatly affect the $RSD(F)$.

Example 8–3. An investigator decides to test a prediction based on a simple model of approach and avoidance gradients. In his experimental situation, the model predicts that a tendency to give responses with direct sexual meaning is partly offset by an avoidance tendency related to guilt and anxieties connected with the area of sex. This condition leads to the expectation that sexual responses will take the form of symbolic responses rather than responses with direct sexual connotations.

A further prediction from the model, one which leads to the experimental test, is that conditions which reduce the anxieties related to sexual responses will permit a greater degree of direct sexual type responses; i.e., the avoidance gradient will be reduced.

In order to test these notions, the investigator uses ambiguous stimuli, similar to the Rorschach test cards, and has independent raters score each subject's responses on the basis of three classes of sexual connotation: direct, secondary, and remote.

One group of 5 subjects takes the test in standard conditions while a second group of 5 subjects is given a quantity of alcohol sufficient to produce some degree of anxiety reduction.

The following data represent the number of responses by each subject judged to be in each of the three response classes.

The analysis of these data requires that one recognize that three scores per subject have been obtained and that a single subject has given scores to only one of the two groups. These features permit us to apply the repeated measures analysis-of-variance technique, in which classes become identified with the correlated or "trials" variable.

The first steps are given without further comment.

$$C = \frac{(434)^2}{30} = 6{,}278.53$$

$$SS_{TOT} = 7{,}286 - 6{,}278.53 = 1{,}007.47$$

$$SS_{BG} = \frac{(211)^2 + (223)^2}{15} - 6{,}278.53 = 4.80$$

$$SS_T = \frac{(117)^2 + (191)^2 + (126)^2}{10} - 6{,}278.53 = 326.07$$

Group	Subject	Classes of Sexual Connotations			Subject Totals
		Primary	Secondary	Remote	
	1	14	21	11	46
I	2	3	12	14	29
(control)	3	8	22	16	46
	4	7	31	15	53
	5	9	18	10	37
	Total	41	104	66	$T_{\text{I}} = 211$
	1	16	14	10	40
II	2	9	10	8	27
(alcohol)	3	15	21	15	51
	4	19	26	14	59
	5	17	16	13	46
	Total	76	87	60	$T_{\text{II}} = 223$
Class Totals		117	191	126	$GT = 434$

To obtain the sum of squares for the interaction of trials (classes) by groups, we first develop the SS_{BC}.

$$SS_{BC} = \frac{(41)^2 + (104)^2 + \cdots + (60)^2}{5} - 6{,}278.53 = 481.07$$

then,

$$SS_{G \times T} = 481.07 - 4.80 - 326.07 = 150.20$$

The subjects within groups terms may be developed next:

$$SS_{Ss/\text{gr. I}} = \frac{(46)^2 + (29)^2 + \cdots + (37)^2}{3} - \frac{(211)^2}{15} = 115.60$$

$$SS_{Ss/\text{gr. II}} = \frac{(40)^2 + (27)^2 + \cdots + (46)^2}{3} - \frac{(223)^2}{15} = 193.73$$

$$SS_{Ss} = 115.60 + 193.73 = 309.33$$

We use (8–4.12) to obtain the correlated error sum of squares

$$SS_{Ss \times T} = 1{,}007.47 - 4.80 - 326.07 - 150.20 - 309.33 = 217.07$$

The division of these sums of squares by the corresponding *df*'s gives the mean-square estimates, from which the three F-ratios are found. The summary table indicates the results of these further calculations.

Source	*df*	*SS*	*MS*	*F*
Groups (G)	1	4.80	4.80	.12
Subjects	8	309.33	38.67	–
Classes (C)	2	326.07	163.04	12.01**
$G \times C$	2	150.20	75.10	5.53*
$Ss \times C$	16	217.07	13.57	–
Total	29	1,007.47	–	–

*$P < .05$ **$P < .01$.

In this experiment, the significance of the classes variable is not of great importance. What is of interest is the shift in the number of "remote" and "secondary" to "direct" responses from the control group to the alcohol condition. The null hypothesis regarding a class by group interaction is rejected at the .05 level of confidence. The observed shifts from control to alcohol conditions in the numbers of responses in the different classes are concluded to be significant.

8–5. The *t*-Test: A Special Case of Analysis of Variance. Historically, the method of testing the hypothesis that two sample means are based on random samples from populations with equal means has been based on the distribution of t, studied in Chapter 5. Usually called the *t-test,* this technique requires the development of a ratio. The numerator of this ratio is the difference between two sample means, which we refer to as D. The denominator of the ratio is the estimated standard deviation of the $RSD(D)$.

The random sampling distribution of the ratio is known to be the t-distribution with $(N - 2)$ degrees of freedom, where N is the total number of scores in the two samples.

Comparison of *t* and *F*. In order to show the close relationship of the t-test and the analysis of variance procedure for testing the same null hypothesis, let us assume that each of two random samples of scores has the same number of scores, n. The assumptions made about the populations in using a t-test are the same as for the analysis of variance in the one-variable case. Under these conditions, the t-ratio used to test the hypothesis of equal population means is

$$t = \frac{M_1 - M_2}{\sqrt{\hat{\sigma}_{M_1}{}^2 + \hat{\sigma}_{M_2}{}^2}} \tag{8–5.1}$$

where the two estimated variances of means are developed separately from each of the two groups (using (5–8.5) or some similar formula).

The same null hypothesis may be tested with an F-ratio by considering the experiment as a single-variable experiment with only two groups. In this case, the resultant F-ratio is based upon 1 and $(N - 2)$ degrees of freedom. The corresponding t-ratio involves a single degree of freedom measure, equal to $N - 2$, which is equal to the denominator df in the F-ratio.

A series of simple algebraic operations, omitted here, will prove that,

$$t^2(df_e) = F(1, df_e) \tag{8–5.2}$$

where df_e refers to the degrees of freedom for the error terms, $N - 2$.

Equation (8–5.2) expresses the fact that, for a two-group experiment, the t-test and the F-test give answers related directly by a square root or squaring operation. If one performs an analysis of variance, the square root of the resultant F will equal the value of t which would be obtained from a t-test on the same data. Conversely, if one were to perform a t-test, the square of the t obtained would be equal to the corresponding F obtained from the data.

This relationship between t and F for a two-group experiment also holds in terms of the probability of occurrence of sample values of these statistics, given the null hypothesis. This is most clearly seen by comparing the .05 and .01 entries of t and F in their respective tables (Tables D and E). The squared value of t for any df_e is equal to the $F(1, df_e)$.

The result of this comparison is that any two-group experiment suitable for analysis of variance methods might just as well be subjected to a t-test. The identical statistical decision about the null hypothesis will be reached using either method. However, why not use the analysis-of-variance test?

Two reasons for using the analysis-of-variance rather than the t-test are offered. The first, a minor reason, is that one need not obtain square roots in the analysis-of-variance procedures. The second reason is much more important. The analysis-of-variance technique easily generalizes to the case of more than two levels of a single independent variable as well as to other more complex designs, as illustrated in the last two chapters. The t-test, given the sampling distribution in Table D, is *appropriate only* for experiments involving *two groups of scores.*

Limitations and misuses of the t-test. The basic limitation of the t-test is expressed above: only two groups of scores may be considered in an experiment.

Misuses of the t-test are numerous in research literature. Probably the most frequent misuse, still found far too often, is in "multiple t-tests." Consider a design involving four groups of subjects. Since there are four sample means, there are six distinct *pairs* of means which might be tested (one member of each pair against the second) by use of a t-test. However, the random sampling distribution of t is based on the assumption that only two sample means have been selected in the experiment. Consider the average difference between the highest and lowest sample means. It should be intuitively clear that this average difference will increase as the number of means increases from which the highest and lowest values are chosen.

The procedure of testing all possible pairs of means with t-tests will result in an increased proportion of rejection of H_0 when H_0 is true, i.e., an increased proportion of type I errors.

Unless special tables of the distribution of t are used, the t-test should never be employed in experiments with more than two groups. The only possible exception is when two particular groups are specified, in advance of running the experiment, as candidates for direct comparison with one another.

The student will certainly encounter numerous correct applications of the t-test in research literature. He might well consider the squared value of t and think in analysis-of-variance terms rather than try to master a technique of such limited scope as the t-test for comparing two sample means.

EXERCISES

1. The following are trial totals, based on the n's given, for different groups in repeated measures designs. Find the SS_{BG}, SS_T, and $SS_{G \times T}$ in each case and give the associated degrees of freedom.

(a)

	Trials		
Groups	1	2	3
1	142.5	169.0	174.3
2	101.1	128.5	135.2
	$n = 6$		

(b)

	Trials				
Groups	1	2	3	4	5
1	16.7	18.9	24.3	27.2	27.0
2	13.6	15.0	16.6	18.4	19.0
3	16.5	14.4	17.0	18.1	18.9
		$n = 5$			

(c)

	Trials	
Groups	1	2
1	268	245
2	341	329
3	301	288
4	315	275
	$n = 10$	

2. Perform a repeated-measures analysis of variance of the data given below.

Groups		Trials 1	2	3	4
1	S_1	15	17	19	22
	S_2	16	18	18	21
	S_3	15	19	21	23
2	S_4	7	10	14	16
	S_5	10	13	18	17
	S_6	11	15	19	20
3	S_7	9	14	15	18
	S_8	10	13	17	20
	S_9	9	12	16	17

3. If each person's total score (summed over trials) in exercise 2 were used as a raw score, analyze the data as a one-variable experiment. Compare the F with the equivalent term in exercise 2.
4. Perform a repeated-measures analysis of variance of these data:

Groups		Reaction Times to Stimuli: Light	Sound
No warning	S_1	.33	.30
	S_2	.47	.35
	S_3	.29	.30
	S_4	.38	.29
Warning	S_5	.21	.19
	S_6	.25	.21
	S_7	.29	.22
	S_8	.26	.18

Applications: Use information from all previous chapters where relevant.

5. The following data give magnitudes of galvanic skin responses (*GSR*) for

Groups		Trials 1	2	3	4	5
1 (1.0-sec. interval)	S_1	0	3	2	6	5
	S_2	1	4	12	7	8
	S_3	3	2	5	5	7
	S_4	0	0	3	4	9
	S_5	2	7	7	5	10
2 (0.5-sec. interval)	S_6	3	2	7	6	12
	S_7	5	4	10	9	13
	S_8	2	5	8	11	10
	S_9	4	8	7	7	11
	S_{10}	3	7	10	10	9

each subject on 5 test trials in a classical conditioning experiment. The two groups represent two different treatments with respect to the interval between the conditional stimulus and the unconditional stimulus on the learning trials. Do the groups differ with respect to mean responding and/or rate of conditioning? The numbers are measured amplitudes of recorded *GSR*'s, in mm.

6. Sixteen subjects are asked to estimate the weights of oddly shaped containers of various sizes. One group of 8 subjects is given the volume of each container before making a judgment while the second group of 8 different subjects is not given this information. Each subject judges 4 randomly selected containers from a set of over 200, but, for some reason, the subjects' orders of judging stimuli are not recorded. The following data are the differences between an estimated weight and the true weight (in oz.). Does the knowledge of volume affect the scores?

Group		Scores			
	S_1	−17	2	5	14
	S_2	3	22	−7	−1
	S_3	10	12	13	2
1	S_4	16	−10	8	6
(knowledge)	S_5	29	13	15	18
	S_6	14	−8	−2	6
	S_7	19	7	16	−1
	S_8	3	.12	9	13
	S_1	29	14	6	15
	S_2	−3	10	13	18
	S_3	16	0	29	30
2	S_4	13	10	−7	12
(control)	S_5	24	11	0	21
	S_6	13	6	18	14
	S_7	9	3	−4	15
	S_8	10	36	21	11

7. A research publication indicates that the means of two groups, each with 16 different subjects, have been compared with a *t*-test. The value of *t* is reported as 2.35; what would be the corresponding value of *F* for the analysis-of-variance test of this hypothesis? If $M_1 = 12.54$ and $M_2 = 15.60$, what is the value of the MS_{WG} in this study?

9

Chi Square

9–1. Frequency Data. **General characteristics.** The type of scores encountered in normal-curve and analysis-of-variance procedures have been termed *variates.* The use of such scores implies that some aspect of a situation, an object, behavior, etc., has been so measured that the magnitude of the score represents the amount, magnitude, or degree of the aspect under study. We now wish to consider the use of numbers to represent only the quality of numerosity. Our earlier discussions of classification and counting serve as a convenient basis for this use of numbers.

If one considers a set of distinct elements, the first use of the numerosity concept is to count the total number of such elements. However, even in this simple case, and certainly in the case of partitioning a set of elements, a fundamental problem is the definition of the set under consideration.

Mathematical or other rational definitions of sets and subsets are usually accomplished in a straightforward manner. But, the definition of sets and subsets of elements representing entities in the "outside" world often requires difficult and arbitrary decisions. For instance, if one wishes to classify behavior in an eyelid-conditioning experiment into two classes, "response" and "no response," certain clear rules of definition must be established. Just how much movement of the eyelid is required to constitute a blink? During what time intervals will a response be counted? Decisions of this sort must be reached before one can reasonably classify and count, on the basis of the characteristics of behavior being considered.

Formal discussion of the technical problems involved in the definition of sets and subsets is not intended in this text, but the topic is of enough importance that the student of statistics must at least be aware of its existence.

The problem of clearly defining the qualifications for membership in a set or subset is of particular relevance to the use of the *chi square*, χ^2, statistic to be discussed in this chapter. Specifically, we wish to define a population set and subsets within or from the population in such a way that complete partitioning of the population will be accomplished. This means that each element to be classified must be assigned to one and only one of the subsets. The definitions must not allow ambiguity with regard to the proper assignment of any element; moreover, all elements must be assignable to some subset.

Within this framework of a complete partitioning of a population of elements, we may now describe the statistical features of sets and subsets in very simple terms. The number of elements in a class is the basic datum derived from this procedure of definition, classification, and counting.

Since the number zero will be used as the origin for counting the number of elements, it may readily be seen that frequency data qualify as a ratio scale.

The comparison of one subset with another subset is most meaningfully handled by use of relative frequencies or proportions. In the same way that the totals of two groups of scores are converted to means for the sake of comparison, the frequencies within two classes are usually converted to relative frequencies in order to make meaningful comparisons. We are often more interested in the proportion of the total number of population elements which belongs to a given class than in the raw frequency *per se*.

For example, a thoughtful coed may be unimpressed with the fact that a school has 1,000 male students, until she finds that this figure is 95 per cent of the total enrollment. As another illustration, a person who has been tossing a coin a number of times may report that the total number of heads is now 15 more than the total number of tails. Such information is difficult to utilize unless one knows the total number of tosses.

The intention here is to illustrate why a single frequency or the simple difference between two frequencies is not as meaningful as a proportion or the difference between two proportions.

It should be added that the total number of elements in a population or sample set is of statistical importance. Just as the standard error of the mean decreases with increased sample sizes, the random sampling variability of the *proportion* of elements in a subset will decrease as the sample size (summed over all subsets) increases.

Partitioning of subjects. In terms of research in the behavioral sciences, the elements in a population are often people or other organisms. Certain characteristics of the subjects in an experiment will be used to classify or partition the subjects. Very often, we think and speak more in terms of classifying characteristics than in terms of classifying the subjects. For instance, an investigator might report 60 per cent responding on trial 5 in an eyelid-conditioning study. In a formal sense, this statement means that 60 per cent of the subjects may be classified in a subset defined in terms of persons making a specific type of response on trial 5.

The first of these methods of reporting and thinking is clearly the more convenient of the two. However, occasional use of the more formal method of thinking has the advantage of alerting one to the high level of abstraction involved in the partitioning of subjects on the basis of one or a few characteristics. When we design an experiment to measure some aspect of behavior, we are usually also designing the experiment so as not to measure countless other aspects of behavior.

The major statistical point to be carefully noted about the partitioning of subjects concerns the "score" a subject is given. In the case of a variate, X, a subject is usually assigned to a condition (subset of the experiment) and his score, X, depends upon his behavior in the experimental situation.

However, in the present case of the partitioning of subjects on the basis of some characteristic, *a person's "score" is always equal to one.* That is, a single person is just one element of a set. *What varies in such an experiment is not a person's score but the subset to which that score of one is assigned.*

In those situations in which the score assigned to a subject is free to vary according to his behavior, one should use the descriptive and inferential statistics discussed in the four previous chapters. In contrast, when a single subject is classified as belonging to a certain subset on the basis of some characteristic (or behavior), and when that subject is simply counted (given a score of one), then we must think in terms of frequency data. The chi square statistic is an important means of answering statistical questions when frequency data are encountered.

9–2. The Chi Square Statistic. We define the statistic chi square, denoted χ^2, with δ degrees of freedom, as

$$\chi_\delta^2 = \sum_{i=1}^{\delta} z_i^2 \tag{9–2.1}$$

where the z_i's are drawn at random from a normal distribution, in standard z-score form. Verbally, one might say that χ^2 is equal to the sum of δ squared z-scores randomly selected from a normal distribution with a mean of 0 and a σ of 1.

It should be clear that the value of χ^2 will be non-negative; the smallest possible value is 0. It should also be noted that, on the average, the value of χ^2 will increase as more and more z-scores are drawn, i.e., as the *df* increases.

At this point, one may reasonably ask, "Since χ^2 is based on the normal distribution, which is a continuous number scale, how can this statistic be useful for frequency data, which clearly give rise to a discontinuous or discrete number scale?" The answer to this question is that the chi square statistic, when used with frequency data, yields an approximation to an exact answer which is laborious to obtain in most problems. The approximation obtained happens to be an exceedingly good one, under certain conditions, of course.

The approximation depends upon the results of the *central limit theorem*, given in Chapter 5. This theorem indicates that when large samples are used, the $RSD(M)$ is a normal distribution. It should be clear that the $RSD(T)$ is also normally distributed, the conversion from means to totals merely being a linear transformation. As will be discussed shortly, the frequency of cases in a subset may be thought of as a total, and when a large sample of elements is used, the random sampling distribution of the subset totals will be normal.

The definition in (9–2.1) will not be used, in the form given, with frequency data. However, the computational formulas to be offered depend upon (9–2.1) and the above approximation of frequency data to a normal distribution.

In a sense, the χ^2 has two assumptions implied by (9–2.1). These are a normal distribution of scores and random sampling of scores from this population. Although we allow a certain latitude in the normal-distribution assumption, the notion of random sampling must not be violated in working with the χ^2 statistic. Let us examine this assumption more closely.

Since the normal distribution has an infinite number of scores, random sampling techniques from the distribution are equivalent to sampling with replacement. The importance of this is that the value of one element selected has no effect upon the value of any other element selected. This concept is usually termed *independence*. This is quite similar to our notion of correlated and uncorrelated scores in the last chapter. If elements are chosen from a population in such a manner that the selection of an element falling into some subset influences the probability of selection of an element for any other subset, then such a procedure violates the notion of independent selection. Such a violation will invalidate the procedures to be given involving the use of χ^2. Since the selection of subjects in an experiment usually constitutes sampling without replacement, we must assume that the population is very large, in order to strictly satisfy this assumption of independence of sampled elements.

9–3. Hypotheses Related to Frequency Data. Prior to illustrating the use of χ^2 on frequency data, we must examine the parameters of the population. This will allow us to formulate and test hypotheses related to frequency data.

Although certain properties of Question-I and Question-II procedures with frequency data are contained in a later discussion, our goal in this chapter is to ask and answer questions about population parameters from samples from the population. We intend, then, to work problems of the sort of Question III.

Population proportions for subsets. The typical use of χ^2 assumes that a population of elements (subjects) is very large or infinitely large. If the method of classifying elements into subsets has been established, then the only remaining description (parameter) of a population is the relative frequency of elements in the various subsets in the population. These relative frequencies will be expressed as proportions of the total number of elements. Once again, we omit the technical definition of a proportion of an infinite number of cases; just think in terms of extremely large numbers of elements.

Let us assume that our partitioning method produces K distinct subsets of the population. We describe this population by giving the K proportions of elements found in these subsets. Using p_k to denote the proportion of elements in class (subset) k, the description of the population is merely a listing of the values for $p_1, p_2, \ldots, p_K$. Since the classes form a complete partition of the population, we find that

$$\sum_{k=1}^{K} p_k = 1 \qquad (9\text{–}3.1)$$

Hypotheses about the population. Since our description of the population takes the form of proportions within subsets, when we desire to formulate hypotheses about a population of elements, the hypotheses generated and tested will take this same form. Clearly, given the definitions for forming subsets, any hypothesis for which

(9–3.1) does not hold cannot possibly be true. We restrict our thinking to a set of proportions which add to one when summed over all K subsets.

To be explicit, let us assume that a population may be partitioned on the basis of the sex of the subjects into two classes, male and female. We may now offer a hypothesis about the population by suggesting that the proportions of males and females, p_m and p_f, are each equal to .50. This hypothesis is consistent with (9–3.1) and takes the form of stating the proportions for each subset in a complete partitioning. Other possible hypotheses are: $p_m = .25$, and $p_f = .75$; $p_m = .07$, and $p_f = .93$; $p_m = .67$, and $p_f = .33$; etc.

9–4. Expected Frequencies. In order to test hypotheses about population proportions, let us develop some expectations about random samples from our hypothesized population. We will then be able to use the χ^2 statistic to judge the seriousness of departures from our expectations about sample data.

Since random samples are the result of our sampling procedure, we may again speak of probabilities. With p_k as the proportion of population elements in a given subset, the probability, P_k, that a single randomly selected element will belong to class k, is equal to p_k. With any size sample, drawing an indefinitely large number of samples will result in an average proportion of p_k elements in the subset k in the samples. We may speak of this average p_k proportion of the sample elements as the *expected proportion* in the sample subset k. It is seen to be equal to P_k, the probability that a single randomly selected element will belong to class k.

In order to calculate a χ^2 statistic using a convenient computational formula, we desire an *expected frequency* in each sample class. If p_k is the expected proportion of sample elements in class k, the *expected frequency* in that class depends upon the total number, N, of sample elements chosen. Using $f_{e,k}$ to denote the *expected frequency* in class k, we have

$$f_{e,k} = Np_k \tag{9–4.1}$$

Equation (9–4.1) will yield an expected frequency for each subset of scores in a single random sample of size N. The p_k's may be interpreted in two slightly different ways. The p_k's may represent the true population proportions. In that case, (9–4.1) will yield a set of "true" expected frequencies for a sample of size N. Another way of describing this result is to say that the set of f_e's obtained gives the average value of the frequencies in each class when an infinite number of samples, each of size N, is considered.

The second method of interpreting the p_k's in (9–4.1) is to use the set of hypothesized values of the population proportions discussed in the previous section. In this case, which is the usual one for hypothesis-testing purposes, use of (9–4.1) will yield a set of expected frequencies *which would be true if our assumptions about the population are true.* Results of a single sample which are widely different from those "expected" will result in rejection of the assumed p_k values.

Restrictions and degrees of freedom. One should carefully note that the K expected frequencies found by using (9–4.1) are not all free to vary independently of one another. Examine the sum of the f_e's:

$$\sum_{k=1}^{K} f_{e,k} = \sum_{k=1}^{K} Np_k = N \sum_{k=1}^{K} p_k = N \qquad (9\text{–}4.2)$$

by use of (9–3.1). The restriction of the set of expected frequencies comes from the fact that we have a set of p_k values which sum to 1. It is certainly reasonable to expect that the total of the expected frequencies in a sample of N scores is N. Otherwise, one would be in the awkward situation of expecting either more or less elements in the whole sample than are known to be in the sample. Another view of the restriction of our sample values expected in each class is to observe that we need to use a portion of the result of the sampling procedure, the sample size N, in order to calculate the set of expected frequencies.

Both of these arguments result in loss of 1 degree of freedom from what might have been thought to be the appropriate *df* for the related chi square test of the hypothesis about p_k's.

The basic data for this type of problem are the numbers of elements in the K classes in a sample of N elements. Each of these K values might vary to some degree from the expected value. The χ^2 derived from the experimental test is based on K squared difference scores (differences between observed and expected frequencies). Even so, we find that only $(K - 1)$ of the expected values may be varied independently, in order that the total of N elements be obtained.

Whenever (9–4.1) is used to obtain expected frequencies, as described above, the related chi square will have $(K - 1)$ degrees of freedom associated with it.

9–5. Applications of Chi Square. The actual computations in a chi square problem are simple and usually far less numerous than those involved in analysis-of-variance problems. The computational formula for chi square may be given as:

$$\chi^2 = \sum_{k=1}^{K} \left[\frac{(f_{o,k} - f_{e,k})^2}{f_{e,k}} \right] \tag{9–5.1}$$

where the $f_{o,k}$'s refer to the observed frequencies, the $f_{e,k}$'s refer to the expected frequencies, and K is equal to the number of subsets of elements. The *df* for the resulting χ^2 is usually equal to $K - 1$.

Verbally, we may describe the operations in (9–5.1) by saying that each expected frequency is subtracted from its corresponding observed frequency, and the difference is squared. Then the squared difference between observed and expected frequencies is divided by the expected frequency. The results of these operations, *done separately for each class,* are then added together to obtain the value of χ^2.

It may already be apparent to the reader that the development of the set of expected frequencies is essential to the use of (9–5.1). Students often have more difficulty in the development of the expected frequencies than in the actual computation of the value of the chi square. It is for this reason that the discussion of the expected frequencies was given priority over the introduction of (9–5.1).

The $RDS(\chi^2)$. In order to assess the degree to which our observed frequencies deviate from those expected under a set of p_k values hypothesized to be correct, we have constructed a statistic, χ^2, which will be zero if all pairs of observed and expected frequencies show no differences. Inspection of (9–5.1) will show that the value of χ^2 increases as expected and observed values grow farther apart. When the set of p_k values under test is correct, i.e., true for the population considered, then the value of χ^2 will tend to be smaller than when the population proportions are actually different from the p_k values tested.

The region of rejection for a χ^2 test will contain the 5 per cent of the values of χ^2 which are larger than the other random sampling values of χ^2. In this procedure, as in earlier hypothesis testing, we will falsely reject a set of true population parameters (p_k's) only .05 of the time. Moreover, when the population proportions differ from those hypothesized, the test value of χ^2 will be in the region of rejection more than .05 of the time, on the average.

The random sampling distribution of chi square is based on the assumption that the expected proportions of cases in the sample are equal to the corresponding proportions in the population. The $RSD(\chi^2)$, then, indicates how frequently selected values of chi square would be obtained when sampling from populations with known parameters (proportions). As indicated at the beginning of this chapter, the average value of χ^2 in such cases does depend upon the number of degrees of freedom associated with the statistic. Figure 9–1 shows some typical random sampling distributions of χ^2 for various degrees of freedom values.

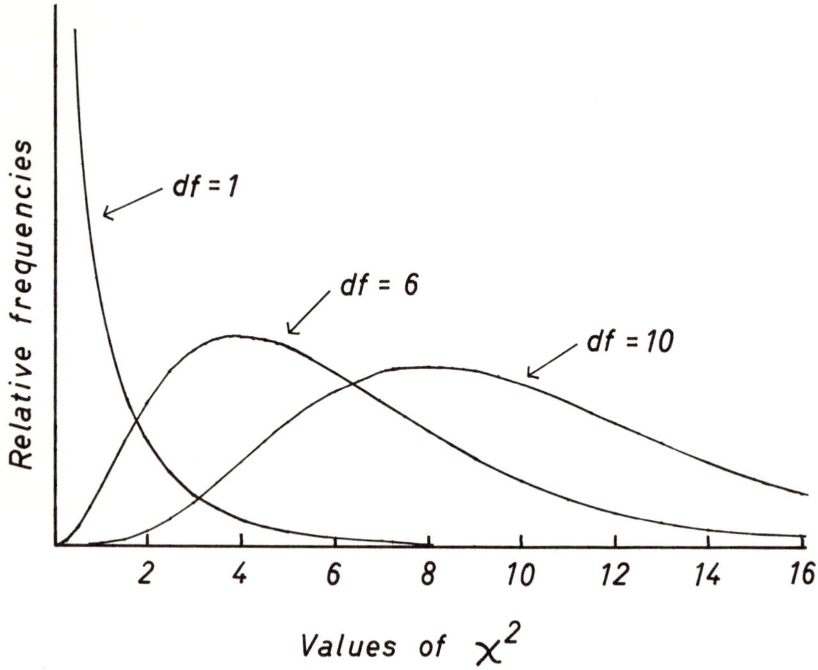

Fig. 9–1. Three random sampling distributions of *chi square* for different degrees of freedom. The curves are drawn to slightly different scales on the *y*-axis.

Table F in the Appendix gives information about the various $RSD(\chi^2)$. The first entry needed to use Table F is the *df* corresponding to a test value of χ^2. Within any row, a number of values of χ^2 are given. These values indicate, for the *df*'s related to that row, the cutting points on the distribution of χ^2 which divide the distribution into two parts. The first part contains χ^2 values less than the one tabled. The second part contains all values of the distribution which are as large or larger than the tabled value. At the top of each column the proportion of the chi square distribution in the second of these two parts is given. For example, consider the random sampling distribution of chi square based on one degree of freedom. Using the first row of Table F, note the number 3.841. This is a value of chi square which divides the distribution into two parts. The column heading above the number 3.841 is .05, which indicates that .05 of all random sample values of chi square in this distribution are equal to or greater than 3.841. The rest of the distribution contains values of χ^2 less than 3.841.

The use of Table F for hypothesis testing is easily accomplished by examining the tabled value of chi square under the .05 column and in the row corresponding to the *df* value. If the test value of chi square as computed from the data is as large or larger than this tabled value, we have a value among the highest .05 of those expected by chance. This is defined as within the region of rejection and permits rejection of the set of p_k values under test.

Example 9–1. In a probability learning experiment subjects are required to predict which one of three lights will light on each trial. They do this by selecting one of three response levers to pull on each trial. The investigator is interested in a mathematical model to describe the learning which occurs when one of the three lights is turned on more often than the others. He needs to make an initial estimate about the probabilities of selecting each of the three levers on the first trial of the experiment. The simple question is then asked, "Are the initial probabilities of selecting a given lever all equal?" If this hypothesis is tenable, the investigator's task is simplified.

The available data come from 120 subjects. By examining only the first response of each subject, we may classify the subjects into three classes depending upon which response each person chose on his first trial. The results are given as the observed frequencies, f_o's.

Response classes	f_o	f_e
1	38	40
2	52	40
3	30	40
	$\Sigma f_o = N = 120$	$\Sigma f_e = 120$

The hypothesis to be tested concerns the population proportions of selecting each of the three levers. We note that this hypothesis requires that the three p_k values be equal. Since the sum of these proportions must equal one (9–3.1), we find that $p_1 = p_2 = p_3 = \frac{1}{3}$. The fraction $\frac{1}{3}$ is more convenient than a decimal number in this case but .3333 could be used just as well.

The set of expected frequencies may be developed for each class by multiplying p_k by N.

$$f_{e,1} = \tfrac{1}{3}(120) = 40., \quad f_{e,2} = \tfrac{1}{3}(120) = 40., \quad f_{e,3} = \tfrac{1}{3}(120) = 40.$$

These expected frequencies are shown next to their corresponding observed frequencies.

The value of chi square for testing this hypothesis may now be quickly obtained.

$$\chi^2 = \frac{(38 - 40)^2}{40} + \frac{(52 - 40)^2}{40} + \frac{(30 - 40)^2}{40} = \frac{248}{40} = 6.20$$

The number of classes is 3 so that the *df* for this problem is equal to 2. Use of Table F shows that .05 of the sampling values of χ^2 are 5.991 or larger. Since our value of 6.20 is among this set of the highest 5 per cent of random sample values we may reject the hypothesis that $p_1 = p_2 = p_3$.

It should be noted that any set of observed values may yield a significant value of χ^2 for some sets of p_k values but not for others. The obtained value of χ^2 depends upon both the observed frequencies and the hypothesis under test. For instance, if the investigator in example 9–1 had reason to believe that p_1 and p_3 were equal to each other but that p_2 was twice as large as either of the others, then the resulting expected frequencies would have been 30, 60, and 30, respectively. The same set of observed values shown in the example would now have given a chi square value of 3.20, which is not significant.

The use of the chi square statistic is not restricted to frequency data resulting from a classification based on a single characteristic. Partitions formed from considering more than one characteristic of the population elements may also be suitable for hypothesis testing through the use of χ^2. If the hypothesis to be tested allows a statement about the p_k values for each possible class, then the procedures just illustrated will permit a test of that hypothesis for a given random sample of data. Example 9–2 shows such an analysis.

Example 9–2. An investigator wishes to examine the distribution of students attending a high school conference on driving education for teenagers. He knows that three schools are involved in the conference and he has the total enrollment figures from each high school, in terms of the numbers of both men and women. For school *A* there are 658 men and 702 women; for school *B* there are 1,357 men and 1,289 women; and school C has 946 men and 878 women. Considering the entire enrollment of all three schools to constitute the population, we can now calculate the population proportions in six different classes. These six classes result from the two-fold classification of subjects on the basis of both sex and school. The total N is found to be 5,830.

Each population frequency is now divided by the total N to obtain the population proportions in each class. The results may be given as follows:

	School		
	A	*B*	*C*
Men	.113	.233	.162
Women	.120	.221	.151

These six proportions are known to be true for the population considered. When we now examine the observed frequencies of students in attendance at the conference in which the investigator is interested, there is a question as to what hypothesis is to be tested. Since the population proportions are known, it is not reasonable to test them. However, the persons in attendance at the conference are certainly not drawn at random from the population in any strict sense of the term. On the other hand, the sample may be considered a random sample from some unknown population of persons attending such conferences. In the first case, we may consider the hypothesis under test to be the randomness of the sample. Since the population values are known, a significant χ^2 would suggest that the sample is not as might be expected when sampling persons at random from the known population. In the second case, the population proportions known merely serve as test values in place of the unknown population proportions. In either case a significant value of χ^2 would force us to reject one or both of the basic assumptions related to the $RSD(\chi^2)$, i.e., the hypothesized proportions and the use of random sampling.

The investigator found the following frequencies in each cell (class) in attendance at the conference. The total of these observed frequencies is 522.

	Obtained Frequencies		
	School *A*	School *B*	School *C*
Men	87	91	45
Women	101	116	82

The development of the set of expected frequencies requires the use of (9–4.1). The N and the p_k values are used to obtain the expected frequencies given next.

	Expected Frequencies		
	School *A*	School *B*	School *C*
Men	59.0	121.6	84.6
Women	62.6	115.4	78.8

The expected frequencies should be summed as a check on one's calculations. Here they do sum to 522.0.

The χ^2 may now be obtained by use of (9–5.1):

$$\chi^2 = \frac{(87 - 59.0)^2}{59.0} + \frac{(91 - 121.6)^2}{121.6} + \frac{(45 - 84.6)^2}{84.6} + \frac{(101 - 62.6)^2}{62.6} + \frac{(116 - 115.4)^2}{115.4} + \frac{(82 - 78.8)^2}{78.8}$$

$$= 13.5 + 7.7 + 18.5 + 23.6 + 0.0 + 0.1$$

$$= 63.4$$

Since the N was the only part of the data in the sample used in the calculation of the expected frequencies, we have placed only one restriction on the data. The degrees of freedom associated with the resulting χ^2 are equal to the number of classes minus one. This gives a *df* of 5 for the present example. *Table F* indicates that a χ^2 based on 5 *df* will equal or exceed 11.07 only .05 of the time by chance. Our value of 63.4 is well into the region of rejection and the hypothesis may be rejected with a high degree of confidence.

The best way of stating our conclusion in this case may well be that the observed sample is certainly not a random sample from the known population considered.

9–6. Restrictions on Chi Square. **Small expected frequencies.** Since the use of chi square with frequency data is based upon the fact that the random sampling distribution of the frequency within a given class approximates the normal distribution, we must use care in applying this statistic to frequency data when the expected frequency in a class is small. The central limit theorem applies to "large" samples. When the expected value for a given class is small, the random sampling distribution of the total for that class departs rapidly from a normal distribution. The χ^2 value obtained from such a condition, when interpreted with Table F, is a poor approximation to the true probability of differences between observed and expected frequencies.

For hypothesis testing purposes, a limitation is placed on the size of any expected frequency in a χ^2 application. All expected frequencies must be 5 or larger in order to use the χ^2 statistic with the distributions given in Table F.

$$f_{e,k} \geq 5, \qquad \text{for all } k. \tag{9–6.1}$$

This restriction applies to *expected frequencies*. The observed values may be zero or any other number without violating this rule.

When expected values do fall lower than 5, χ^2 should not be used. One of two possible courses is taken in such cases. Either the exact probability value is calculated, by methods shown in Chapter 10, or the classification scheme is so modified that (9–6.1) is true. The latter technique should be used with care so that the purpose of the hypothesis-testing procedure is not destroyed or greatly changed. An illustration of this recourse is given later in this chapter.

Correction for continuity. We have noted in an earlier chapter that a discrete function often may be considered as an approximation to a continuous function. In this regard, it was noted that the apparent limits of a score or range of scores are slightly in error if the intention is to approximate a continuous distribution. The real limits of a score or range of scores are formed by considering a discrete value to represent a number which has been rounded to some extent. For instance, if a frequency score of 10 is to be used in relation to a continuous distribution like the normal curve, we would consider that the score of 10 was actually a range of values from 9.5 to 10.5 on the normal curve.

If a normal-curve problem was encountered in which the proportion of cases equal to a frequency score of 10 or greater was desired, the appropriate portion of the normal distribution would include all values equal to 9.5 or greater.

Consider an application of χ^2 which tests the hypothesis that a coin is *unbiased;* i.e., that the population proportions of heads and tails are equal. If a sample of 15 elements is selected, the expected number of heads (and tails) is 7.5. Clearly, the best possible outcome of such an experiment is an 8 to 7 split between the two types of events. The use of (9–5.1) will yield a value of χ^2 slightly greater than zero, even though the results are as close as possible to the expected frequencies! However, if we consider the frequency score 8 to include all values from 7.5 to 8.5 and the score of 7 to be 6.5 to 7.5, then we see that obtained scores include their respective expected scores.

The result of such reasoning is the recognition that the probability values related to the $RSD(\chi^2)$ are slightly in error when frequency data is used. Fortunately, the degree of error is very small when two or more degrees of freedom are related to the χ^2.

When a *chi square* is based on frequency data and has only one degree of freedom, the magnitude of error is sufficient to require a correction. Since the cause of such error is that the frequency scale is not continuous, the correction is called a *correction for continuity.* This correction is merely the use of a "real" limit for an observed score rather than the integer value itself. The following formula is a modification of (9–5.1) which should always be used for an application of chi square with one degree of freedom when discrete data are analyzed:

$$\chi^2(1\ \text{df}) = \sum_{k=1}^{K} \left[\frac{(|f_{o,k} - f_{e,k}| - .5)^2}{f_{e,k}} \right] \qquad (9\text{–}6.2)$$

In (9–6.2), the absolute value of the difference between an observed and expected frequency is reduced by $\frac{1}{2}$ unit. This means that the resulting χ^2 value in each class or *cell* is less than would have been obtained without the correction. An exception is when the observed and expected values are equal. Such cases are trivial since the hypothesis under test certainly would not be rejected.

It may be seen that the correction given in (9–6.2) is equivalent to squaring the difference between the expected frequency and the real limit of the observed frequency which is closest to the expected value.

Example 9–3. In an experiment, 47 rats are run in a T-maze. On the first trial of the experiment, 28 rats make a right turn and 19 make a left turn. Test the hypothesis that there is no difference between the probabilities of turning right or left at the start of the experiment.

The hypothesis to be tested concerns the population of rats which might have been tested in this experiment. We assume that the animals used are a random sample of this population. The hypothesis is that the probability of a right turn on the first trial, p_r, is .50, and the same for a left turn, p_l.

The total N in the sample is 47, so that the expected frequencies are equal to 23.5.

Since only two classes are possible in the partitioning, the degrees of freedom related to the χ^2 to be developed are equal to one. The correction for continuity should be used. The χ^2 according to (9–6.2) is

$$\chi^2 = \frac{(|28. - 23.5| - .5)^2}{23.5} + \frac{(|19. - 23.5| - .5)^2}{23.5}$$

$$= \frac{(4.0)^2 + (4.0)^2}{23.5} = 1.362$$

Table F indicates that a one-degree-of-freedom χ^2 must be 3.841 or greater for rejection at the .05 level of confidence. Therefore, the observed frequencies do not differ from the expected frequencies sufficiently to reject the hypothesis tested.

In summary, we have two restrictions on the use of (9–5.1) for obtaining a χ^2: (a) all expected frequencies should be 5 or greater, and (b) when one degree of freedom is associated with the χ^2, (9–6.2) should be used to obtain the statistic; i.e., a correction for continuity should be used.

9–7. Contingency Chi Square. When two (or more) characteristics of elements are used as the basis of classification, it is often of interest to test the hypothesis that population proportions for one characteristic do not vary as a function of the other characteristic. This hypothesis is *similar* to the interaction of two independent variables studied in the analysis of variance procedures. As a simple illustration, let us consider the population of persons in the United States at the moment. In order to classify these people let us examine both the sex of a person and his or her height. Let us simply divide the population into persons over 5 ft. 10 in., and persons 5 ft. 10 in. or less. Combined with sex, we now have four classes which we may call "tall men, short men, tall women, and short women." In this problem we are not interested in the proportion of men or women, or even in the proportions of tall and short persons. What is of interest is whether the proportions of tall (and short persons), *within each sex,* are equal. Consider just the men for a moment. Suppose that .51 of the *men* are in the tall class. Now suppose that of the women, .37 are in the tall class and .63 in the short class. Clearly the classification on the basis of height is related to the sex variable, although a cause and effect relationship should *not* be inferred.

If the population proportions of tall and short persons, considering each sex separately, are the same for both men and women, then the characteristics are unrelated. We may state this as: *The proportions of elements found in the different conditions of one characteristic do not depend upon or are not* contingent *upon which class of a second characteristic is considered.*

The test of a hypothesis about such a lack of contingency should be carefully distinguished from our earlier hypotheses about frequency data. In the present hypothesis, the exact proportions within any class of elements is not specified in advance. What is specified is that the proportion of the men who are "tall," whatever that might be, is equal to the proportion of women who are "tall."

Expected frequencies. In order to develop expected frequencies for a *contingency* χ^2 problem, we need estimates of the population proportions within each class of a given characteristic, ignoring the other characteristic. For instance, in terms of the previous illustration, we wish to estimate the proportions of men and women in the population, and the proportions of tall and short persons in the population. The χ^2 to be obtained should not be a test of these proportions in any way, since our interest is not in these values at all but in the equality of such average values over each subdivision.

In order to avoid a test of the average proportions within the classes of a single characteristic, we use the observed frequencies in the entire sample as exactly equal to the expected frequencies for each characteristic. For example, if the observed numbers of men and women were 85 and 15, respectively, ignoring the other characteristic, we proceed as if the population had .85 men and .15 women. Similarly, if we were to obtain 36 tall subjects and 64 short ones, we assume that the population consists of .36 tall persons and .64 short ones, ignoring sex.

Once these average values are assumed, the expected frequencies *within* each class of the two-fold array may be obtained. Using the hypothetical data given above, it is clear that the 85 men should be divided into .36 tall persons and .64 short persons. The same division should be true of the 15 women if the two characteristics are unrelated. This would suggest that $(85)(.36) = 30.6$ tall men are to be expected. Similarly we would expect 54.4 short men, and 5.4 tall women and 9.6 short women.

In order to calculate the expected frequencies for a contingency χ^2 in an easy manner, let us arrange the possible classes into a row by column table, with the different levels of one characteristic arranged along the rows and the levels of the other characteristic arranged along the columns. Table 9–1 shows a general outline of this arrangement and indicates the method of obtaining the expected frequencies. In this table, we denote each cell (class) with a double subscript, the first subscript indicating the row and the second the column. The expected frequencies may then be denoted as $f_{e,i,j}$, where the i indicates the row and j the column. Thus, $f_{e,3,2}$ would stand for the expected frequency in the cell found in the third row, second column.

The total frequency in each row and column is denoted as f_{Ri} and f_{Cj}, respectively. We may now give a formula for finding each expected frequency, using only the observed data:

$$f_{e,i,j} = \frac{(f_{Ri})(f_{Cj})}{N} \tag{9–7.1}$$

where the N is the total number of cases in the sample.

Table 9–1 illustrates the use of (9–7.1).

TABLE 9–1

The Expected Frequencies in a Contingency Chi Square Problem[a]

Characteristic I Levels	Characteristic II Levels						Totals
	1	2	. . .	j	. . .	J	
1	$\frac{(f_{R1})(f_{C1})}{N}$	$\frac{(f_{R1})(f_{C2})}{N}$	. . .	$\frac{(f_{R1})(f_{Cj})}{N}$	. . .	$\frac{(f_{R1})(f_{CJ})}{N}$	f_{R1}
2	$\frac{(f_{R2})(f_{C1})}{N}$	$\frac{(f_{R2})(f_{C2})}{N}$	. . .	$\frac{(f_{R2})(f_{Cj})}{N}$	. . .	$\frac{(f_{R2})(f_{CJ})}{N}$	f_{R2}
.	.	.	. . .	.	. . .	.	.
.	.	.	. . .	.	. . .	.	.
.	.	.	. . .	.	. . .	.	.
i	$\frac{(f_{Ri})(f_{C1})}{N}$	$\frac{(f_{Ri})(f_{C2})}{N}$	. . .	$\frac{(f_{Ri})(f_{Cj})}{N}$	. . .	$\frac{(f_{Ri})(f_{CJ})}{N}$	f_{Ri}
.	.	.	. . .	.	. . .	.	.
.	.	.	. . .	.	. . .	.	.
.	.	.	. . .	.	. . .	.	.
I	$\frac{(f_{RI})(f_{C1})}{N}$	$\frac{(f_{RI})(f_{C2})}{N}$	. . .	$\frac{(f_{RI})(f_{Cj})}{N}$	. . .	$\frac{(f_{RI})(f_{CJ})}{N}$	f_{RI}
Totals (observed)	f_{C1}	f_{C2}	. . .	f_{Cj}	. . .	f_{CJ}	N

[a] The entries in each cell indicate the method of calculating the expected frequencies for that cell. See the text for further discussion.

Degrees of freedom. The development of expected frequencies for the contingency χ^2 application requires a number of restrictions to be placed on the observed data. Although we obtain RC observed cell totals (where R and C are the number of levels of the two characteristics) and the same number of expected frequencies, the restrictions on the observed data are more extensive than in the earlier applications of χ^2. Not only do we use (and hence restrict) the N in the sample, we also use the row and column totals from the observed data.

First restricting N to the observed total permits us to fix $R - 1$ of the row totals at their observed values. The remaining row total is determined, in a sense, since the sum of row totals is already restricted to N. A similar analysis of the column totals allows us to count the number of restrictions as $(R - 1) + (C - 1) + 1$. This total number of restrictions on the cell totals reduces the number of cell totals which may vary freely from RC to $RC - (R - 1) - (C - 1) - 1$.

Simple algebra reveals that the total degrees of freedom for a *contingency* χ^2 takes the same form as the *df*'s in an interaction term from analysis of variance techniques, i.e.,

$$df = (R - 1)(C - 1) \tag{9–7.2}$$

The student may gain some insight into this concept of degrees of freedom by fixing the row and column totals for a row-by-column table and actually filling in cell frequencies one at a time. In so doing, he will find that $(R - 1)(C - 1)$ such numbers may be filled in arbitrarily without violating the restrictions on row and column totals; however, the remaining cell totals will now be determined exactly by the restrictions.

Calculations. The RC cell totals, one observed and one expected for each cell, now form the basis for computing the χ^2 as a test of the "no-contingency" hypothesis. The procedure is now exactly the same as in our earlier applications. The two limitations for the use of χ^2 also hold: no expected frequency should be less than 5, and a 1 *df* problem requires the correction for continuity (such a case will arise only with a 2-by-2 table). Equation (9–5.1) or (9–6.2) is applied, depending upon the number of degrees of freedom, with $K = RC$.

Example 9–4. There would appear to be no reason biologically that the sex of an individual would be related to his hair color. However, there is a rather pronounced tendency for women to modify hair color. One might suspect that the proportions of persons with various "apparent" hair colors would be different for the two sexes. In order to test such a hypothesis, let us imagine that a college population is used and 280 subjects are selected at random from the population. Each person is classified according to both sex and hair color. The following are possible observed frequencies resulting from such an experiment.

Sex	Hair Color				Totals
	Brown	Black	Blond	Red	
Men	65	67	22	9	163
Women	18	34	52	13	117
Totals	83	101	74	22	280

We wish to test the hypothesis that the proportions of men with various hair colors are the same as for women with these hair colors, in the population. This is a contingency problem since no prior proportions are offered for either characteristic *per se.*

Note that a χ^2 statistic should be appropriate here, since the data are frequencies resulting from a partitioning of the population, and sampling is random from this population.

The use of (9–7.1) gives the following set of expected frequencies for each cell, with the observed values given once again in the parentheses.

Sex	Hair Color				Totals
	Brown	Black	Blond	Red	
Men	48.3 (65)	58.8 (67)	43.1 (22)	12.8 (9)	163
Women	34.7 (18)	42.2 (34)	30.9 (52)	9.2 (13)	117
Totals	83	101	74	22	280

All of the expected frequencies are greater than 5.

The *df* is 3, $(2 - 1)(4 - 1)$, so we will use (9–5.1) to find the test value of χ^2.

$$\chi^2 = \frac{(16.7)^2}{48.3} + \frac{(8.2)^2}{58.8} + \frac{(-21.1)^2}{43.1} + \frac{(-3.8)^2}{12.8} + \frac{(-16.7)^2}{34.7} + \frac{(-8.2)^2}{42.2} + \frac{(21.1)^2}{30.9} + \frac{(3.8)^2}{9.2}$$

$$= 5.77 + 1.14 + 10.33 + 1.13 + 8.04 + 1.59 + 14.41 + 1.57$$

$$= 43.98$$

The random sampling distribution of χ^2 with 3 *df* (Table F) shows this answer to be highly significant. We may reject the hypothesis of no relationship between the characteristics of sex and apparent hair color.

Since the hypothesis tested in this contingency problem is related to proportions, it is advisable to actually look at the observed proportions of hair color within each sex. Dividing the observed frequencies of the men by the total number of men (163) we obtain .40, .41, .13, and .06 for the various hair colors in the order used earlier. The corresponding values for the women (using 117 as the denominator) are .15, .29, .44, and .11. The largest discrepency between the pairs of values is in the "blond" category, a result which is hypothetical but probably not unrealistic.

Another method of looking at these data, which is equivalent to the same contingency hypothesis, is to examine the proportions of each sex found in a given hair-color class. Giving the men first in each case, we find .78 and .22 for brown, .66 and .34 for black, .30 and .70 for blond, and .41 and .59 for red. Looking at the four proportions for the men, we find that the men predominate in this total sample in both the brown and black categories but women are more frequent in the blond and red classes.

There is essentially no difference between these two methods of viewing the data in a contingency problem, but often one way of describing the data is more easily understood. Both may be used in order to gain insight into the experimental interpretation to be given the results.

9–8. Goodness of Fit. A common application of χ^2 is to test a distribution of scores against the expected values derived from some specified distribution such as the normal curve. This type of application is a Question-III hypothesis-testing procedure. It is quite similar to our earlier applications involving K observed and expected frequencies, but the development of the expected p_k values is more complex. In this regard it is similar to the contingency χ^2 in that formula (9–5.1) or its modification may be used *once the expected values are obtained.*

The *goodness-of-fit* test is different in only two ways from our previous applications. First, the elements in the population and sample are scores, X's, which have been classified into a frequency distribution. The frequencies in each score class may now be used just as the frequencies of elements in the various classes of any partitioning procedure. The second difference is that the degrees of freedom related to the χ^2 may be less than $K - 1$, as in the contingency problem.

Since a goodness-of-fit test asks a question about the shape of a frequency distribution of scores, one would set the observed and expected values of certain parameters of the distribution in question equal to one another, as was done with the row-and-column values in the contingency test. It is this operation that reduces the number of degrees of freedom from $K - 1$.

Let us examine only a single goodness-of-fit test, the test for normally distributed scores. Such an application requires that a set of observed scores be grouped into a frequency distribution. The method of grouping the scores is quite arbitrary. Care should be used so as to determine the class intervals used for grouping without regard for the actual data to be tested. Without such a precaution, one runs the risk of allowing one's desire about the outcome of the test to influence the grouping procedure.

In testing a set of observed frequencies for their correspondence to a normal distribution, three restrictions are placed on the data. The first is the usual one that the observed and expected N's be equal. The other two restrictions stem from the need to set the mean and standard deviation of the expected set of X's equal to the similar values of the observed scores. Therefore, if we use K groups of score values (usually a range of X's in each group), the df for the resulting χ^2 will be equal to $K - 3$.

If the expected number of cases in some class (usually at the tails of the distribution) is less than five, a common practice is to combine such a class with an adjacent one. This is illustrated in Example 9–5. However, this technique is not always legitimate in a strict sense of scientific methodology. Another approach is illustrated in Example 9–6 (for the same data as in Example 9–5). This second method overcomes the problem of ever getting expected values less than 5 and also simplifies the calculation of the χ^2.

Example 9–5. In the first stage of an experiment on stimulus generalization, an investigator conditions 116 subjects to a tone. The unconditioned stimulus is the application of an ice cube to the subject's wrist. The response measured is the reduction of blood volume in the finger tips, and is measured by sensitive instruments on a polygraph. During this conditioning stage, the latency of the first conditioned response for each subject is recorded in seconds and hundredths of a second. The investigator wishes to test whether or not the distribution of 116 latency measures can be considered normal in form.

The calculation of the mean and s of the data yields values of 35.24 sec. for the mean and 10.239 sec. for the standard deviation. The data are placed into a frequency distribution as given below and the expected proportion and expected frequencies in each class are determined by using the mean and s from the data (as well as N) and a table of the normal distribution.

In the analysis of these data, the real limits between each range of scores is changed to a z score as shown in columns III, IV, and V. The p_k values result from examination of the normal curve tables in z-score form. For example, if we wish to find the proportion of scores on a normal curve for the range 55 to 51, we simply examine the z-scores which form the limits of this range. These are 1.98 and 1.49. The proportion of cases in a normal distribution with values between 1.98 and 1.49 is found by subtraction as illustrated in Chapter 5.

The resulting p_k values are those expected from a normal distribution having a mean and standard deviation equal to the set of scores being examined. The expected frequencies are obtained as usual by multiplying each p_k by the N of 116.

We find that a χ^2 cannot be meaningfully calculated from the observed and expected cases as shown above, because four of the expected frequencies are less than 5. In order to overcome this problem, let us simply combine those classes of low expected frequencies with an adjacent class. Note that the combination of the classes 65–61 and 60–56 would still give an expected fre-

I Class	II f_0	III Real limits	IV $x = X - M$	V $z = x/s$	VI p_k	VII f_e
65–61	2				.0068	0.8
		60.5	25.26	2.47		
60–56	4				.0171	2.0
		55.5	20.26	1.98		
55–51	5				.0442	5.1
		50.5	15.26	1.49		
50–46	5				.0906	10.5
		45.5	10.26	1.00		
45–41	10				.1463	17.0
		40.5	5.26	.51		
40–36	30				.1830	21.2
		35.5	.26	.03		
35–31	27				.1892	21.9
		30.5	−4.74	−.46		
30–26	16				.1517	17.6
		25.5	−9.74	−.95		
25–21	10				.0962	11.2
		20.5	−14.74	−1.44		
20–16	4				.0481	5.6
		15.5	−19.74	−1.93		
15–11	2				.0190	2.2
		10.5	−24.74	−2.42		
10– 6	1				.0078	0.9
	$N = 116$					$\Sigma f_e = 116$

quency less than 5. We must include these two classes with the next class, 55–51, in order to obtain the desired goal. The final classification, observed frequencies, and expected frequencies are given next. The χ^2 obtained from each class is also given. These have been computed as usual by use of (9–5.1). The total of these eight terms is 12.43.

Class	f_o	f_e	χ^2
over 50	11	7.9	1.22
50–46	5	10.5	2.88
45–41	10	17.0	2.88
40–36	30	21.2	3.65
35–31	27	21.9	1.19
30–26	16	17.6	.15
25–21	10	11.2	.13
under 21	7	8.7	.33
			$\chi^2 = 12.43$

The evaluation of the total χ^2 requires the consideration of the degrees of freedom related to the test value. In this case, we have used 8 classes of scores, but have placed three restrictions on the expected frequencies: they have been developed so that the sum of the f_e's is 116, the mean is 35.24 and the standard deviation is 10.239. The *df* for the χ^2 is then $8 - 3$, or 5. The .05 value of a χ^2 with 5 *df* is 11.07. Since the obtained χ^2 is larger than that, we may reject the hypothesis that the obtained scores are a random sample from a normally distributed population.

The data from Example 9–5 could have been tested for their goodness of fit to a normal distribution in a number of slightly different ways, depending upon the classes selected for grouping the data. Certainly some methods would be more sensitive to departures from the normal curve in some portions of the range of scores than other methods. Unless there is particular interest in some segment of the distribution over other segments, there is no general rule which will allow a less than arbitrary decision to be reached concerning the class intervals used.

One method of grouping the data for a χ^2 goodness-of-fit test which has some merit in terms of computational ease is given in Example 9–6. In this method, all expected frequencies are set equal to one another in advance, and the problem of expected values less than 5 is eliminated at the outset.

Example 9–6. For this example, we use the same data and problem as in Example 9–5. Prior to classifying the data, however, let us construct a classification system in which the classes each have the same expected p_k's. For instance, let us use $\frac{1}{10}$ of the normal distribution in each of ten classes. Dividing the distribution at the mean and taking consecutive *z*-scores such that each range contains .10 of the normal curve, the *z*-scores may then be transformed to their raw score equivalents by use of the mean of 35.24 and *s* of 10.239, taken from the observed data.

z-Scores	$x = (z)(s)$	$X = (+x) + M$	$X = (-x) + M$
0.0	0.0	35.24	–
±0.26	±2.66	37.90	32.58
±0.52	±5.33	40.57	29.91
±0.84	±8.60	43.84	26.64
±1.28	±13.11	48.35	22.13
over |1.28|		over 48.35	less than 22.13

We may now construct the ranges of observed and expected scores. The p_k's are each .10 so that the expected frequency in each group will be (116)(.10), or 11.6. The χ^2 may now be computed with a common denominator for each cell. This requires only one division. The raw scores are not given here but we assume that each one is known to 2 decimal accuracy.

Since we have 10 classes, and have made 3 restrictions on the data, the *df* for this χ^2 is 7. The distribution of χ^2 with 7*df* indicates that our obtained value is significant beyond the .01 level of confidence. As in Example 9–5,

Range of Scores	f_o	f_e	$f_o - f_e$	$(f_o - f_e)^2$
Over 48.35	14	11.6	2.4	5.76
48.35–43.84	6	11.6	−5.6	31.36
43.83–40.57	6	11.6	−5.6	31.36
40.56–37.90	9	11.6	−2.6	6.76
37.89–35.24	23	11.6	11.4	129.96
35.23–32.58	15	11.6	3.4	11.56
32.57–29.91	15	11.6	3.4	11.56
29.90–26.64	9	11.6	−2.6	6.76
26.63–22.13	8	11.6	−3.6	12.96
under 22.13	11	11.6	−0.6	.36
				248.40

$$\chi^2 = \frac{248.40}{11.6} = 21.41$$

one would reject the hypothesis that these data are realistically described by the normal distribution. The fact that the obtained probability of the present χ^2 is less than that obtained in Example 9–5 is coincidental. Neither method of grouping the data is consistently "better" than the other.

9–9. Approximations of Exact Probabilities. The application of the χ^2 statistic to frequency data has already been cited as an approximation. As we shall see in Chapter 10, any particular set of observed frequencies in K classes has an exact probability of happening, given the true proportions of scores in the population. The probability value associated with a given χ^2 is an approximation of the sum of a number of exact probabilities, specifically those probabilities associated with all outcomes which are as deviant or more so from expectation as the observed frequencies. The χ^2 takes into account, so to speak, deviations from the expected values in both the direction observed and in the opposite direction. For instance, if one were to test the hypothesis that the probabilities of heads and tails are both .50 in a coin-tossing experiment, using 10 tosses of the coin would yield expected frequencies of 5.0 for both outcomes. If we obtain 9 heads and 1 tail in an actual experiment, the χ^2 (corrected for continuity) may be interpreted as a probability which equals the sum of the probabilities of the following exact outcomes: 10 heads and 0 tails, 9 heads and 1 tail, 1 head and 9 tails, and 0 heads and 10 tails.

In cases where the χ^2 is not appropriate, such as with small expected frequencies or in the evaluation of the probability of some set of outcomes not equal to the two "tails" of the distribution of outcomes, one must resort to the use of the exact probabilities. Certain portions of Chapter 10 will indicate how this is to be accomplished.

EXERCISES

1. For each situation given, find the expected frequencies for the different classes implied.
   (a) $N = 120$, $p_1 = .40$, $p_2 = .15$, $p_3 = .45$.
   (b) Tossing a true die 300 times.
   (c) You are given the hypothesis that accidents happen in proportion to the populations of communities. The populations for 4 cities which had a total of 1460 accidents are, $A = 1{,}000{,}000$, $B = 200{,}000$, $C = 25{,}000$, and $D = 475{,}000$.
   (d) If an instructor claims to assign grades on a curve which is .08 *A*'s, .24 *B*'s, .40 *C*'s, .22 *D*'s, and .06 *E*'s, and claims that men and women do equally well, what would be expected from a class of 210 men and 180 women?
2. Test each hypothesis in Exercise 1 with the corresponding observed frequencies given below.
   (a) $f_{o,1} = 65, f_{o,2} = 10, f_{o,3} = 45$
   (b) f_o's: 38, 61, 52, 47, 59, and 43
   (c) $f_{o,A} = 801, f_{o,B} = 245, f_{o,C} = 83, f_{o,D} = 331$
   (d)

	Grades				
	A	B	C	D	E
Men	21	63	81	42	3
Women	16	54	65	32	13

3. In (d) of Exercises 1 and 2, the instructor claims that the class is somewhat better than usual. Do his grades allow you to reject his previous curve as a hypothesis? Are the grade distributions shown related to the sex of the students?
4. A person says that he is unlucky in card games. In support of his claim, he counts the number of bridge hands played during an evening, which is 28, and finds that on 13 of these he was not dealt a single ace. The probability of being dealt a bridge hand not containing an ace is .3038. Can you reject the hypothesis that .3038 is a true probability for this person?
5. For the following frequency data, can you reject the hypothesis of equal probabilities of passing the test for the two grade levels shown?

	Pass	Fail
Freshmen	95	19
Sophomores	49	1

Applications; use information from all previous chapters where relevant.

6. An experimenter has a theory from which he can predict the mean and standard deviation for the normally distributed response measures of each subject in a concept identification task. He uses 15 subjects to test his

theory. For each, he finds a z-score which describes the degree of departure from prediction for that subject. Given these independent z-scores, can you reject the hypothesis that subjects behave according to his theory? (Hint: the sum of independent χ^2's is distributed as χ^2 with df equal to the sum of the separate df's)

−1.20 −.67 .35 .33 −.44 .14 .05 .12
−.43 .43 −1.01 −.55 2.02 .88 1.48

7. In a pursuit-rotor experiment, the length of hits on the target is recorded for each subject. These are classified into short, medium, and long hits by a predetermined rule. Two practice conditions are used, with 10 subjects in each condition. The frequencies of short, medium, and long hits are tabulated on the last trial for each subject. How would the data be analyzed?
8. The frequencies of accidental deaths in two cities are recorded for three successive years. How would you test the hypothesis that changes in the probability of such deaths over the years considered are independent of which city is involved?
9. The I.Q. scores from a certain intelligence test are known to be normally distributed with $\mu = 100$, and $\sigma = 15$.
   (a) If a random sample of 25 college students shows a mean I.Q. of 112, can you reject $H_0: \mu = 100$?
   (b) If the 25 college students are distributed with 5 below an I.Q. of 100 and 20 above this value, can you reject the hypothesis that they were randomly selected from the original population?
10. The results of a standardized test in English are taken from 15 different people in each of three different colleges. For each college, 5 people were tested in September, 5 in January, and 5 in May. The scores are given below. Analyze the data and indicate why the results might be difficult to interpret.

College	Month of testing		
	September	January	May
	67	81	61
	92	87	93
X	48	89	88
	75	52	82
	69	68	78
	66	70	58
	69	65	65
Y	43	91	71
	88	80	68
	71	74	83
	82	90	93
	95	86	68
Z	81	72	97
	78	84	81
	75	78	86

10

Probability

In this chapter we shall examine a number of types of statistical problems which differ from one another, at least slightly. They have in common a general method of solution called probability theory. The formal aspects of probability theory will be drastically condensed in our discussions; however, the procedures suggested for problem analysis and solution are based on the theorems derived in mathematical probability treatments. Although the coverage of formal derivations is quite limited, a broad spectrum of elementary probability problems is considered. The student may have difficulty with many of the elementary probability problems. A good deal of practice with such problems seems to be required before any degree of general competence is acquired. However, in an effort to offset this discouraging state of affairs, a method of analyzing a probability problem in terms of a set of relevant characteristics is offered. This method should allow the student to recognize and solve those problems illustrated in this treatment. Moreover, it is hoped that such an analysis will permit the student to recognize problems which are more complex or of different types than the ones given.

With respect to our general types of statistical questions, all of the probability problems relate directly to the Question-I case, in which we have or assume knowledge of the population and develop answers about particular outcomes in random samples from the population. The other types of statistical questions may arise in this context, but the means of answering these questions will depend upon the appropriate solution of the Question-I type.

10–1. General Concepts. Let us structure our thinking about probability by using the terminology of set theory. As in previous work, we begin by considering a population of N_* distinct elements. This population set forms the basis for the definition of an *experiment*, a *sample space*, an *event*, and the *probability of an event*.

The experiment. We define an *experiment*, in the context of probability theory, as the random selection of a sample of r elements from the population set. The random selection procedure may be either with or without replacement.

It should be noted that the term experiment does not imply that a person has direct or even indirect control over the selection procedure; one may simply be an observer of "natural" events. Such happenings may or may not be appropriately analyzed according to the laws of probability. This is an empirical question in the same way that the application of any mathematical formulation to the real world may or may not be useful. One merely forms an analogy between our formal notion of random sampling and the events in the "outside" world. As in other scientific applications of logic and mathematics, the usefulness of such an approach has been remarkable.

The sample space. The outcome of an experiment is described by giving information about the population elements contained in the sample. Each different outcome of an experiment which is logically possible is called a **sample point**. We will restrict our attention to cases in which each sample point is equally likely to occur. In practice, this usually requires an examination of the elements selected in the samples, *including their order of selection*. If we now find that each sample point, as defined by our description of the sample, contains an equal number of all possible samples which differ in any respect, then we may assume that each sample point is equally likely. Usually, the sample points may be so defined that each possible sample that is different from any other sample in any respect constitutes a sample point.

For example, if the population consists of the 52 cards in a standard deck of playing cards, and if we select a random sample of 2 cards with replacement, then the sample points may be defined as all of the different possible *ordered pairs* of elements. In this case, the selection of the ace of spades followed by the king of diamonds in one sample would constitute a single sample point. The selection of the king of diamonds and then the ace of spades would be a different sample point.

In a sense, the definition of sample points forms the basis for a complete partitioning of all possible outcomes of the sampling experiment. We define the *sample space* as the set of all possible sample points. The sample space will also be called the **universe** set, because we will always restrict our attention to those outcomes of an experiment which are included in the sample space. The sample points are clearly members or elements of the universe set. Let us keep in mind the restriction that each such element of the sample space is assumed to have an equal likelihood of being selected.

An event. In practice, we may not wish to distinguish between all possible sample points. For instance, in the above illustration of the selection of two cards from a deck of cards, our interest may be in the color of the cards in the sample. For this purpose, the universe consists of elements which are too finely described. One might be interested in an outcome of such an experiment which may be described as the selection of two red cards. This outcome will be termed an *event*. This particular event may be described in terms of the sample points by listing all of the many sample points which have two red cards. The set of sample points which satisfies our requirements of some event is called the *event set* or simply the *event*.

Notation. We shall refer to sets by the use of capital letters. The sample space or universe will be denoted as U. Any event within the universe will be given by some other letter, such as A or B. The definition of a set A is accomplished when rules have been set down which either allow the assignment of any element in U to set A or indicate that the element does not belong to set A.

The measure of a set. The concept of the *measure of a set* which we will use is a much restricted version of the mathematical concept by the same name. We will define the measure of a set as the number of elements in that set. Let us refer to this measure of a set A as

$$m(A) = \text{measure of set } A = \text{number of elements in set } A \qquad (10\text{–}1.1)$$

As a special case, let us label the measure of the universe U as N:

$$m(U) = N \qquad (10\text{–}1.2)$$

The $m(U)$ should not be confused with the number of elements in the population, N_*. The measure of U may be far different than N_*, either larger or smaller.

It should be clear that the measure of a set, A, may not exceed N, since the definition of the event A is limited to sample points, each of which is an element of U.

10–2. The Probability of an Event. We now define the probability of an event, A, denoted as $P(A)$:

$$P(A) = \frac{m(A)}{m(U)} \tag{10–2.1}$$

The definition (10–2.1) must be understood as given, i.e., it is a definition and not a derivation based on other postulates or theorems. As such, this definition may or may not be consistent with our earlier concepts of probability as discussed in Chapter 4. Although our present definition is consistent with the mathematical limits notion of probability, given certain assumptions, it would take us far afield to demonstrate the consistency.

Since the $m(A)$ is less than or equal to $m(U)$, we may state that the $P(A)$ is less than or equal to unity. However, in order to include a zero probability from (10–2.1), we must introduce the concept of an empty set.

The null set. We define the *null set*, $\emptyset$, as a set which contains no elements. This null (or empty) set is formally considered to be a subset of any set of elements. We may then say that the universe set, U, contains, as a subset, $\emptyset$. Any definition of a set A which contains none of the possible sample points of an experiment is therefore a permissible exception to our general rule of defining an event in terms of sample points. The $m(\emptyset)$ is defined as zero, for obvious reasons.

We may now include a probability of zero for an event A which might be defined in such a way as to include none of the possible sample points in U.

The definition given for $P(A)$ indicates the general method of solving any probability problem. If one can define and measure the universe and the event in question, (10–2.1) permits the probability of event A to be obtained directly. The difficulties encountered in probability problems arise in the definition of sample points and the measurement of U and the event in question.

10–3. Complex Events. We are often faced with the need to describe an event in terms of other events. When such a description is made, we may consider the event of interest to be a complex event, because our definition is made in terms of other events. Two particular methods of forming such a complex event will be introduced. Since our events are sets of elements (sample points), we refer to these two types of complex events as *intersection sets* and *union sets*.

In order to study these complex sets, let us define a set which includes all elements in the universe which do not belong to some given set. If A is a set of elements in U, $\bar{A}$ is called the **complement** of A. The complement of A is defined as a set of all elements which do not belong to A. It is clear that all N elements in U are thus partitioned into two classes, A and $\bar{A}$.

Intersections. The *intersection* of two sets A and B is itself a set. We denote this intersection set as $A \cap B$. $A \cap B$ is a set of elements such that each element in the intersection set is also an element in both sets A and B. Any element in an intersection of two sets must be a member of each of the two sets considered. Figure 10–1 illustrates this type of relation in a form called a *Venn diagram.* In the figure, all elements in U which belong to a given set are imagined as placed together so that a single circle can be drawn around the set of elements. Perceptually, the intersection of A and B is now indicated by an overlapping area in the diagram.

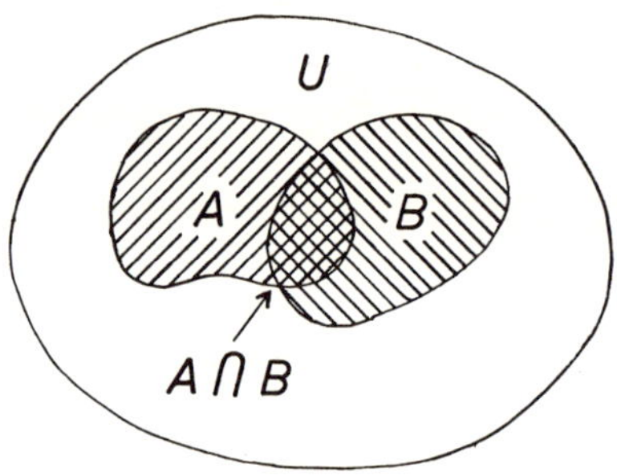

Fig. 10–1. A Venn diagram representing the intersection of two sets.

Suppose that a population of the 52 cards in a standard deck is sampled by drawing a single card at random. Then the universe of sample points will be the 52 possible cards which might be selected. If we define event A as all red cards and event B as all face cards (kings, queens, and jacks), then the intersection of these two sets will contain six elements: king of hearts, king of diamonds, queen of hearts, queen of diamonds, jack of hearts, and jack of diamonds. In terms of the measures of each set, we see that $m(U)$ is 52, the $m(A)$ is 26, the $m(B)$ is 12 and the $m(A \cap B)$ is 6.

Let us consider the intersection of a set A with its complement, $\bar{A}$. By definition, an element in set A cannot belong to $\bar{A}$, and *vice versa.* Therefore, the intersection of A and $\bar{A}$ contains no elements, i.e., is $\emptyset$.

One may define two sets in such a way that their intersection contains no elements. Such sets are called *mutually exclusive* sets. In terms of probability theory, if two events, A and B are so defined that their intersection is the $\emptyset$ set, we call the two events mutually exclusive:

$$A \text{ and } B \text{ are mutually exclusive if } A \cap B = \emptyset \qquad (10\text{–}3.1)$$

Figure 10–2 illustrates two mutually exclusive sets in terms of a Venn diagram.

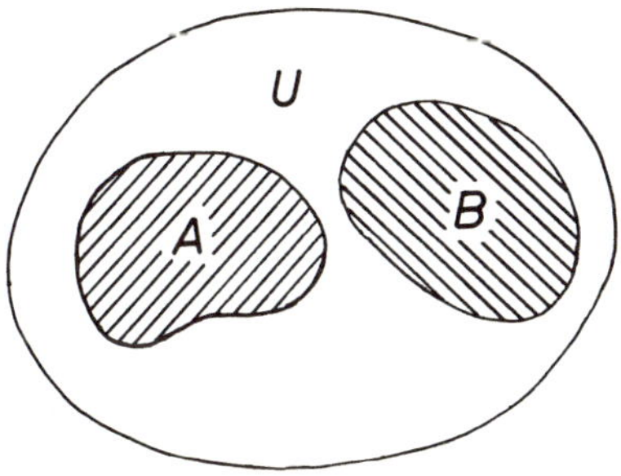

Fig. 10–2. A representation of two mutually exclusive sets.

It should be noted that a set, A, may be a subset of another set, B. For example, A may be all of the hearts in a deck of cards and B may be all of the red cards. It should be clear that the intersection of A and B would now contain all of the elements of A. Ordinarily, we will not have great interest in this set relationship. But the intersection of two sets has great importance in the solution of many probability problems.

Unions. The second complex set to be considered is called a *union*. The *union* of two sets, A and B, will be written as $A \cup B$ and is defined as a set of elements, each of which is either a member of set A or a member of set B (or of both). Figure 10–3 shows a diagram of a union set composed of the simple sets A and B. Note that the union of two sets may well contain more elements than either of the two sets forming the union.

Let us now consider two partly overlapping sets and describe three major parts of the array. Figure 10–4 illustrates the configuration of interest. If A and B are the two primary sets which have some elements in common, we may consider three types of elements. First, we have those elements which belong to set A but do not belong to set B. Second we have those elements which belong to B but not to A. Finally, some elements belong to the intersection of A and B.

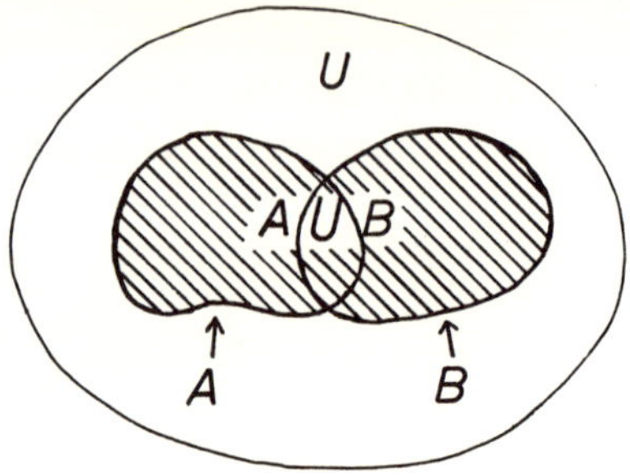

Fig. 10–3. The union of sets A and B, illustrated by the entire shaded area shown.

Note that these subsets are each mutually exclusive of one another. We may describe the first of these sets as the intersection of A with the complement of B. This would be written as

$$A \cap \bar{B}$$

Here, an element in this complex set must belong to A and *not* belong to B. In the same fashion, the second case includes a set of elements which is described as

$$B \cap \bar{A}$$

The final set is the intersection of A and B, $A \cap B$.

This method of viewing the two sets A and B and their relationship allows us to express the union of A and B in terms of three parts which are mutually exclusive of one another. Hence

$$A \cup B = (A \cap \bar{B}) \cup (\bar{A} \cap B) \cup (A \cap B) \qquad (10\text{–}3.2)$$

In (10–3.2) we have simply extended the definition of a union to include three sets rather than two. It is true that each of the three

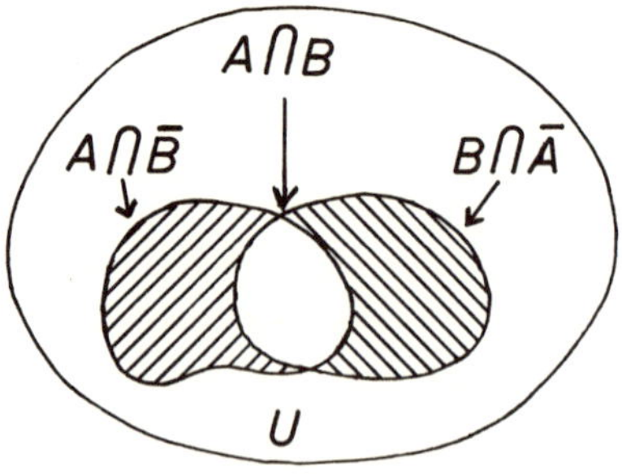

Fig. 10–4. The representation of the union of sets A and B by the combination of three mutually exclusive sets.

sets joined together in the union is an intersection of two others, but (10–3.2) may be understood by the use of Fig. 10–4.

Probabilities of complex events. The probability of an intersection or a union event may be expressed by use of definition (10–2.1). In the case of an intersection, we have

$$P(A \cap B) = \frac{m(A \cap B)}{m(U)} \tag{10–3.3}$$

Let us illustrate by using the experiment of drawing a single card from a deck. If we define A as the red cards and B as the set of 12 face cards, then the $m(U)$ is 52 and the $m(A \cap B)$ is 6, as given earlier. Hence the probability of selecting a card which is both red and a face card (the intersection) is equal to $\frac{6}{52}$ or about .115.

The probability of a union event may be described in a similar way as

$$P(A \cup B) = \frac{m(A \cup B)}{m(U)} \tag{10–3.4}$$

The usefulness of (10–3.2) may now be demonstrated if we note that the measure of the union of two mutually exclusive events is equal to the measure of one event plus the measure of the second.

$$m(A \cup B) = m(A) + m(B) \quad \text{(if } A \text{ and } B \text{ are mutually exclusive)}$$

The use of this property and (10–3.2) will permit a modification of (10–3.4):

$$P(A \cup B) = \frac{m(A \cap \bar{B}) + m(\bar{A} \cap B) + m(A \cap B)}{m(U)} \tag{10–3.5}$$

This modification is possible because the three parts of the union as shown in (10–3.2) are mutually exclusive of one another.

We will now attempt a further modification of (10–3.5). In so doing, note that the union of $(A \cap \bar{B})$ with $(A \cap B)$ is simply the set A, and that these two sets are mutually exclusive. Thus, $m(A \cap \bar{B}) + m(A \cap B)$ is equal to the $m(A)$, and a similar result holds for $(\bar{A} \cap B)$. We now add and subtract $m(A \cap B)$ in the numerator of (10–3.5) and obtain

$$P(A \cup B) = \frac{m(A \cap \bar{B}) + m(A \cap B) + m(\bar{A} \cap B) + m(A \cap B) - m(A \cap B)}{m(U)}$$

$$P(A \cup B) = \frac{m(A) + m(B) - m(A \cap B)}{m(U)} \tag{10–3.6}$$

The numerator of (10–3.6) must be equal to the measure of the union of A and B as seen from (10–3.4). We may reason this equality out verbally by considering the addition of the measure of A and the measure of B. It may be seen that the number of elements in the union of A and B may be smaller than this sum because any element in the intersection of A and B will have been counted twice, once because the element belongs to A and again because it belongs to B. The subtraction of the number of elements in the intersection of A and B is just sufficient to make allowance for the duplication of counting elements in the intersection.

The division of each term in the numerator of (10–3.6) by the $m(U)$ yields the following important formulation:

$$P(A \cup B) = P(A) + P(B) - P(A \cap B) \qquad (10\text{–}3.7)$$

Let us work an example using (10–3.7).

Example 10–1. What is the probability of rolling a pair of dice and obtaining at least one 6 on the two faces?

In this problem, we will consider the population as the digits 1, 2, 3, 4, 5, and 6. Each of these elements (numbers here) represents a possible face on a die which may result from rolling that die.

The experiment suggested may be construed as the selection of two elements from the population. Since there are two separate dice, we must allow for the same result on each selection, i.e., we consider sampling with replacement.

The sample space is most easily described by keeping the two dice separate. This may be likened to keeping track of which number is selected first in a given sample and which is selected second. A complete description of the universe is then given by listing all possible ordered pairs of the numbers 1 through 6. The measure of the universe here is found to be 36. This number may be obtained by simply enumerating the possible ordered pairs, i.e., 11, 12, 13, 14, 15, 16, 21, etc. We define event A as the collection of sample points in which a 6 occurred on the first selection. B is the set of points in which a 6 occurred in the second selection. In each case, there are six points included. Hence,

$$P(A) = \tfrac{6}{36}, \qquad P(B) = \tfrac{6}{36}$$

The problem asks for the probability of the event which consists of at least one six. This includes a six on the first selection, or a six on the second selection, or a six on each selection. In terms of the events A and B, our required event is the union of A and B. We then desire

$$P(A \cup B) = P(A) + P(B) - P(A \cap B)$$

A simple examination of the sample space will show that the number of points in the intersection of A and B is 1. The pair 66 is the only case in which both A and B are true simultaneously. Hence the $P(A \cap B)$ is $\tfrac{1}{36}$.

We may now obtain

$$P(A \cup B) = \tfrac{6}{36} + \tfrac{6}{36} - \tfrac{1}{36} = \tfrac{11}{36} = .306$$

It is interesting to note that the probability is less than the sum of $P(A)$ and $P(B)$. One intuitive approach to this problem often encountered is that the probability of a six on one die is $\frac{1}{6}$ and the same on the second, hence the probability of one or the other must be $\frac{2}{6}$. The error in this analysis is that on some trials of the experiment, both dice give sixes, but the event has still occurred just one time and not twice. This "duplication of effort" is analogous to the intersection and must not be counted twice, so to speak, in the calculation of the union event.

10–4. Independent Events. Two events are said to be *independent* of one another if the probability that one event will occur remains the same whether or not the other event has occurred. We will not attempt to define independent events in a formal fashion, but certain characteristics of such events must be considered.

In order for two events to be independent, it is necessary that either event may occur without the other and that both may occur simultaneously. Another way of stating the latter is that the intersection of the two events must not be $\emptyset$.

However, just the fact that the intersection of the two events has a non-zero probability is not enough to produce independence. The degree of overlap between the two must be precisely enough to permit the following equation to be true.

$$P(A \cap B) = P(A)P(B) \tag{10–4.1}$$

Equation (10–4.1) is offered without proof. It follows from a formal definition of the independence of the $P(A)$ with respect to the occurrence of event B.

Intuitively, one should suspect that two real events are independent if there is no physical relationship between them. For example, if two dice are rolled, the outcome on one die is said to be independent of the outcome of the second, because there is no physical way suspected in which the *outcome* on one die may affect the outcome on the second. Naturally, (10–4.1) is based upon mathematical reasoning and postulates. The question of whether or not (10–4.1) actually applies to two given dice is a matter for testing. However, to the author's knowledge, even with biased dice, no one has ever found two dice whose outcomes were not independent of each other. For instance, it may well be that the probability of a 6 on one die is quite high (loaded die), say .45. Even so, this probability remains at .45 no matter what may happen on the other die on a given toss.

Let us return to the problem in Example 10–1. The probability of the intersection of A and B was found to be equal to $\frac{1}{36}$. This was accomplished by actually counting the measure of $(A \cap B)$. Use of

(10–4.1) will give the same answer. Since event A is the occurrence of a 6 on one die, and B the same on the second, we may consider the two events to be independent of one another. The probability of their *joint occurrence*, the intersection, is then found by multiplying the $P(A)$ by the $P(B)$. This gives $(\frac{6}{36})(\frac{6}{36})$, which reduces to $\frac{1}{36}$.

Independent trials. A kind of special case of independent events concerns the performance of an experiment (in the probability sense) a number of times. We refer to each experiment as a *trial* in this context. By a redefinition of the sample space, the possible sequences of outcomes on the different trials can be handled in the same way as any probability problem. However, it is often more convenient to use (10–4.1) in finding the probability of a particular sequence of events.

Certainly in many cases, the outcome of a trial can have no effect upon the probability of an outcome on a later trial. We may then state that the trials are independent of one another and use (10–4.1) to find the probability of the joint occurrence of particular outcomes on two (or more) successive trials.

Example 10–2. Find the probability of drawing cards at random, one at a time with replacement, and obtaining an ace, a king, and a queen, in that order.

We may consider this problem to consist of independent trials, three of them. Since replacement is used and random sampling employed on each trial, there is no reason to suggest that the outcome on any one trial will affect other trial outcomes. The event of an ace on trial 1 will be called A. Let B represent a king on trial 2 and C a queen on trial 3. On each trial, the sample space consists of 52 points and the measure of the event in question is 4 in each case. Hence, we see that $P(A) = P(B) = P(C) = \frac{1}{13}$. The problem requires that each of the three events takes place. This is simply the intersection set in some more elaborate sample space. The answer is obtained by use of (10–4.1). The extension of (10–4.1) to three independent events is quite easy. Simply consider the first two trials alone. $P(A)P(B)$ will give the $P(A \cap B)$. Now consider that $A \cap B$ is an event and reapply (10–4.1) with $(A \cap B)$ representing one event and C the second.

The solution may be given as

$$P(A \cap B \cap C) = P(A)P(B)P(C) = \tfrac{1}{13} \cdot \tfrac{1}{13} \cdot \tfrac{1}{13} = .00046$$

Let us summarize the general result of the independent trials case. If k independent trials are considered, the probability of the joint occurrence of event A_1 on trial 1, A_2 on trial 2, etc., written as $P(A_1, A_2, \ldots, A_k)$, is given as:

$$P(A_1, A_2, \ldots, A_k) = P(A_1)P(A_2) \cdots P(A_k) \qquad (10\text{–}4.2)$$

Special forms of $P(A \cup B)$. We now have two special cases in which the probability of the union of two events may be found from only the knowledge of $P(A)$ and of $P(B)$. The general solution of the union problem given in (10–3.7) requires the use of the probability of the intersection of A and B. The two cases in which this intersection probability may be obtained without specific examination of the sample space are: (1) when A and B are mutually exclusive, and (2) when A and B are independent. In the first case, if A and B are mutually exclusive events, we know that $P(A \cap B)$ is zero, from the definition of such events. This would allow us to find the $P(A \cup B)$ from (10–3.7) as

$$P(A \cup B) = P(A) + P(B) \text{ (where } A \text{ and } B \text{ are mutally exclusive)} \quad (10\text{–}4.3)$$

For example, if A represents obtaining a 6 on a die and B represents a 5 on the die, and if we consider a single toss as the experiment, then the probability of getting either a 5 or a 6 is equal to $\frac{1}{6} + \frac{1}{6}$ or $\frac{1}{3}$.

In the second case, two independent events occur jointly with a probability equal to $P(A)P(B)$. This allows (10–3.7) to be given as

$$P(A \cup B) = P(A) + P(B) - P(A)P(B) \quad \text{(where } A \text{ and } B \text{ are independent)} \quad (10\text{–}4.4)$$

In nearly all further work, the union of two events will be considered only when one of these two cases applies. The calculation of the union probability in other cases is often impossible with information at nand.

Gambler's fallacy. A classic misconception about the "laws of chance" may now be dispelled. When a single event occurs over and over in a series of independent trials, people have a strong tendency to expect an alternative event to occur next. For instance, if an unbiased coin is tossed a number of times, the probability of obtaining a head on a given toss is .50, ignoring the case in which the coin rests on its edge. If the trials do constitute independent trials, then the past history of events has no bearing on the outcome on a given trial. The fact that tails may have occurred 10 times in a row does *not* alter the probability of heads on the next trial!

In spite of the fact that the "long run" expectation is for .50 heads and .50 tails, the previous history of independent trial outcomes does not affect the future trial outcomes. One may understand this by analogy, perhaps. The probability of making a million dollars starting from nothing is quite small. However, if one already has \$999,999, then the probability of getting to one million dollars is not bad at all. The probability of obtaining 11 heads in a row starting from trial 1 is quite small, $(\frac{1}{2})^{11}$ by use of (10–4.2). However, given 10 heads in a row, the probability of having 11 in a row after the next toss is $\frac{1}{2}$.

One may also ask, "What change in the world takes place when 10 heads in a row are obtained which will have an effect on the coin the next time it is tossed?" The laws of probability are not derived from metaphysics or mysticism. Moreover, a number of persons have learned from hard experience that the real world corresponds to the mathematical realm more closely than to their personal intuition or hunches.

In fact, if one obtains 10 heads in a row at the start of a series of coin tosses, the laws of chance predict that heads will be 10 ahead of tails at any future point, on the average, of course. The expectation of .50 heads and .50 tails is not really violated by this result, because a difference of only 10 in total frequencies becomes vanishingly small when the total number of trials increases and raw frequencies are converted to proportions.

A final point of some relevance here concerns the possible lack of correspondence between the real world and the probability analysis. If a coin of questionable balance actually came up heads 10 times in a row, would one reasonably argue that tails are more likely than heads? The expectation for an increased probability of tails after 10 heads runs contrary to any analysis of independent trial outcomes.

Perhaps one cause of such a fallacy is an assumption, perhaps unconscious, by people that sampling is taking place without replacement! If one knew that a bag contained 50 red and 50 black balls, then the selection (without replacement) of 10 red balls in a row would greatly increase the probability that the next ball chosen would be black. This type of problem will be considered later.

10–5. Types of Sampling Experiments. As indicated at the beginning of the chapter, a number of slightly different problems are all encompassed by the area of probability theory. In order to arrive at a correct solution of these problems, it is imperative that one make those distinctions between problem types which lead to different methods of solution.

We have already considered some of these distinctions, such as whether or not events are mutually exclusive or independent. Let us now set up a series of characterizations of probability problems by asking a series of questions. The pattern of answers to these questions may then be used as a sort of diagnosis of the problem type under consideration. Following this diagnostic scheme, we may consider the methods of solving a number of the possible types of problems.

Relevant distinctions. First, we consider those distinctions which are important in probability problems and then we will develop a shorthand notation system for describing these possible distinctions.

One important consideration is the number of elements sampled from the population. This number, r, if greater than one, will force us to make subsequent decisions about the definition of the universe, sample space. If the number of elements selected is merely one, then the sample space will be most easily described as equivalent to the population of elements.

If more than one element is to be selected from the population, a very important difference in approach to the problem depends upon the method of sampling used. If sampling is with replacement, the universe set will be quite different in most cases from the set if sampling is without replacement. A failure to make such a distinction will often lead to the use of formulas which are inappropriate to the problem.

In some problems, the definition of the population of elements is made in such a way that the elements are not all distinct from one another. The type of analysis used in these cases is different from the usual case in which all elements are distinctly different from each other. We will need to examine this possibility prior to attempting a problem solution.

Another important distinction, which is relevant when the number of elements selected from the population is greater than 1, is whether or not the order of selection of elements is important or unimportant. For example, if one were to ask for the probability of tossing two coins and obtaining one head and one tail, the requirements of the event in question are rather different from those in a problem specifying which was obtained first and which second. The application of certain formulas to be given later depends upon the existence of a problem in which the order of selecting elements is not important.

Finally, in the case where more than one trial of the experiment is used, we must determine whether events are independent, mutually exclusive, or neither of these.

Method of analysis. Let us now restate the relevant distinctions given above in the form of questions. Each major answer to the questions will be given a number and a letter. We may then describe a particular problem with this notation scheme.

I. How many elements are selected from the population in a single sample?

  I–a. One element.

  I–b. More than one element.

II. How many trials (samples) are considered in the experiment?

  II–a. One trial.

  II–b. More than one trial.

III. Are the elements of the population so defined as to be distinct from one another?

  III–a. Yes, they are distinct.

  III–b. No, they are not all distinct.

IV. Is the order of selecting elements important or not?

  IV–a. Yes, order is important.

  IV–b. No, order is not important.

V. Is sampling done with or without replacement?

  V–a. With replacement.

  V–b. Without replacement.

VI. Are the component events mutually exclusive or independent?

  VI–a. Yes, they are mutually exclusive.

  VI–b. Yes, they are independent.

  VI–c. No, neither of these.

A particular problem may be described by giving the set of numbered answers true for that problem. It should be noted that some answers preclude the asking of certain questions. For instance, if I–a is true, only one element is drawn, then IV and V are not relevant. Hence many problems may be described without listing all six possible answers.

Recall the problem in Example 10–1. In that example, we asked for the probability of obtaining at least one 6 when rolling a pair of dice. Let us use the set of questions given above to characterize this problem. The method used in the example was to consider a single trial on which two elements were selected. We would then have I–b and II–a as the first two answers about this problem. The elements (1 through 6) are so defined that III–a is also true.

The answer to IV concerning the order of selection of the elements must be answered in terms of the sample space used. Although order is not important in the verbal statement of the problem, the sample points are ordered pairs of elements. In this case, the answer IV–a is correct. It might be mentioned here that many problems can be solved in more than one way, depending mainly on the definition of the universe elements. In the present example, the method used is clearly one in which the order of selecting elements is part of the information contained in the sample points.

Item V is answered with V–a since the selection of a given outcome on one die does not prevent the same outcome on the second die. Also, VI–b is true since we have reason to think that the two dice operate independently.

The usefulness of this sort of characterization of a probability problem will become apparent shortly as we examine methods of counting the number of elements in different types of sample spaces.

Use of this check list of characteristics of a probability problem permits one to select the correct method of counting the elements in the event set or sets and in the universe set. Once this has been accomplished, our basic definitions of the probability of a simple or complex event may be used.

In order to illustrate the need for special methods of counting these sample points, reconsider the problem from Example 10–1. Rather than asking about the probability of at least one 6 in rolling a pair of dice, suppose we wish the probability of at least one 6 when rolling 5 dice. If we define the sample space as the set of all points possible when considering the order of selection of the elements in the population, one would quickly find that enumeration of the possibilities is very extensive. The sequences 11111, 11112, 11113, 11114, 11115, 11116, 11121, 11122, etc., are all sample points; however, the complete listing of all sample points would take a great deal of time, and allow for a number of errors. The actual measure of the universe here is 7,776.

Our basic problem is to develop efficient methods of shortening the counting of elements in such sets. We will find that the correct methods of counting elements change somewhat as the answers to the questions about sampling experiments change.

10–6. Methods of Counting. Permutations. These are rearrangements of a finite number of things in all possible orders. Any single permutation of all those possible is different from all other permutations.

One way of thinking of permutations is in terms of numbers. For instance, if we allow all 10 digits to be used in each position of a two-digit number, there are 100 possible permutations. These consist of the numbers 00 through 99. It is usual, when enumerating permutations, to make changes from one permutation to the next in a systematic way. This reduces the chance of error. The above-mentioned 100 permutations might be constructed by starting with 0 in the first position and varying the second number: 00, 01, 02, . . . , 09. Then, placing a 1 in the first position, we start the sequence over again in the second location: 10, 11, 12, . . . , 19, and so forth, until all possibilities are exhausted.

One must keep in mind that permutations are sequences; i.e., the order of the numbers or elements is important. For example, 67 and 76 are two different permutations since the same two digits occur in different orders. Later we will consider the concept of *combinations,* in which case 67 and 76 will be considered as the same, because order is not considered.

Let us illustrate permutations formed by sampling without replacement. Let us assume that three letters, a, b, and c, are to be placed in sequences of three letters, using each letter only one time. The result would be six sequences: *abc, acb, bac, bca, cab,* and *cba.*

The two illustrations of permutations given above are both easy enough that the number of permutations may be found without special formulas. However, in general, it will be necessary to count the total number of permutations by means of special formulas.

Let us first consider the number of permutations which may be formed by *sampling with replacement.* In this type of counting formula, we will consider a population of N_* distinct elements. Let us generalize the notation by using the letter n instead of N_*. From the n distinct elements we wish to sample r elements using a replacement technique. The problem is to count the number of possible permutations, with the order of selecting elements as a relevant factor.

It may be seen that the first selection may give rise to n possible different elements. Since replacement is used, the second element selected may also be any one of n different things. Now consider the possible sequences which might result from just these two selections. Any one of the n first elements may be followed by each of the n second selections. In other words, each way of starting the sequence is followed by n different ways of continuing the sequence. The result will be $n \times n$ or n^2 permutations.

If we now consider a third selection, each of the n^2 sequences of two elements may again be continued in n different ways. This would yield n^3 permutations for the first 3 selections. The general solution of the problem may be given as:

$$n^r = \text{number of permutations, with replacement} \qquad (10\text{–}6.1)$$

where n is the number of elements possible on each selection and r is the number of elements sampled.

One should note that r may be greater than n for (10–6.1), because the population is never exhausted.

Let us illustrate (10–6.1) by finding the number of permutations possible when rolling five dice. This is best thought of as rolling a single die five different times. The population consists of 6 elements, the numbers 1 through 6. Since a number is not excluded from subsequent selection when it is used on a given selection, we find that the requirements of (10–6.1) are met. The answer is then 6^5, or 7,776.

Now we must examine the case in which permutations are formed by selection *without replacement.* Again let n be the number of distinct population elements and r the number of selections made. Since we remove the element from the population on each selection, it is clear that n is the length of the longest sequence which may be formed. If one attempts to form sequences larger than n, he will find the $n + 1$ selection impossible; there will be no elements remaining in the population.

As before the first selection may result in n different results. The second selection is limited to $n - 1$ different ways of continuing the sequence, because only $n - 1$ elements remain in the sample when we make the second selection. Each of the n starting points may be continued in any of the $n - 1$ ways which can result on the second selection. This means that $n(n - 1)$ is the total number of sequences from the first two selections. Again, on the third trial, we have one less population element available than on the last selection. The general solution for the number of permutations without replacement is:

$$(n)(n-1)(n-2)\cdots(n-r+1) = \text{number of permutations without replacement} \qquad (10\text{–}6.2)$$

where r is less than or equal to n.

We may illustrate (10–6.2) by asking for the number of possible permutations which may be formed when dealing two cards from a standard deck. By dealing cards, we imply that replacement is *not* used. The population consists of 52 different cards and r is equal to 2. Equation (10–6.2) yields 52 as the first factor, n. The last factor is $n - r + 1$, or $52 - 2 + 1$. This shows that the second factor, 51, is the last in the series. The answer is then (52)(51), or 2,652.

The form of (10–6.2) may be changed if we introduce the concept of *factorials,* which will be useful in later formulations as well. A factorial of a number, X, is equal to the product of integers 1 through

X, inclusive. We denote such a product as $X!$; the exclamation sign means the *factorial* of X and is not simply a punctuation mark.

$$X! = (X)(X-1)(X-2)\cdots(2)(1) \tag{10–6.3}$$

Let us illustrate with a few concrete examples:

$$\begin{aligned} 1! &= 1 \\ 2! &= (2)(1) = 2 \\ 3! &= (3)(2)(1) = 6 \\ 4! &= (4)(3)(2)(1) = 24 \end{aligned}$$

As a special case, we *define* 0! as equal to 1, not zero. Thus it is possible to divide a quantity by 0!, or 1.

In terms of the notation of factorials, (10–6.2) may be rewritten as

$$\frac{n!}{(n-r)!} = \text{number of permutations without replacement} \tag{10–6.4}$$

This equation is formed from (10–6.2) by multiplying and dividing (10–6.2) by $(n-r)!$. The numerator then consists of

$$n(n-1)(n-2)\cdots(n-r+1)(n-r)!$$

which is equal to $n!$.

Example 10–3. If we have 5 different people to choose from, in how many different ways may we seat 3 people on a sofa?

We assume that there are only 3 seats available on the sofa. In this case, a seating arrangement is an ordered sequence, counting from left to right (or right to left). Clearly, our selections must take place without replacement and we desire to make three selections. The number of population elements is 5.

Use of (10–6.4) will yield

$$\frac{5!}{(5-3)!} = \frac{(5)(4)(3)(2)(1)}{(2)(1)} = \frac{120}{2} = 60$$

Equation (10–6.2) will give the same result:

$$(5)(4)(3) = 60$$

In this method, note that the last term is $n - r + 1 = 3$.

There are two methods for calculating the number of permutations of n distinct population elements, one for replacement techniques and a different one for sampling without replacement. We see that question V in our diagnosis scheme is especially important for correct counting of permutations in a sample space.

Combinations. Many probability problems concern the collection of elements found in a sample, but are not concerned with the order of sampling the elements. An analogy to our earlier work may be made here. The mean of a set of numbers depends upon the particular numbers in the set and certainly not on their order within the set. A *combination* consists of a set of r elements. The characteristics of the set are merely the particular elements in the set, ignoring order.

As stated earlier, the sequence 67 is different from 76, but the set with the numbers 6 and 7 is the same, in terms of combinations, no matter in what order the numbers may be given.

We restrict our attention to combinations formed from n distinct elements taken in samples of size r, without replacement. The calculation of the number of combinations possible when replacement methods are used is too difficult for consideration in this text.

First, note that the number of combinations formed from selecting all n population elements is equal to 1. When we select less than n elements, the number of combinations increases because there are numerous ways of forming the sets of selected elements. When r is equal to 1, we may see that the number of sets of single elements formed is n. For intermediate values of r the number of combinations may be either greater than or equal to n.

In order to obtain a formula for counting the number of different combinations which may be formed in sampling from the population, let us consider the relation between combinations and permutations. When we restrict our attention to sampling without replacement, any combination of r elements, all of which are different from one another, includes a number of permutations. For instance, when sampling from the set of 10 digits without replacement, a particular combination of 3 digits, such as 583, may be compared to the number of permutations of the same 3 digits. The number of permutations which may be formed using these 3 digits is $r!$, since we now require that the 3 digits be selected with order considered. This may be likened to a new problem in which the population consists of the r particular digits in the combination being considered and the number of elements selected is the same, r. It may be seen in our example that 6 permutations are possible using the digits 5, 8, and 3.

The same type of analysis is true for any three digits which are all different. In general, the number of permutations which can be formed from any combination of r different elements is $r!$.

If we knew the number of different combinations possible, and multiplied this number by $r!$, we would then have the number of permutations possible. Let us denote the number of combinations formed from selecting r elements from n distinct population elements, without replacement, as ${}_nC_r$. We have found that this number multiplied by $r!$ is equal to the number of permutations possible under the same restrictions. We have the number of permutations given by (10–6.4), so the following equation must be true:

$$({}_nC_r)(r!) = \frac{n!}{(n-r)!}$$

Dividing both sides of this equation by $r!$ will give the desired number of combinations.

$$_nC_r = \frac{n!}{r!(n-r)!} \tag{10–6.5}$$

where r is less than or equal to n.

It can be readily seen that $r + (n - r)$ is equal to n, so that the two numbers forming factorials in the denominator must sum to the number forming the factorial in the numerator.

Let us find the number of combinations possible when selecting 3 elements without replacement from 5 population elements. Use of (10–6.5) gives

$$_5C_3 = \frac{5!}{3!(5-3)!} = \frac{(5)(4)(3)(2)(1)}{(3)(2)(1)(2)(1)} = \frac{(5)(4)}{2} = 10$$

This result may be illustrated concretely by taking the population elements as the letters *a, b, c, d, e.* The 10 combinations possible are: *abc, abd, abe, acd, ace, ade, bcd, bce, bde,* and *cde.*

The requirements which must obtain for use of the different permutation formulas and the combination formula may now be given in terms of our diagnostic scheme. In general, these formulas are used when more than one element is sampled (I–b) and are applied to a single trial (II–a). Each requires distinct population elements for use in probability problems (III–a).

Then IV and V are used to discriminate between the three basic formulas, i.e., (10–6.1), (10–6.2), and (10–6.5). With replacement used, V–a, we must use (10–6.1) and think in terms of permutations. This is true even when the event in question does not imply a particular order of selecting elements. The number of combinations possible when sampling with replacement is too difficult to find for efficient problem solution. Such problems can be solved by use of permutations. With the combination of V–a and IV–a true, we will use (10–6.1).

With order important (IV–a) and sampling without replacement (V–b), use of (10–6.2) (or 10–6.4) is indicated.

Finally, with the conditions IV–b and V–b both holding, one can use (10–6.5), the combinations formula.

At the end of the chapter, a summary table is given in which the various characteristics of problems possible are shown along with the general methods of solution. For the moment, let us abstract just the characteristics of the permutation and combination problems and set them down together.

1. If we have: I–b, II–a, III–a, IV–a, and V–a, use (10–6.1).
2. If we have: I–b, II–a, III–a, IV–a, and V–b, use (10–6.2).
3. If we have: I–b, II–a, III–a, IV–b, and V–a, use (10–6.1).
4. If we have: I–b, II–a, III–a, IV–b, and V–b, use (10–6.5).

The use of the above guide lines will permit one to obtain the measure of U for the corresponding probability problems. Once the sample space is defined and counted by the use of a permutation or combination formula, the measure of the event set must be found by using the same sample space. The student should not fall into the trap of defining and measuring the universe in terms of permutations and then counting the number of combinations in the event.

The measure of the event is often found by use of the same formula employed for the $m(U)$. In so doing, the population of elements is reduced to only those elements which are permitted in samples belonging to the event. For example, if an event is hands of 3 cards which contain only hearts, the 13 hearts in the deck become a "new" population with n' equal to 13. The permutations possible without replacement may then be counted in the event by sampling 3 cards in all possible permutations from this reduced population of elements.

Use of VI will permit the combination of the probabilities of various simple events into the appropriate complex event when required. In a similar manner, II, when answered with II–b, usually requires the calculation of probabilities of an event for each of a number of trials. Then the independent trials formula may be useful.

The examples in the next section are intended to illustrate these procedures. Each example should be studied carefully, but it should be noted that many problems may be done in alternate ways, with the same answer obtained, of course.

10–7. Examples of Problems Using Permutations and Combinations. Unless otherwise specified, each of the following problems involves a single trial on which more than one element is selected from a population of n elements, all distinct from one another.

Example 10–4. The numbers 1 through 8 are arranged in a random order. What is the probability that the digits 1 and 2 appear as neighbors in the order given?

Our first approach to this problem is to structure the sampling procedure and the population. Clearly, we have 8 distinct population elements. Since the entire 8 elements would occur in an "arrangement," the number of elements chosen is 8. The statement of the problem implies that only one sample is considered, "a random order" indicates this feature. This pattern of characteristics is I–b, II–a, and III–a and allows a combination or permutation definition of the sample space.

Since sampling proceeds without replacement, but order is important in the event asked for (IV–a and V–b), we may work with permutations without replacement, using (10–6.2) or (10–6.4).

The measure of the universe is

$$m(U) = \frac{n!}{(n-r)!} = \frac{8!}{0!} = \frac{8!}{1} = 40{,}320 \text{ (permutations)}$$

This answer indicates that there are 40,320 possible permutations, each of which is equally likely to be obtained on a single sample.

The problem is now to obtain the measure of the event. Since the digits 1 and 2 must occur in that order, let us examine the possible positions in a series of 8 numbers in which these two digits might be found. Numbering the positions 1 through 8 allows us to describe the possibilities more easily. Clearly the digits in question could be in positions 1 and 2. Also they might occupy positions 2 and 3, 3 and 4, 4 and 5, 5 and 6, 6 and 7, and, finally, 7 and 8. Counting these possibilities, we find 7 different pairs of positions which may be used for the event. However, merely specifying the positions of two of eight numbers is not a specification of a permutation. We must now count the number of permutations possible for *each* of these seven placements. This is a small problem in itself.

However, we may proceed as follows: For each placement of the digits 1 and 2, there are six other positions to be filled with the remaining six digits in various arrangements. This is equivalent to the number of permuations of six elements taken six at a time, without replacement. Here $n' = 6$, and $r' = 6$. The solution, using (10–6.2), is 6!. There are then 6! permutations possible for each of the 7 placements of the digits 1 and 2. This gives the measure of the event as

$$7(6!) = (7)(720) = 5{,}040 \text{ (permutations)}$$

The probability of the event is then found from our basic definition as

$$P(A) = \frac{m(A)}{m(U)} = \frac{5{,}040}{40{,}320} = .125$$

Example 10–4 is by no means an easy problem, but was given first in order to illustrate the application of the diagnosis scheme and to illustrate the common difficulty in such problems, which is to obtain the measure of the event or events in question. Usually, one must think about the sample points involved in the event and count them by use of one of the methods for counting permutations and combinations. As in this example, it is often necessary to break up an event in separate parts and then combine the answers found for each part.

The next example is somewhat more straightforward than Example 10–4.

Example 10–5. Find the probability of selecting 3 aces in a row when cutting a deck of cards 3 different times.

Here, the idea of cutting a deck of cards is essentially the same as sampling with replacement. It is assumed that the deck is reshuffled after each selection. The population of 52 cards indicates the n. Since only a single trial seems to be implied in the statement of the problem, we once again find the requirements for either a combination or permutation problem. However, since sampling is done with replacement, (V–a), the permutation method is called for rather than our combinations formula.

The measure of the universe is found by calculating the number of possible permutations of 52 elements taken 3 at a time, with replacement. By (10–6.1), this is

$$m(U) = 52^3 = 140{,}608 \text{ (permutations)}$$

The measure of the event, 3 aces in a row, may also be found by counting permutations. Since each selection in the event set must be an ace, the total possible permutations may be found by considering the selection of 3 cards (r') from a "population" of 4 cards (the aces), with replacement being used. The number of permutations thus formed is again obtained by (10–6.1).

$$m(A) = 4^3 = 64 \text{ (permutations)}$$

The desired probability is

$$P(A) = \frac{m(A)}{m(U)} = \frac{64}{140{,}608} = .000455$$

A probability this small indicates that one should not count heavily upon the occurrence of such an event.

Once more, the counting of the event set was more difficult than obtaining the measure of the universe. This is typical of many probability problems, but there are numerous cases in which the event set is small and readily counted. The next example is one such case.

Example 10–6. Among 10 people in a class are 3 fraternity brothers. What is the probability that all 3 would be selected, at random, in the formation of a 3-person group for a class demonstration?

The selection of 3 people implies that no replacement is allowed and that more than one of the distinct elements (persons) are selected. Also, a single trial experiment is implied. This allows either permutations or combinations to be used, depending upon the further analysis of the problem. Since the event does not specify the particular order of selection, but only that 3 particular persons be chosen, the combinations formula is appropriate. Here we see that just one combination is included in the event. If any other person than the three fraternity brothers is included in the sample, the event has not occurred. Therefore, the measure of the event, A, is

$$m(A) = 1 \text{ (combination)}$$

The universe now consists of sample points which are combinations of 3 people. Using (10–6.5) we may obtain the total number of such combinations:

$$m(U) = {}_{10}C_3 = \frac{10!}{3!\,(10-3)!} = \frac{10!}{3!\,7!} = \frac{(10)(9)(8)}{3!}$$
$$m(U) = 120 \text{ (combinations)}$$

Notice how 7! cancels most of the 10! in the numerator. This is usually possible in combination problems, because all quantities are usually factors of the expression.

The desired probability of event A is now

$$P(A) = \frac{m(A)}{m(U)} = \frac{1}{120} = .0083$$

Problems which require the probability of complex events are usually dealt with by finding the probabilities for the component events and then using the appropriate equations for unions or intersections, as the case may be.

The next example is one in which we combine the combinations notion with our earlier concepts about complex events.

Example 10–7. A box contains 12 balls numbered consecutively 1 through 12. If 3 balls are withdrawn at random, what is the probability that the sample will contain either the number 7 ball or 3 odd-numbered balls?

The preliminary analysis of the problem indicates that a single sample of 3 elements is chosen from a population of 12 distinct elements. These conditions allow us to think in terms of permutations or combinations.

The "either-or" statement in the problem should be interpreted to mean that a complex event is under study. In all such cases, the either-or statement allows the combination of *both* simple events to satisfy the requirements as well.

Let us define event A as being the selection of the number 7 ball in the sample, and define event B as being the selection of 3 odd-numbered balls. We wish to find the probability of the union of these two events. From (10–3.7) we find the problem broken down into three parts.

$$P(A \cup B) = P(A) + P(B) - P(A \cap B)$$

First examine event A. Since the order of selection is not specified, let us use combinations. The $m(U)$ will be the number of combinations possible when selecting 3 elements without replacement from 12 population elements.

$$m(U) = {}_{12}C_3 = \frac{12!}{3!\,9!} = \frac{(12)(11)(10)}{(3)(2)(1)} = 220 \text{ (combinations)}$$

In order to count the number of combinations which include the 7 ball, we limit our selections to the 11 other balls, assuming the 7 ball to be already in the sample. The question, then, reduces to the asking of how many different pairs of balls may be placed with the number 7 ball. Since order is disregarded and sampling is without replacement, the combinations formula is again used, just as in the $m(U)$. Here we select 2 balls to be placed with the seven ball. These are drawn from the other 11 balls.

$$m(A) = {}_{11}C_2 = \frac{11!}{2!\,9!} = \frac{(11)(10)}{2} = 55 \text{ (combinations)}$$

The probability of A is, therefore,

$$P(A) = \tfrac{55}{220} = .25$$

The probability of event B is found in much the same fashion. Using combinations again, we find the measure of U is the same as found above. The sets of balls which are all odd numbered may be counted by reducing our population of permissible elements to the 6 odd-numbered balls. The number of different sets of 3 drawn from this population is

$$m(B) = {}_6C_3 = \frac{6!}{3!\,3!} = \frac{(6)(5)(4)}{6} = 20 \text{ (combinations)}$$

Then we obtain $P(B)$ as

$$P(B) = \tfrac{20}{220} = .0909$$

The next event which must be considered is the intersection of A and B. First, note that the two events are not mutually exclusive, because a set of 3 odd-numbered balls (B) might well contain the 7 ball (A). Since the two events are not mutually exclusive, we should inquire as to whether or not they are independent. Does the occurrence of event A change the probability of obtaining event B? In this case, one should see that it well may. If A has occurred, we already have one odd-numbered ball. The selection of three odd-numbered balls is thus aided. This reasoning does not conclusively prove that the intersection of A and B does not follow Equation (10–4.1) which gives $P(A \cap B) = P(A)P(B)$, but the obvious connection between the two events suggests that one should attempt to measure the intersection event directly.

The samples (combinations) which are part of the intersection of A and B must have 3 odd-numbered balls, one of which must be the number 7 ball. We may measure this intersection by finding the number of ways of forming such samples. Again assume that the 7 ball is included in the set of 3 balls. The number of ways of selecting the remaining 2 balls in the sample is equal to the number of ways 2 elements may be chosen from the remaining set of 5 elements which are desired, i.e., there are only 5 odd-numbered balls in the population if the 7 ball is already in the sample.

This reasoning permits

$$m(A \cap B) = {}_5C_2 = \frac{5!}{2!\,3!} = \frac{(5)(4)}{2} = 10 \text{ (combinations)}$$

Again, the measure of the universe is the same as found earlier, 220 combinations:

$$P(A \cap B) = \tfrac{10}{220} = .04545$$

Once these three event probabilities are found, the final answer is obtained quickly:

$$\begin{aligned} P(A \cup B) &= P(A) + P(B) - P(A \cap B) \\ &= .25 + .0909 - .04545 \\ &= .29546 \end{aligned}$$

As a check on our reluctance to regard A and B as independent events, we may find $P(A)P(B)$. This is .0227, which is not equal to $P(A \cap B)$ as required of independent events.

It may be well to reemphasize that the examples given here are not easy ones. The more elementary problems usually contain very few sample points in the event set. As may be gleaned from the previous examples, the calculation of $m(U)$ is typically rather straightforward. However, even elementary problems may require a good deal of labor when large factorials are needed. The next example illustrates a problem which is simple conceptually but difficult in terms of the arithmetic.

Example 10–8. Find the probability of being dealt all 13 spades from a standard deck in a single deal of a bridge hand.

Here we must know that a bridge hand consists of 13 cards. Our problem requires that they all be spades. The population may best be regarded as the 52 different cards and a single deal is specified.

Sampling takes place without replacement so we may consider a combinations approach to the problem. A single combination (all 13 spades) is required, so that the measure of the event is 1 combination.

The measure of the universe is the number of different combinations of 52 elements taken 13 at a time,

$$m(U) = {}_{52}C_{13} = \frac{52!}{13!\,39!}$$

The actual computation of this number is quite laborious. A number of cancellations between numerator and denominator may be made, but the answer is still very large. Without giving the intermediate calculations, the correct answer is

$$m(U) = 635{,}013{,}559{,}600$$

The probability of the event is slightly less than 1 in 635 billion, an answer we shall not bother to change to a decimal.

It is true that combination problems of the type shown here may be done with the permutation formula (10–6.4). The reason that combinations are usually preferred when possible is that the number of permutations is larger than the number of corresponding combinations. Also the designation of the event set is often perceptually easier using combinations. In cases where one has difficulty with the combination method, the problem can, however, be restructured to one involving permutations.

As the final example in this section, we consider a type of problem which may be done in a number of ways. The choice of method does not change the answer, of course.

Example 10–9. Five men check their hats with a hat-check girl, who finds later that the identification slips have been removed from the hats. Two of the men return to the counter together and the girl decides to simply pick 2 hats at random and put them on the counter. What is the probability that the 2 correct hats are placed on the counter? We assume that each man will pick his own hat if the correct 2 are placed before them.

The 2 men reduce the event in question to the selection of some particular pair of the 5 population elements. The sampling takes place without replacement and the order of selection is not important. These characteristics allow us to use the combinations technique. The measure of the universe is

$$m(U) = {}_5C_2 = \frac{5!}{2!\,3!} = 10 \text{ (combinations)}$$

Only one of the possible pairs of hats will satisfy the requirements of the event, hence, the $m(A)$ is one.

$$P(A) = \tfrac{1}{10} = .10$$

An alternative solution to this problem is obtained with permutations. Since sampling is without replacement, $m(U)$ may be found, with (10–6.2), as 20 permutations.

The possible permutations in the event consist of those ordered pairs of hats which are correct for the two given men. Let us arbitrarily say that the men require hats numbered 1 and 2. The 2 permutations which contain these numbers are 1,2 and 2,1. Then the $P(A)$ is

$$P(A) = \tfrac{2}{20} = .10$$

Still another method of solving the problem in Example 10–9 involves a type of reasoning not yet considered. View the selection to the two hats to be two trials. If one finds the probability of "success" on the first trial and then finds the probability of "success" on the second trial, the multiplication of these two probabilities will yield the probability of the joint occurrence of success on each trial.

This technique may sound similar to the independent trials or events idea expressed in (10–4.2), but it is really quite a different method. The difference is that the calculation of the probability of the event on the second trial is performed under the assumption that the first event has taken place. This concept is called **conditional probability.**

The first hat selected by the girl in this example may be either of the two required. Success on this trial will be called A; $P(A)$ is then $\frac{2}{5}$ or .40.

The calculation of $P(B)$, a correct second selection, is done on the assumption (or is *conditional* upon the assumption) that event A has taken place on the first trial. If this is true, then of the remaining 4 hats, only one remains which belongs to the required pair. The probability of the second selection being correct, given that the first was correct, is now $\frac{1}{4}$ or .25.

Multiplying .40 by .25 we obtain the .10 found in the other two methods of solving this problem.

One should carefully note that the probability of success on the second trial would be $\frac{2}{4}$ or .50 if the first selection were unsuccessful. This dependency of $P(B)$ upon the event A or $\bar{A}$ indicates that the two events are not independent. Although a trial by trial type of analysis of probability problems is often useful for both independent and non-independent sequences of trials, it is well to note that the latter requires the *conditional probability* line of reasoning.

10–8. The Binomial Expansion. A most useful method for finding the probabilities of certain types of events is based upon the *binomial expansion*. Let us first specify a class of events which is appropriate for this method of analysis. The characteristics of such events are most easily specified by considering a population of elements of only 2 kinds. In our previous analysis scheme, this would mean that item III would be answered to the effect that the elements are *not* all distinct from one another. For instance, rather than consider each card in a deck as distinct from every other card, we might simply use the color of the cards in specifying the elements. In that case, we find that there are 26 red cards and 26 black cards, and that no further distinction is made among the population elements.

The sampling method suitable for the use of the binomial expansion may be specified by I–a, II–b and V–a, that is, only one element is selected, but more than one trial is used, and the replacement method is used (between trials).

Since we now have only 2 types of elements in the population, the outcome of each trial may be given by stating which type of element was chosen. Let us arbitrarily call the 2 types of elements A's and B's. With a single element chosen, the sample space on each trial is identical to the population. It may be seen that the probability of selecting an A element, called event A hereafter, is equal to the number of A's divided by N_*. For simplicity we will use

$$p = P(A), \qquad \text{and} \qquad q = P(B)$$

where B is the selection of a B type element.

A and B are mutually exclusive and they exhaust all possible outcomes. We find the important relationship, then, that

$$p + q = 1.00 \tag{10–8.1}$$

Since an element selected on one trial is always replaced prior to the next trial, p and q remain the same on all trials. The events (A or B) on the different trials are then independent events.

A single outcome. The information given above has specified all the answers to our diagnostic scheme with the exception of that for IV, "Is order important?" This question is irrelevant if viewed with respect to population elements, because we have assumed that only 1 element is selected on each trial. However, the use of Equation (10–4.2) for finding the probability of a particular sequence of independent events requires that those events take place in a specified order. The binomial expansion, used in a probability question, is intended to apply to the *total number* of times events A and B occur without regard to any particular order.

For example, if a coin is tossed 10 times, what is the probability of obtaining 4 heads and 6 tails? Considering this to be a series of 10 independent trials, there are a number of permutations of the outcomes of the 10 trials all of which will result in 4 heads and 6 tails.

Let us follow this example through more completely. We may consider the population to consist of any even number of elements, $\frac{1}{2}$ of which represent heads and the other half tails. Letting the event heads be A, we see that $p = q = .50$.

Let us now take some particular series of the 10 independent trials which yields 4 heads and 6 tails: $AAAABBBBBB$. By use of (10–4.2), we find that the probability of this particular sequence of events is $ppppqqqqqq$, or p^4q^6, which is equal to $(.5)^{10}$. This is not the probability of getting 4 heads and 6 tails in such an experiment, because there are other sequences of these events which would also yield the desired number of each of the events. For example, $ABABABABBB$ and $BBBBBBAAAA$ are possibilities, each of which has a probability of $(.5)^{10}$ as well! It may be seen that any 2 particular sequences of events, *viewed as events themselves*, are mutually exclusive. Any replication of the experiment can result in only one particular sequence.

All such sequences which have the same number of A's and B's (heads and tails here) will have equal probabilities of occurrence. Since they are all mutually exclusive sequences, the probability of the *union* of all the desired sequences will be the probability of the desired number of A and B events. This reasoning permits the following statement to be made for sequences of independent events of the type under discussion.

For two types of mutually exclusive events, A and B, with probabilities p and q (where $p + q = 1$), (1) any particular sequence of n independent trials which yields k events of type A and $(n - k)$ events of type B has probability equal to p^kq^{n-k}; and (2) the probability that n independent trials result in exactly k events of type A and $(n - k)$ events of type B is *equal to the number of sequences of such events which yield this outcome multiplied by* $p^k q^{n-k}$.

Returning to our coin tossing example, we may now see that $(.5)^4(.5)^6 = (.5)^{10}$ must be multiplied by the number of sequences which yield 4 heads and 6 tails.

The solution of this problem may be accomplished if we use the notion of combinations to count the permutations. This rather odd technique results from considering the restriction of 4 heads and 6 tails. All of our desired permutations (sequences of events) contain 4 heads and 6 tails, but in *distinguishably* different orders. Let us consider the 10 different *positions* in the sequences, numbered 1 through 10, as being a population of elements. We now consider the sets of such positions in which the heads may be located. Once this is accomplished, for a given sequence, the tails will fill the remaining locations. Since all heads are alike, we must find the different combinations of locations in which to put the heads.

The number of heads to be placed is 4, so that we wish to find the number of combinations of the 10 locations taken 4 at a time:

$$_{10}C_4 = \frac{10!}{4!\,6!} = 210$$

If one were to consider selecting positions for the tails first (why not?), the same answer is obtained. With 6 locations needed for the tails, we have

$$_{10}C_6 = \frac{10!}{6!\,4!} = 210$$

This reasoning generalizes to all such sequences of independent events. First, let us examine one more particular case: Assume that an unbiased die is rolled 4 times. We define event A as the outcome of the face numbered 1 on a given roll. Event B is any other outcome. Here $p = \frac{1}{6}$ and $q = \frac{5}{6}$. Let us now find the probability of obtaining exactly 3 A's (and 1 B) among the 4 outcomes.

The probability of any particular sequence of 3 A's and 1 B is $(\frac{1}{6})^3(\frac{5}{6})^1$ which is 5/1,296. Here we may consider the number of possible sequences directly. The single B may clearly be placed in any 1 of 4 positions, the remaining positions being taken by A's. We have $AAAB$, $AABA$, $ABAA$, and $BAAA$.

The use of the combinations formula gives the same number of sequences. Selecting 3 positions for the A's gives

$$_4C_3 = \frac{4!}{3!\,1!} = 4$$

and the same is true if we consider placing the single B before the A's.

In general, if one event takes place exactly k times, in n trials, the number of sequences which are included, is given below.

$$_nC_k = \text{number of sequences with } k \text{ events of one type} \qquad (10\text{–}8.2)$$

where n is the number of trials and k the number of events of one of two types.

The solution to our general problem may now be stated. For the independent trials problem, with only two possible events,

$$P(k\ A\text{'s and } n - k\ B\text{'s}) = ({}_nC_k)\, p^k q^{n-k} \qquad (10\text{–}8.3)$$

where p is the $P(A)$, and q the $P(B)$.

All possible outcomes. We have now found a method for obtaining the probability of exactly k events of one type and $(n - k)$ of a second type in n independent trials. There are a number of such possible outcomes, when we ignore the particular sequences and use only the total number of each type of event.

In order to study all possible outcomes, we may now examine the binomial expansion in its entirety, (10–8.3) being the general term of this expansion. From algebra, it is known that

$$(a + b)^2 = a^2 + 2ab + b^2$$
$$(a + b)^3 = a^3 + 3a^2b + 3ab^2 + b^3$$

and, in general,

$$(a + b)^n = a^n + \frac{n!}{(n-1)!\,1!}\, a^{n-1}b + \frac{n!}{(n-2)!\,2!}\, a^{n-2}b^2 + \cdots + \frac{n!}{(n-k)!\,k!}\, a^{n-k}b^k + \cdots + b^n \qquad (10\text{–}8.4)$$

The expression in (10–8.4) should be viewed with respect to p and q rather than a,b as general terms. When this is done, we see that $p + q$ raised to any power, n, is still equal to one. If p and q, as defined in this section, are substituted into (10–8.4), each term of the expansion will be seen to be a case of (10–8.3). Each term of the binomial expansion, when p and q are used, gives the probability of some exact distribution of the number of events A and B for n independent trials. The $(n + 1)$ terms which are found in the expansion exhaust the possible outcomes of the independent-trials experiment. For instance, if 4 trials are used, we may obtain 4, 3, 2, 1, or 0 events of type A (or B). The expansion of $(p + q)^4$ will give 5 terms, each of which gives the probability of one of the 5 different outcomes. It may be noted in (10–8.4) that the coefficients of the a and b products are of the form $_nC_k$, as required in (10–8.3).

If each term in the binomial expansion is interpreted as a probability of some exact outcome of the sampling experiment, we have given the random sampling distribution of such outcomes. In this case, the statistic being considered is the number of trials on which one of the two types of events may take place. Such a distribution is called the *binomial distribution.*

The shape of the binomial distribution depends upon two parameters, the number of trials, n, and the value of p (or q). Let us examine the binomial distribution for certain specific values of these parameters. Assume that the event A is a 6 when rolling a die; p is then $\frac{1}{6}$ and q is $\frac{5}{6}$, representing the probability of any other outcome. Let us use 4 trials and find the probabilities of all possible outcomes by expanding $(p + q)^n$.

$$\left(\frac{1}{6} + \frac{5}{6}\right)^4 = \frac{4!}{4!\,0!}\left(\frac{1}{6}\right)^4\left(\frac{5}{6}\right)^0 + \frac{4!}{3!\,1!}\left(\frac{1}{6}\right)^3\left(\frac{5}{6}\right)^1 + \frac{4!}{2!\,2!}\left(\frac{1}{6}\right)^2\left(\frac{5}{6}\right)^2 + \frac{4!}{1!\,3!}\left(\frac{1}{6}\right)^1\left(\frac{5}{6}\right)^3 + \frac{4!}{0!\,4!}\left(\frac{1}{6}\right)^0\left(\frac{5}{6}\right)^4$$

$$= \left(\frac{1}{6}\right)^4 + (4)\left(\frac{1}{6}\right)^3\left(\frac{5}{6}\right)^1 + (6)\left(\frac{1}{6}\right)^2\left(\frac{5}{6}\right)^2 + (4)\left(\frac{1}{6}\right)^1\left(\frac{5}{6}\right)^3 + \left(\frac{5}{6}\right)^4$$

$$= .00077 + .01543 + .11574 + .38580 + .48225$$

The sum of these 5 terms is .99999, which is equivalent to 1 when errors of rounding are considered. This feature is a necessary condition of the algebra and is also required by the addition of the probabilities of a set of exhaustive, mutually exclusive events. We interpret the 5 terms above as being the probabilities of getting exactly 4, 3, 2, 1, and 0 of the event with probability p, respectively. Note that the most likely outcome is zero 6's.

Let us examine other distributions of a similar form and then graph each distribution. Assume that event A includes both a 5 and a 6 when rolling the die. This union event has $p = \frac{1}{3}$, hence, $q = \frac{2}{3}$. Keeping the number of trials at 4, we now expand $(\frac{1}{3} + \frac{2}{3})^4$ and obtain

$$(\tfrac{1}{3} + \tfrac{2}{3})^4 = .01235 + .09877 + .29630 + .39506 + .19753$$

Examination of these 5 terms shows that the most likely outcome is now 1 of event A. The distribution has changed because p has changed.

Let us return to the distribution of the number of sixes, but use an n of 6 rather than 4.

$$(p + q)^6 = p^6 + 6p^5q + 15p^4q^2 + 20p^3q^3 + 15p^2q^4 + 6pq^5 + q^6 \quad (10\text{–}8.5)$$

Substituting $p = \frac{1}{6}$ and $q = \frac{5}{6}$, we obtain

$$(\tfrac{1}{6} + \tfrac{5}{6})^6 = .00002 + .00064 + .00804 + .05358 + .20094 + .40188 + .33490$$

These seven terms, in the order given, are the probabilities of getting exactly 6, 5, 4, 3, 2, 1, and 0 sixes when rolling a true die 6 times.

As a final binomial distribution, let us use 6 trials and a p of $\frac{1}{3}$ and a q of $\frac{2}{3}$. The general expansion shown in (10–8.5) is again used, but substitution of these new single event probabilities gives

$$(\tfrac{1}{3} + \tfrac{2}{3})^6 = .00137 + .01646 + .08230 + .21948 + .32922 + .26337 + .08779$$

In Fig. 10–5, the two binomial distributions formed from the 4 trial examples are plotted. Figure 10–6 shows similar distributions for the two 6-trial problems. These figures illustrate that the binomial distribution changes as a function of both parameters, the number of trials and the value of p. In each case, a related distribution of the number of times event B, with probability q, might occur is possible, but these distributions are found quickly from the distribution of event A, because of the restriction that A and B are mutually exclusive and exhaustive.

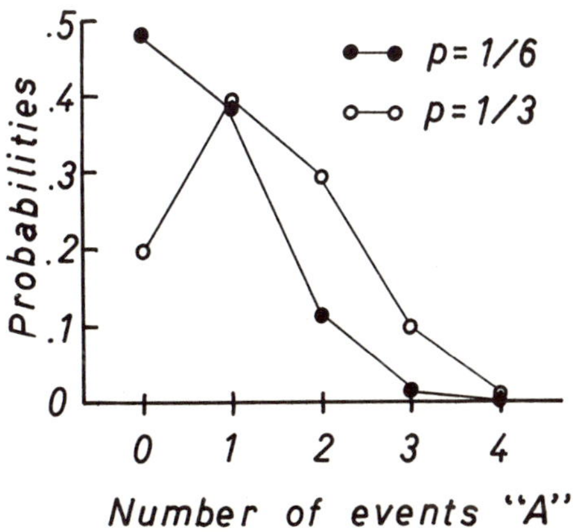

Fig. 10–5. Two binomial distributions which differ with respect to the value of p. The n is 4 in each case.

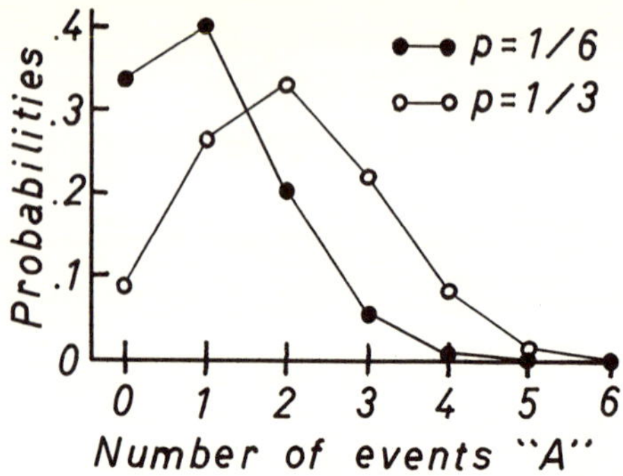

Fig. 10–6. Two binomial distributions with n's of 6 but with different values of p.

We would then have

$$P(n - k \text{ of event } B) = 1 - P(k \text{ of event } A) \qquad (10\text{–}8.6)$$

The normal approximation to the binomial. We may think of the sampling experiment related to the binomial expansion as the selection of n scores (with replacement) from a population consisting of a proportion, p, of scores which are equal to 1 and a proportion, q, of scores which are equal to 0. The outcome of the sampling experiment is then described by the *total* (T) of the set of n elements chosen. For example, if event A occurs on 6 of 8 trials, we represent this as the selection of 6 scores of 1 and 2 scores of 0. The total of the sample of scores is then 6. From the central limit theorem, we see that the $RSD(T)$ becomes normally distributed as the sample size becomes large.

This reasoning permits us to conclude that the binomial distribution should be nearly a normal distribution if the number of independent trials in the experiment is large enough. When p is neither very small nor very large, the binomial distribution is well fit by the normal curve when n is over 20. Even an n as small as 10 is suitable if the value of p is near .50.

Another method of viewing this problem stems from the discussion of the chi square statistic in Chapter 9. In that chapter, the expected number of times an event might occur was defined as equal to np, where n is the total number of elements chosen and p the probability of the event on each selection. We restricted the use of chi square, in that chapter, to cases in which np was at least as large as 5, for the sake of *hypothesis testing*. In order that the binomial distribution be well fit by a normal curve, we may use a similar restriction here.

If both np and nq are 10 or greater, the corresponding binomial distribution may be described by the normal distribution. It may be shown that the mean of the binomial is

$$np = \text{mean of the binomial distribution} \qquad (10\text{–}8.7)$$

where it is understood that the number of events with probability p is the statistic used.

Equation (10–8.7) may be referred back to one of the distributions shown in Fig. 10–5. Using the case in which $p = \frac{1}{6}$ and $n = 4$, we found that the probabilities of 4, 3, 2, 1, and 0 events of the type counted were .00077, .01543, .11574, .38580, and .48225. In order to find the mean of the distribution directly, we simply multiply the value of the score (the number of events) by the proportion of times (probability) the score occurs and add up these products for each possible score. This is equivalent to finding a mean from a relative frequency distribution. This gives

$$\begin{aligned} \text{mean} &= (4)(.00077) + (3)(.01543) + (2)(.11574) + (1)(.38580) \\ &\qquad + (0)(.48225) \\ &= .66665 \end{aligned}$$

This value should be the same as that obtained from (10–8.7):

$$\text{mean} = np = (4)(\tfrac{1}{6}) = .66667$$

The difference between the two values is due merely to errors in rounding.

One should note that the binomial distributions in Figs. 10–5 and 10–6 are quite skewed. In such cases, the mode and mean are not equal to one another. This may also be stated as a difference between the expected number of events (the mean) and the most likely exact outcome (the mode).

Equation (10–8.7) is true for all binomial distributions whether or not they are well fit by the normal distribution.

It is also known that the standard deviation of the distribution of outcomes called the binomial may be found from the parameters of the distribution. Let us denote this standard deviation of the number of events as σ_{bi}. The appropriate computation is then

$$\sigma_{bi} = \sqrt{npq} \qquad (10\text{–}8.8)$$

The use of (10–8.8) and (10–8.7), when the normal curve serves as an approximation to the binomial, is accomplished in the same fashion as any normal-curve hypothesis test. However, since the binomial distribution is discrete rather than continuous, the real limits of the discrete score should be used. Example 10–12 illustrates this technique.

Example 10–10. An investigator wishes to assign each of 8 subjects to one of his two experimental conditions. He does this by flipping a coin for each assignment. What is the probability that he will obtain exactly 4 subjects in each of the two groups?

We assume that the coin is unbiased and that the 8 tosses are independent trials. Since there are two decisions or outcomes possible on each trial, the binomial expansion is then appropriate. Here, a specific outcome is required, 4 events of the first type and 4 of the second. The n is 8 and k is 4. Using (10–8.3), we obtain

$$P(4\ A\text{'s and }4\ B\text{'s}) = {}_8C_4(\tfrac{1}{2})^4(\tfrac{1}{2})^4 = 70(\tfrac{1}{2})^8 = \tfrac{70}{256} = .274$$

We may note that this split is likely about $\frac{1}{4}$ of the time. This is true, even though a four to four split is more likely than any other split, and even though the mean number of heads is 4 ($np = (8)(.5) = 4$). If the investigator wishes to divide the subjects at random, with the assurance that there will be 4 subjects in each group, he should use a random sampling technique without replacement.

Example 10–11. In a four-alternative multiple-choice test, what would be the probability, by a person who is guessing (by means of chance), of getting the first 3 questions all correct or all incorrect?

We may note that the division of each trial into 2 outcomes, correct or incorrect, suggests that this is a binomial problem. Since chance is specified as the technique used to answer each question, the trials may be considered independent.

The population of elements on each trial is a set of four alternatives, one of which is the correct type of element. Therefore, p is $\frac{1}{4}$ and q is $\frac{3}{4}$. The problem requires the probability of an either-or outcome. Since the possible outcomes in the binomial expansion are mutually exclusive, we find the probability of each of the two required events and add the probabilities.

The first possibility is all correct on the first three trials, n is 3 and k is 3. This term of the binomial expansion is best illustrated in (10–8.3):

$$P(3\ A\text{'s and }0\ B\text{'s}) = p^3 = (\tfrac{1}{4})^3 = .0156$$

The similar use of this equation for the second contingency gives

$$P(0\ A\text{'s and }3\ B\text{'s}) = q^3 = (\tfrac{3}{4})^3 = .4219$$

The probability of the union of these two is .4375.

Example 10–12. A student is given a 75-question multiple-choice test in which each question has four alternatives. The instructor has been careful to use questions which are not of such type that a little knowledge will lead a student to the wrong answer. The student in question scores 26 correct on this test. Can you reject the hypothesis that the student simply guessed at random on each question?

In this problem, we assume that the hypothesis is true and develop the probability of a result as deviant from that expected, or more so than was actually obtained. This is the general form of the hypothesis-testing use of statistics. Here, we have no reason to expect that a "significantly" low total of correct responses is anything other than a freak. However, if the student knows at least some of the material covered in the test, he may well get more questions correct than might be expected.

Again, if the hypothesis is true as specified, the expected distribution may be obtained by use of the binomial expansion. However, the result of 26 events of the correct type *and all those more deviant than this result,* would require that one calculate the probabilities of 50 terms of the binomial expansion. Without access to a high speed computer, such calculations would not usually be undertaken.

However, since the n is large, we may use the normal approximation to the binomial if np and nq are not small. Here, by assumption, the value of a correct response on a given trial may be taken as $\frac{1}{4}$ (p) and q is then $\frac{3}{4}$. The expected mean number of correct responses is np or 18.75. Let us denote this as μ_{bi}.

The value of σ_{bi} is found as

$$\sqrt{75(\tfrac{1}{4})(\tfrac{3}{4})} = \sqrt{\tfrac{225}{16}} = \tfrac{15}{4} = 3.75$$

The use of the normal curve requires the calculation of a z-score and its interpretation in the normal-curve table. Here we use the real limit of the frequency score of 26 (25.5) and deviate this value from the mean of the distribution. Division by the standard deviation gives the required z-score.

$$z = \frac{X - \mu_{bi}}{\sigma_{bi}} = \frac{25.5 - 18.75}{3.75} = 1.80$$

Table C shows that the area of the normal curve between the mean and a z-score of 1.80 is .4641. This means that .0359 of all of the scores on this normal distribution are as high or higher than the obtained value.

The conclusion would depend upon the region of rejection chosen prior to making this test. If one were to define this region of rejection as the highest 5 per cent of the random sampling distribution, we would reject the hypothesis. If, rather, one were to define the region of rejection as the highest 2.5 per cent of the scores, our test value would fall short of significance. In either case, the student's performance probably would have been judged as failing. If we wish to set the probability of a type-I error at .05, but still wish to reject only with high deviations from the mean, the highest 5 per cent of the distribution would be the region of rejection.

10–9. Summary of Problem Types. In Table 10–1, a summary of various kinds of probability problems is presented along with the general methods of solution for each problem type. It must be emphasized that Table 10–1 is intended as an adjunct, not a substitute, to a thoughtful analysis of a probability problem. Use of such a table for a systematic analysis would usually be discontinued when one has gained experience with the area of probability.

It should also be mentioned that Table 10–1 is far from all-inclusive, with respect to the possible types of probability problems. However, one advantage to the use of this table is that problems which are of a different type than covered in this chapter may be recognized as such.

TABLE 10–1

A Diagnostic Method of Solving Probability Problems

<table>
<tr><th colspan="5">Questions to be answered[a]</th><th rowspan="2">Method of solutions</th></tr>
<tr><th>I</th><th>II</th><th>III</th><th>IV</th><th>V</th></tr>
<tr><td rowspan="5">a</td><td>a</td><td>–</td><td>–</td><td>–</td><td>Simply obtain the number of population elements in the event and divide by N_*.</td></tr>
<tr><td rowspan="4">b</td><td rowspan="2">a</td><td rowspan="2">–</td><td>a</td><td>Sequences of independent trials. Possibly a binomial problem.</td></tr>
<tr><td>b</td><td>A trial-by-trial conditional probability problem.</td></tr>
<tr><td rowspan="2">b</td><td rowspan="2">–</td><td>a</td><td>Usually a binomial problem.</td></tr>
<tr><td>b</td><td>Redefine the population as distinct elements.</td></tr>
<tr><td rowspan="10">b</td><td rowspan="5">a</td><td rowspan="4">a</td><td rowspan="2">a</td><td>a</td><td>A permutations problem, use (10–6.1).</td></tr>
<tr><td>b</td><td>A permutations problem, use (10–6.4).</td></tr>
<tr><td rowspan="2">b</td><td>a</td><td>Must be done as a permutations problem with (10–6.1).</td></tr>
<tr><td>b</td><td>Use combinations, (10–6.5).</td></tr>
<tr><td>b</td><td>–</td><td>–</td><td>Redefine as 1 element per trial with more than one trial.</td></tr>
<tr><td rowspan="5">b</td><td rowspan="4">a</td><td rowspan="2">a</td><td>a</td><td>Use (10–6.1) on each trial, and then combine trials as a union or intersection.</td></tr>
<tr><td>b</td><td>Use (10–6.4) on each trial. Combine trials with union or intersection methods.</td></tr>
<tr><td rowspan="2">b</td><td>a</td><td>Use (10–6.1) on each trial, and combine trials as required.</td></tr>
<tr><td>b</td><td>Use (10–6.5) on each trial; then use union or intersection methods.</td></tr>
<tr><td>b</td><td>–</td><td>–</td><td>Redefine population as distinct elements.</td></tr>
</table>

[a] The answers represented in the table are: I–a, one element per trial; I–b, more than one element per trial; II–a, one trial (sample); II–b, more than one trial; III–a, distinct population elements; III–b, indistinct population elements; IV–a, order is important; IV–b, order is not important; V–a, elements chosen with replacemant; V–b, elements chosen without replacement.

The knowledge that one does not know how to solve a problem is often far superior, in the long run, than would be the misapplication of a procedure that one thinks is correct.

A number of the alternatives given in Table 10–1 have not been illustrated in this chapter. These are included for the sake of completeness.

Table 10–1 may be used by asking and answering each of the first five diagnostic questions given in this chapter. Asking these questions in turn, one will be led from left to right in the table, moving up or down as a function of the answers to each question, and one will arrive at a suggested method of analysis. In a number of cases, it is suggested that the problem be restructured, such as by so defining population elements that they are distinguishable from one another, rather than indistinct. After this kind of restructuring, the problem may be analyzed once again.

In problems calling for the probability of a complex event, it is usually advisable to use the union or intersection method first. *Each* component event may then be analyzed by use of Table 10–1. Item VI, concerning mutually exclusive or independent events, is applicable to all problems which require the analysis of complex and/or repeated events. As such, VI is used in conjunction with the equations given in Section 10–3.

EXERCISES

1. Find the measure of U, i.e., the number of elements in the sample space, for each of the following:
   (a) Selecting 3 elements, with replacement, from a population of 12 distinct elements.
   (b) Rolling a die 4 times.
   (c) Selecting all 6 of the population elements, in order.
   (d) Selecting one of two alternatives on each of 5 trials.
   (e) Dealing 3 cards from a standard deck.
2. Find the probability of drawing 3 random digits all of which are even numbers.
3. Find the probability that two randomly selected digits will sum to 15.
4. Four people are chosen at random from a group of 6. Find the probability that 3 particular people will be included in the selected group.
5. A drawer contains 4 blue socks, 3 red socks, and 2 green socks:
   (a) Find the probability of drawing, at random, 2 blue socks or 2 green socks in a sample of 2.
   (b) If 2 socks are chosen at random, find the probability that they will both be the same color.
6. A deck of 26 cards, 13 hearts and 13 spades, is used to deal a hand of 5 cards. What are the chances of being dealt all spades?
7. A multiple T-maze has 7 choice-points (with two alternatives each) with only one correct choice at each position. If selections were made at random, what would be the probability of a perfect run on the first attempt at the maze?
8. If one selects 3 numbers at random, without replacement, from a set of ten numbers, 0 through 9, what is the probability of each of the following events?
   (a) Obtaining a 6, 7, and 8, in that order.
   (b) Obtaining a 6, 7, and 8, in any order.
   (c) Obtaining a 6 followed by a 7, on some pair of selections.

9. In rolling a pair of dice 2 times, what is the probability of getting at least one roll which totals 4 on the faces?

10. In drawing 6 cards at random from a standard deck, with replacement, what is the probability of each of the following?

    (a) 4 red cards and 2 black cards.

    (b) 4 hearts.

    (c) no spades.

11. Find the random sampling distribution of the number of 6's when rolling 5 dice at a time.

12. If 60 dice are rolled at once, what is the approximate probability of getting 16 or more 3's on the dice?

Applications; use information from all previous chapters where relevant.

13. Following pretraining, an investigator hypothesizes that each person in an experiment will choose a specific response with a probability of .25:

    (a) If 20 of his 50 subjects make the predicted response, should he reject his hypothesis?

    (b) In a separate group, only 2 of 15 subjects choose the predicted response. If $p = .25$, express the probability, without simplification, of so few or fewer subjects selecting this response.

14. Subjects are each given a 3-choice situation, with only one choice being correct. If subjects are assumed to be guessing and are allowed to continue guessing until the correct choice is made, the number of choices (T) needed by each person may be 1, 2, or 3. If the obtained results were: $T = 1$, 43 subjects; $T = 2$, 29 subjects; and $T = 3$, 28 subjects, could you reject the guessing hypothesis?

15. Two groups of 8 monkeys each are given 15 two-choice trials. The numbers of correct responses for each animal are given below. Do the two groups differ significantly in their average probabilities of correct responding?

    Group 1: 7, 13, 12, 10, 10, 10, 9, and 11

    Group 2: 14, 8, 13, 13, 12, 11, 14, and 12

11

Correlation

Throughout our discussion, any population or sample of elements has been described by a single score for each element. Clearly, any population or sample of real things has a number of characteristics associated with each element. If one were to describe a set of people in terms of the weights of these people, the description would certainly be incomplete. In fact, giving any finite number of characteristics of this group of people would still leave some aspects unmentioned. We will not attempt to suggest a method for complete description of any set of real things; however, in this chapter, we shall extend our description of elements by considering two characteristics of each element.

In general, we shall associate two scores with each member of a set. One of these scores, which we shall call X, will be descriptive of one of these characteristics, and the second score, which we may denote as Y, will describe the other characteristic.

As in previous work, the total number of elements in a set will be called N or N_*. It must be understood that each of the elements has two scores associated with it, both an X and a Y. The set of scores resulting from this situation may be thought of as a set of N (or N_*) ordered pairs of numbers. Using subscripts to indicate these paired numbers, a general pair in such a set of scores could be written as (X_i, Y_i). Although there are $2N$ numbers in such a set, we will consistently refer to the N pairs; i.e., the number N refers to the number of elements, or the number of pairs, or the number of X's, or the number of Y's, but never to the total number of scores under study.

The pairing of the numbers, one an X and one a Y, is of crucial significance in this chapter, because the primary feature of the numbers is the relationship between the partners thus formed.

We shall often refer to the characteristics represented by the scores as variables, the X variable and the Y variable. This terminology arises from the fact that X's will typically differ from pair to pair, and so will the Y's. If such variation does not occur for either of these types of scores, one would find the discussions in this chapter without meaning.

The concept of variability of scores has been explored quite fully in previous chapters. The present chapter will stress a concept called *covariation* or *correlation*. Another term for this concept is *concomitant* variation. The idea may be expressed informally first. Assume that John and Mary are two high-school students who are going steady. If one were to record the locations of both students at a number of different time periods, these pairs of observations would reflect the obvious; not only is each person changing location from time to time, but they are going around together. There is a co-relationship or correlation between the two sets of measurements.

There are numerous variables which vary together or show covariation. On the average, tall people weigh more than shorter people. The amount of clothing worn by people varies as the temperature of the environment changes. And, we are assured by some, the height of hem lines on skirts in America varies as a function of similar measurements taken in Paris. Many of these concomitant variations are obvious to anyone, but the statistician is interested in a more precise measurement of these relationships and certainly in the possibility that some "obvious" relationships might actually be spurious, i.e., might only be true for some atypical sample of observations.

The first question we shall attack, then, is how to describe the degree of covariation between two sets of paired scores by means of a statistic.

11–1. Requirements for a Correlational Statistic. As a means of illustrating some desirable characteristics that a measure of correlation should possess, let us consider a set of paired observations of identical aspects taken from a set of elements. Assume that the elements are 10 different cities in this country and that at a specific time the temperature in each city is measured in two different units, degrees Fahrenheit and degrees Centigrade. Each city is then given two scores, an X and a Y, reflecting the units in the order given above.

Our two variables, X and Y, represent the same physical variable, the temperature. It is certain that the two variables, in terms of the physical world, have the highest degree of correlation possible, because they are representing the same thing. However, as noted in an early chapter, the two different measurement scales used tend to obscure this identity of the variables. For this case of identical physical variables, and all like it, we would wish to have a correlational statistic which takes on its maximum value. Since our two sets of scores differ, on the average, in both their respective means and standard deviations, the obvious way of seeing the identity of the pairs of scores would be to convert them to z-scores. Then it will be possible to define a statistic which is not sensitive to the irrelevant differences between the scores, i.e., differences in means and standard deviations.

This feature of the *correlation coefficient,* which will be defined shortly, is similar to the idea that dispersion is independent of the magnitude of the mean of a set of scores. The correlation coefficient we desire is independent of both the means and the standard deviations of the two variables being studied. We may then state that the degree of correlation between two variables does not change if one of the variables (or both) are modified by use of a linear transformation. This is, of course, the description of the relationship between our Centigrade and Fahrenheit temperatures.

However, let us note a case in which a linear transformation of one set of measurements produces scores that are a bit different in kind than we have been considering. Suppose that each Fahrenheit measurement were changed to a new score by multiplying the score by -1. This is a linear transformation but the new scale of temperature would be inversely related to the original one. The high temperatures in degrees Fahrenheit would now be paired with low numbers on the new scale, and low Fahrenheit values would be paired with the higher numbers on the new scale. It is still true that the *degree* of correlation should be perfect, but it would be nice if the statistic would reflect the difference between these two cases. The first case is called a direct relationship between X and Y, whereas the second would be termed an indirect relationship.

The correlation coefficient we will develop will have a maximum value of unity. For cases in which the degree of relationship is still perfect but high scores on one variable are paired to low values of the second variable and *vice versa,* the correlation coefficient will take on a negative value. For a perfect degree of correlation, with the inverse relationship between the number scales, a value of -1 will be obtained.

A third situation of relevance for the characteristics of a correlational statistic is the condition where absolutely no relationship between X and Y scores is present. This lack of correlation between variables is the kind of situation which should occur, on the average, when the Y scores are paired at random with the X scores.

The lack of any correlation may be meaningfully assigned a value of zero for the corresponding statistic. This is what is done in the correlation coefficient to be discussed.

In summary, we shall define a statistic which reflects the degree of correlation between two standardized sets of measurements, using a zero value to indicate no relationship, and a plus or minus one to indicate perfect correlation, the minus value indicating an inverse relationship between the two number scales used.

11–2. Prediction of One Score from Another. In the case of a perfect relationship between the X variable and the Y variable, it is possible to predict a single score on one variable from knowledge of its partner score. Even when the correlation is not perfect, a prediction of one score from knowledge of its partner may be made which is "better than a chance guess." One method of illustrating the relationship between two sets of scores and of illustrating the prediction of one score from another is a graphic presentation of the data.

Graphic analysis. We may use the rectangular coordinate system to represent two sets of scores. Let the X values be given on the x-axis of such a graph and the Y values on the y-axis. Any pair of XY scores may now be represented by a *single* point in such a graph. The location of any single point, with respect to the x-axis, indicates the value of X for the pair of scores represented. The location of the point relative to the y-axis gives the corresponding value of Y.

For illustration of this method, the data given in Table 11–1 will be used. In this table, artificial data are given which might represent the measurements taken on two variables for each of 30 people. A graph representing these data, which is called a *scatter diagram*, is shown in Fig. 11–1. Note that the figure contains 30 points, each of which indicates the values of one of the 30 pairs of numbers in Table 11–1. For example, the fifth subject shown in the table has an X of 27 and a Y of 36. This pair of numbers is indicated by the point in the upper right corner of Fig. 11–1. This point is above the number 27 on the x-axis and across from the number 36 on the y-axis. Figure 11–1 shows a clear tendency for the values on the x-axis to vary along with the values on the y-axis. Notice that the points on the figure fall within an envelope which is long and narrow. This envelope may be used to illustrate the notion of prediction of one score from its partner.

Select some score on the X variable. For example, let us inspect the figure with the X value of 19. If you know that an individual

TABLE 11–1

Artificial Data for Thirty Subjects on Two Variables

Subject	X	Y	Subject	X	Y
1	14	18	16	21	28
2	7	10	17	17	15
3	5	8	18	19	24
4	22	29	19	17	21
5	27	36	20	11	16
6	12	18	21	10	15
7	13	21	22	9	12
8	14	25	23	7	11
9	10	19	24	14	20
10	8	12	25	18	26
11	11	14	26	5	9
12	13	15	27	14	20
13	19	22	28	12	11
14	19	25	29	15	18
15	9	13	30	16	18

had this score of 19, what statement might be made about his Y score? If one moves directly up from the score of 19 on the x-axis, one finds 3 possible scores for a partner on the Y variable. In general, one would suggest that any of the values within the envelope above the score of 19 might occur as a partner score on Y. Since this envelope is long and narrow, we find that the range of possible scores on Y is considerably smaller than the total range of the Y variable. In short, with the data in Fig. 11–1, the knowledge of a person's X score would permit a guess about his Y score which is restricted to a narrower range than if one did not know the value of X.

If the degree of relationship between X and Y were perfect, one would find that the corresponding scatter diagram would have all data points falling along a straight line, i.e., the envelope surrounding the points could be made with a width of zero, a straight line.

The direction of this straight line or the envelope will indicate the direction of the correlation, either a direct or *positive correlation,* or an indirect or *negative correlation.* If the line or envelope has a positive slope, as shown in Fig. 11–1, the correlation is positive. The negative slope type of diagram indicates a negative correlation.

In those cases where two variables are uncorrelated (have zero correlation) the scatter diagram will contain points which are scattered rather evenly over a rectangular shaped zone. A line drawn around the outside of this cluster of points will have no tendency for either a positive or negative slope.

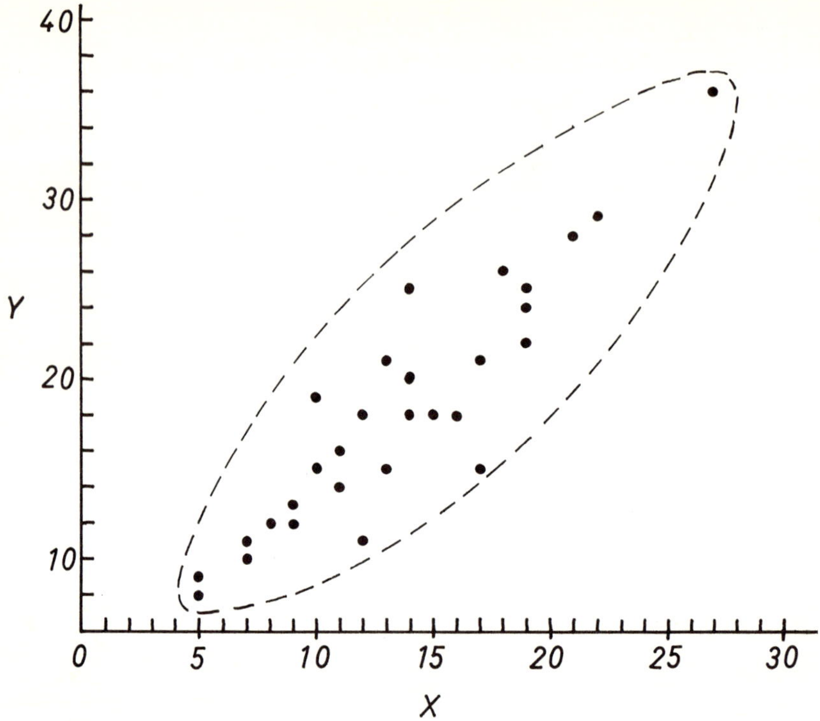

Fig. 11–1. A scatter diagram showing the values of 30 pairs of X,Y scores. Two pairs had identical values and are represented by a single point. A high degree of positive correlation is illustrated.

A somewhat more useful method of plotting a scatter diagram is shown in Fig. 11–2, in which the data from Table 11–1 again are used. This time the mean and standard deviation for X have been computed, $M = 13.6$ and $s = 5.15$, and each X score changed to a z-score. We denote this set as the z_x's. The same conversion has also been done on the set of Y's, $M = 18.3$ and $s = 6.51$. The resulting z-scores will be called the z_y's.

Since we now have pairs of z-scores rather than pairs of raw scores, the resulting scatter diagram is scaled in z-score units on each axis. Figure 11–2 looks quite similar to Fig. 11–1, however, the general case, in which the means and standard deviations of the two variables are not similar, would often show subtle but clear differences between two such graphs. There will usually be a decrease or an increase in the slope of the envelope around the data points.

The scatter diagram of a set of paired points, when scaled in the respective z-score units, has regularities not found in a raw score scatter plot. For example, in the case of a perfect correlation, not only do the points all fall on a straight line, but the *slope* of that line in a z-score scatter plot will always be 1, either plus or minus. In terms of the numbers, one would find that a given z-score on X would be identical in absolute value with its partner, scaled in z_y.

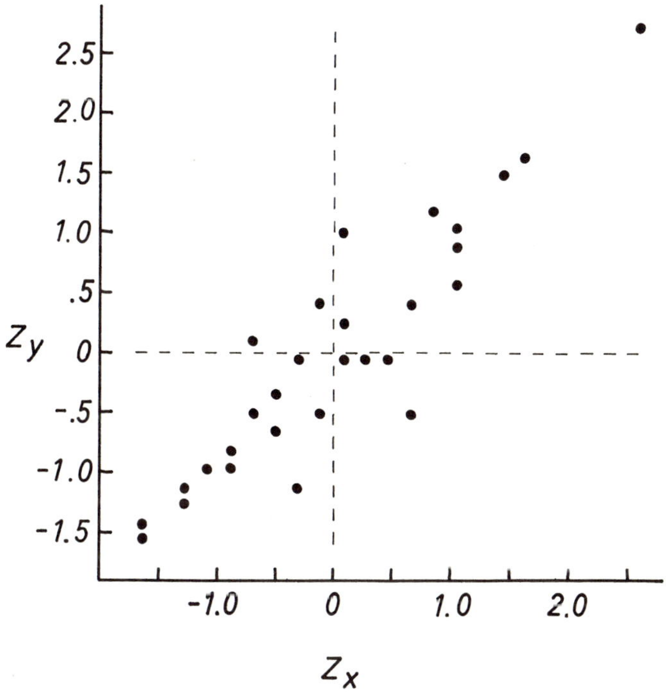

Fig. 11–2. A scatter diagram of the data from Fig. 11–1 scaled in terms of *z-scores* for both X and Y.

Since the slope of the line upon which all points fall, in a case of perfect correlation, is plus or minus one, prediction of a z_y from its corresponding z_x may be easily described as

$$\tilde{z}_y = 1z_x \qquad \text{or} \qquad \tilde{z}_y = -1z_x \tag{11–2.1}$$

where $\tilde{z}_y$ is used to denote a *predicted* score.

We shall now attempt to describe a method of prediction of one z-score from its partner which is of the same linear form as (11–2.1), but will be useful in cases in which the correlation between X and Y is not perfect.

The prediction equation. Although we will restrict our discussion to z-scores on the two variables for the present, it should be clear that any equation written in z-score form may be modified to raw-score form by the formulas for z-scores.

The type of correlational statistic, and the characteristics resulting from its use, which we are discussing, may be termed *linear* correlation. This term is appropriate for a number of reasons, but the primary reason is the method of prediction on which the statistic is based.

Since we have determined that (11–2.1) is appropriate for a perfect correlation, our best case is seen to be a straight-line relationship between variables on a scatter diagram. Even when a correlation is not perfect, in this sense, we shall still attempt to describe whatever degree of relationship exists by means of a straight-line prediction equation.

In our previous discussion of Fig. 11–1, the X value of 19 allows a prediction of Y within a certain range. Let us now restrict ourselves to making an estimate of Y which is exact, i.e., a single prediction rather than a range.

In order to make a prediction of a linear form, as in (11–2.1), we need to know the z_x score and the constants of multiplication and addition from the equation for a straight line. The use of z-scores permits us to call the constant of addition zero in such a straight-line equation. This results from the fact that the mean z-score is always zero for a set of numbers. The constant of multiplication alone would permit us to predict a value of z_y from a value of z_x with this straight line type of prediction.

We therefore define the prediction scheme, which is also a method of describing the relationship between the pairs of numbers, as

$$\tilde{z}_y = \beta z_x \tag{11–2.2}$$

In (11–2.2), we assume that the value of β is to be used as a single number for all possible z_x's in a set of data. The value of β is known to be equal to plus or minus one when the correlation is perfect. Our problem is now to find a method of obtaining the *best* value of β for any given set of data.

Before attempting the solution to this problem, which will yield the correlation coefficient we have been working towards, it should be indicated explicitly that the prediction equation (11–2.2) may be used, with changed subscripts, to predict the values of X from Y rather than the other way around. This feature results directly from the use of a straight-line prediction scheme rather than some other method.

11–3. Derivation of r. Our problem, then, is to determine a value of β which is best for any given set of data. Again, as in Chapter 5, the evaluation of the word "best" into statistical terms allows more than one outcome. However, the criterion used in this case, which is called the **least-squares criterion**, is consistent with our selection of the mean as somehow better for our purposes than the median or mode.

The *least-squares criterion* permits us to define "best" as a system that yields a sum of squared numbers which is as small as possible. For example, if one were to examine the sum of squared deviations of a set of numbers from some arbitrary value, one would find that this sum is always larger than the SS, which is based on the mean. In that sense, one might say that the mean is the "best" value for calculating dispersion.

In our current problem, we should worry about the correspondence between a predicted z_y score for each pair of numbers and the actual or observed z_y being predicted. The difference between the observed and predicted z_y scores is an *error in prediction:*

$$z_y - \tilde{z}_y = \text{the error in prediction} \qquad (11\text{–}3.1)$$

The average error in prediction will always be zero when (11–2.2) is used. This may be seen by summing these errors over all N scores:

$$\Sigma\,(z_y - \tilde{z}_y) = \Sigma\,(z_y - \beta z_x) = \Sigma\,z_y - \beta\,\Sigma\,z_x = 0 - \beta 0 = 0$$

This is true for any finite value of β. However, the least squares criterion requires that we examine the *squared error in prediction.* If we symbolically square each error in prediction and then add these together, the least-squares criterion requires that this sum be made as small as possible. In doing this, we shall change to the average squared error in prediction, just as a SS is changed to an average by changing it to the variance:

$$(z_y - \tilde{z}_y)^2 = (z_y - \beta z_x)^2 = z_y{}^2 - 2z_y\beta z_x + \beta^2 z_x{}^2 \qquad (11\text{–}3.2)$$

Now (11–3.2) represents the squared error in predicting a z_y for any person or pair of numbers in a set, and is expressed in terms of z_x, z_y, and β. Let us now add these squared errors in prediction over all N pairs:

$$\Sigma\,(z_y - \tilde{z}_y)^2 = \Sigma\,z_y{}^2 - 2\beta\,\Sigma\,z_x z_y + \beta^2\,\Sigma\,z_x{}^2 \qquad (11\text{–}3.3)$$

We now convert to an average sum of squared deviations from prediction by dividing each side of (11–3.3) by N. We may call the result the variance of errors in prediction of z-scores, denoted as $\sigma_{U_z}{}^2$.

The subscript U indicates the term *unpredicted*, which is another way of speaking about an error.

$$\sigma_{U_z}{}^2 = \frac{\Sigma z_y{}^2}{N} - 2\beta \frac{\Sigma z_x z_y}{N} + \beta^2 \frac{\Sigma z_x{}^2}{N} \qquad (11\text{–}3.4)$$

However, we found in Chapter 3 (3–5.1) that the sum of all the z-scores in any set of numbers is equal to N. This permits simplification of (11–3.4) to:

$$\sigma_{U_z}{}^2 = 1 - 2\beta \frac{\Sigma z_x z_y}{N} + \beta^2 \qquad (11\text{–}3.5)$$

For ease in the next step, define

$$A = \frac{\Sigma z_x z_y}{N}$$

and rewrite (11–3.5) as

$$\sigma_{U_z}{}^2 = 1 - 2\beta A + \beta^2 \qquad (11\text{–}3.6)$$

We now add and subtract an A^2 term in (11–3.6) in order to complete the square in the last three resulting terms of (11–3.7).

$$\sigma_{U_z}{}^2 = 1 - A^2 + A^2 - 2\beta A + \beta^2 \qquad (11\text{–}3.7)$$

The last three terms may now be factored to obtain

$$\sigma_{U_z}{}^2 = 1 - A^2 + (A - \beta)^2 \qquad (11\text{–}3.8)$$

We now substitute the value of A back into (11–3.8)

$$\sigma_{U_z}{}^2 = 1 - \left(\frac{\Sigma z_x z_y}{N}\right)^2 + \left(\frac{\Sigma z_x z_y}{N} - \beta\right)^2 \qquad (11\text{–}3.9)$$

Equation (11–3.9) is the desired form of the sum of squared deviations from prediction, divided by N. We desire to make the value of (11–3.9) as small as possible. Note, however, that all of the terms in (11–3.9) are fixed in value for any set of data with the exception of β. This comes directly from the prediction equation (11–2.2) in which only the value of β was in question. We are now in a position to define a β which will yield the "best" predictions in terms of the least-squares criterion. The third term in (11–3.9) is a squared term, and, as such, can be no smaller than zero. It will be seen that (11–3.9) will attain its smallest possible value when the third term is zero. This is possible only when β is equal to the first term within the squared term considered. Not only does this define the best value of β for our prediction equation, but this value of β is also the correlation coefficient we have been seeking.

$$\beta = \frac{\Sigma\, z_x z_y}{N} \tag{11–3.10}$$

We will now change the notation β to its more usual symbol, r, which is the correlation coefficient.

$$r = \frac{\Sigma\, z_x z_y}{N} \tag{11–3.11}$$

The statistic r is defined in (11–3.11) as the average cross product of pairs of z-scores. This statistic is known as Pearson's *product-moment correlation coefficient,* or simply Pearson's r. Pearson, of course, was the person most responsible for devising this measure of relationship between two variables.

Certain features of (11–3.11) may be quickly illustrated. For example, if each z-score on X is equal in value to its partner on Y, then the numerator of (11–3.11) is seen to be equal to the sum of squared z-scores, or N. This will produce the maximum correlation of 1 of which we spoke earlier. If one scale were inverted, the z-score pairs would simply have opposite signs and the correlation would come out to be minus one, $-N/N$. Although difficult to demonstrate, the value of r will be zero if no relationship exists, in our sense of a linear function.

The prediction formula may now be rewritten by replacing β by r.

$$\tilde{z}_y = r z_x \tag{11–3.12}$$

It must be stressed that the calculation of r from (11–3.11) requires that each z_x be multiplied by its corresponding z_y prior to the application of the summation sign. The subscripts of z_x, etc., have been omitted for convenience. It should also be noted that the signs of the products must be kept, in order to obtain the correct value of r.

Example 11–1. Let us find the value of r for the following set of data.

Subjects	X	Y
1	6	18
2	8	14
3	9	20
4	10	22
5	12	26

In using (11–3.11), we must convert each set of scores to z-scores. The data given permit this to be done easily because both means and standard deviations are convenient numbers. The M_x is 9, and M_y is 20. The deviations from the respective means and their squares are found as:

Subject	x	y	x^2	y^2
1	−3	−2	9	4
2	−1	−6	1	36
3	0	0	0	0
4	1	2	1	4
5	3	6	9	36
Totals	0	0	20	80

The standard deviation of the X set, s_x, is found to be 2, whereas the corresponding term for Y, s_y, is 4. These values are divided into the appropriate deviation scores in order to give the two sets of paired z-scores.

Subject	z_x	z_y	z_xz_y
1	−1.5	−0.5	0.75
2	−0.5	−1.5	0.75
3	0	0	0
4	0.5	0.5	0.25
5	1.5	1.5	2.25
Totals	0	0	4.00

The cross products of the paired z-scores are also shown. We must now add these products for all 5 subjects. The result here is 4.00. Use of (11–3.11) will now give the value of r:

$$r = \frac{\Sigma z_xz_y}{N} = \frac{4.00}{5} = .80$$

In order to show the method of prediction which is inherent in the value of r, let us predict each person's z_y by use of (11–3.12). Multiplying each person's z_x by .80, we obtain his predicted z_y. The actual z_y is given with each of the predicted values.

Subject	$\tilde{z}_y$	z_y
1	−1.2	−0.5
2	−0.4	−1.5
3	0	0
4	0.4	0.5
5	1.2	1.5

One may note that the predictions are not perfect. They could only be perfect if the correlation coefficient were plus or minus 1. For these data, the value of .80 is the best prediction coefficient. The interested reader may try other values in the prediction equation in order to study the increase in squared errors of prediction that would be obtained.

Example 11–1 illustrates a property of the prediction scheme involved in linear correlation. Examine again the predicted z_y scores with respect to the corresponding z_x for each person. Notice that the predicted z-scores are always closer to the mean (zero) than the z_x scores on which they were based. This is always true if the correlation is not perfect. This effect is termed *regression toward the mean.* The use of correlation for prediction purposes is often referred to as *regression analysis.*

In the special case of an r of zero, all predicted z-scores are zero, the mean of the predicted variable. When we have no correlation, then, the best single guess is that a person has a score equal to the mean on the variable predicted. It should be clear in this case that the sum of squared errors in prediction would be equal to the sum of squared z-scores, or N. This is true because each z-score minus the predicted value of zero is still the original z-score. This point will be explored more fully later.

11–4. Deviation and Raw Score Formulas. Equation (11–3.11) is a basic definition of r which is very useful for understanding the meaning of the statistic and also for theoretical work with the correlation coefficient. However, (11–3.11) is not very useful for calculating the size of r in most problems. When the means of the two variables are not round numbers, the conversion to deviation scores is quite laborious; moreover, the conversion of the deviations to z-scores requires $2N$ divisions. Let us develop other formulas for finding r which are easier to apply, in the general case.

The first variant of (11–3.11), which is less cumbersome, may be called the deviation-score formula. To obtain this formula from (11–3.11), we simply use the fact that $z = x/s$.

$$r = \frac{\Sigma\, z_x z_y}{N} = \frac{\Sigma\, (x/s_x)(y/s_y)}{N} \qquad \text{or} \qquad r = \frac{\Sigma\, xy}{N s_x s_y} \qquad (11\text{–}4.1)$$

The calculation of r from (11–4.1) does require deviation scores to be used, but does not require that each deviation score be converted to a z-score. Let us apply (11–4.1) to some data.

Example 11–2. Find the correlation between the X and Y variables in the data given.

Subject	X	Y
1	7	30
2	8	33
3	10	18
4	11	20
5	14	12
6	16	10
	$\Sigma X = 66$	$\Sigma Y = 123$
	$M_x = 11$	$M_y = 20.5$

We find the mean X to be 11 and the mean Y to be 20.5. Let us now convert each raw score to a deviation score by subtraction of the appropriate mean. The squared values of these deviation scores are used to obtain s_x and s_y, while the cross products of the pairs of deviations are used in the calculation of the numerator of (11–4.1). A summary of these operations may be given as:

Subject	x	y	xy	x^2	y^2
1	−4	9.5	−38.0	16	90.25
2	−3	12.5	−37.5	9	156.25
3	−1	−2.5	+2.5	1	6.25
4	0	−0.5	0	0	0.25
5	3	−8.5	−25.5	9	72.25
6	5	−10.5	−52.5	25	110.25
Totals	0	0	−151.0	60	435.50

The standard deviations may now be found from the sums of squared deviations:

$$s_x = \sqrt{\tfrac{60}{6}} = 3.16$$
$$s_y = \sqrt{\tfrac{435.5}{6}} = 8.52$$

The sum of the cross products is −151.0. This value is combined with the standard deviations and the N of 6 to obtain the size of r from (11–4.1):

$$r = \frac{\Sigma\, xy}{Ns_xs_y} = \frac{-151.0}{6(3.16)(8.52)} = \frac{-151.0}{161.54}$$
$$= -.93$$

The absolute value of r indicates the degree of correlation present, which is fairly close to the maximum possible. The sign, negative in this case, indicates the direction of the relationship, either direct or indirect. In this example, the negative correlation indicates that high scores on the X variable are paired with low scores on the Y variable and *vice versa*. Inspection of the raw data shows this tendency quite clearly.

When the means of X and Y are not convenient numbers, even (11–4.1) is not too easily employed. A formula using the raw scores without conversion to deviations or z-scores may be derived directly from (11–4.1). The actual derivation is somewhat long and will be omitted; however, the substitution of $x = X - M_x$, $y = Y - M_y$, and use of the raw score equations for s_x and s_y is the method used. The additional steps are algebraic simplifications of the expression. One form of this raw score equation, which is quite suitable to use with a calculator, is:

$$r = \frac{N \Sigma XY - (\Sigma X)(\Sigma Y)}{\sqrt{[N \Sigma X^2 - (\Sigma X)^2][N \Sigma Y^2 - (\Sigma Y)^2]}} \qquad (11\text{–}4.2)$$

Note that (11–4.2) requires the use of raw scores only. The basic term in this equation which reflects the covariance of the variables is the sum of cross products of paired raw scores.

Although (11–4.2) is convenient in certain respects, as indicated above, the numbers, especially within the radical, may become very large. Care must be taken in use of (11–4.2) because errors of large magnitude are not as readily apparent as when finding means, deviation scores, and standard deviations. In the latter case, appropriate to the deviation score formula, large-magnitude errors may often be detected by inspection of the data.

Example 11–3. In a problem-solving experiment, subjects were required to make repeated attempts at solving a concept-identification problem. The investigator recorded both the total number of trials each person needed before reaching the criterion of solution used, and the total number of errors which were made by each subject. A subject might be correct on a given trial without having the correct solution in mind, hence these two measures can differ to some degree. One would certainly expect that subjects who required more trials would also make more errors in these trials than would faster-solving subjects.

Let us describe the degree of relationship between trials to criterion and errors to criterion for the data given below.

Subject	Trials (X)	Errors (Y)	Subject	Trials (X)	Errors (Y)
1	42	17	11	17	10
2	13	7	12	26	16
3	21	12	13	18	6
4	31	18	14	5	2
5	26	10	15	27	15
6	22	11	16	19	8
7	28	18	17	23	14
8	35	14	18	32	15
9	9	3	19	17	9
10	41	24	20	15	6

Use of (11–4.2) with these data requires finding the sum of the raw scores and sum of the squared raw scores for each variable. These results are

$$\Sigma X = 467 \qquad \Sigma Y = 235$$
$$\Sigma X^2 = 12{,}737 \qquad \Sigma Y^2 = 3355$$

The sum of cross products is obtained by finding the XY value for each subject ($42 \times 17 = 714$, etc.) and adding these together over all 20 subjects. This total is

$$\Sigma XY = 6{,}428$$

Use of (11–4.2) then gives, with $N = 20$:

$$r = \frac{128{,}560 - 109{,}745}{\sqrt{[254{,}740 - 218{,}089][67{,}100 - 55{,}225]}}$$
$$= \frac{18{,}815}{\sqrt{(36{,}651)(11{,}875)}} = \frac{18{,}815}{\sqrt{435{,}230{,}625}} = \frac{18{,}815}{20{,}862}$$
$$= .90$$

The high positive correlation suspected is verified with our obtained r.

The regression equation. When prediction of a score on one variable from the other variable is desired, (11–3.12) may be used when z-scores are used. But our usual interest in a prediction use of correlation leads back to raw-score units. The prediction of a raw score on Y from a raw score on X may be given when substitutions are made into the z-score prediction equation. We first note that a predicted z-score may be converted as follows:

$$\tilde{z}_y = \frac{\tilde{Y} - M_y}{s_y} \qquad (11\text{–}4.3)$$

where $\tilde{Y}$ is a predicted raw score. From (11–3.12), we may set this expression equal to rz_x and change to raw score terms:

$$\frac{\tilde{Y} - M_y}{s_y} = rz_x = r\,\frac{(X - M_x)}{s_x}$$

Solving for $\tilde{Y}$, the predicted raw score, we obtain

$$\tilde{Y} = (X - M_x)(r)\left(\frac{s_y}{s_x}\right) + M_y \qquad (11\text{–}4.4)$$

Equation (11–4.4) shows the method of predicting a raw score on Y from its partner raw score on X. Note that one needs both mean and standard deviation of each variable as well as the correlation coefficient.

To illustrate (11–4.4), let us use the data from Example 11–2. The first subject in that example had a raw score on X of 7. The prediction of Y for this score of 7 requires that one substitute the quantities from that example into (11–4.4). The computations are shown below:

$$\begin{aligned}\tilde{Y} &= (7 - 11)(-.93)\left(\frac{8.52}{3.16}\right) + 20.5\\ &= (-4)(-.93)(2.7) + 20.5\\ &= 30.54\end{aligned}$$

which corresponds closely to the observed Y of 30. In general, there is little that can be said about the degree of correspondence of a single predicted score and the obtained value. As we shall see later, the squared error in prediction, *on the average,* can be calculated from the knowledge of r.

11–5. Significance Tests. Our study of correlation thus far has been descriptive. We must now consider the estimation of population values for the correlation coefficient and the problem of hypothesis testing about such population values.

Let us denote the population correlation coefficient with the Greek letter equivalent of r, which is ρ.

We may first consider an unbiased estimate of ρ. The solution is simply r, assuming random sampling, of course. Our best single estimate of the population correlation is the value obtained in a random sample.

The descriptive relationships and calculations already encountered are quite valid for *any* type of population distributions. However, in order to test hypotheses about population correlation from random samples, we must make some rather stringent assumptions about the population of X and Y variables. We assume in the rest of this section that both the X's and Y's are normally distributed in the population.

The first hypothesis test which one might consider is whether or not the population correlation is different than zero:

$$H_0: \rho = 0 \qquad (11\text{–}5.1)$$

The $RSD(r)$ has been studied when this null hypothesis is true, and tables are available in other sources which allow one to test the hypothesis by use of a table. However, we shall use a test of this null hypothesis which is more involved, because the same procedures will allow other types of hypotheses to be tested. For instance, we may know that a correlation is .65 for some particular population. If a sample from some other population is used, we may wish to test that our sample is drawn from a population with this same .65 correlation. This test will be done in the same way as the test of a zero correlation in the population.

We may generalize our hypothesis H_0 to include any possible population value, ρ_H, which is desired.

$$H_0\colon \rho = \rho_H \qquad (11\text{–}5.2)$$

where ρ_H is any hypothesized population correlation.

Fisher's Z transformation. Fisher has developed a statistic which is a non-linear transformation of r. For each value of r, a type of logarithmic function is used to obtain a single value of Z. Rather than studying the random sampling distribution of r, it is more efficient, for testing hypotheses, to study the corresponding $RSD(Z)$. This sampling distribution turns out to be essentially normal for moderate-size N and nearly any value of ρ which may be true for the parent population. The mean of the $RSD(Z)$ is equal to the value of Z which corresponds to the population correlation ρ.

The standard error of Z, which is the standard deviation of the $RSD(Z)$, denoted as σ_Z, is given by the following approximation.

$$\sigma_Z = \sqrt{\frac{1}{N - 3}} \qquad (11\text{–}5.3)$$

where N is the number of pairs of scores in the sample.

The values of Z which correspond to different values of r, or ρ, are given in Table G in the Appendix. In this table, only positive values of Z and r are given. A negative value of r corresponds to the negative value of the Z taken from the table. For example, from Table G, an r of .550 is seen to have a corresponding Z of .6184, hence an r of $-.550$ would have a corresponding Z of $-.6184$.

We now have sufficient information to make a test of the H_0 given by (11–5.2).

Testing for a specified population correlation. Since the $RSD(Z)$ is approximately normal, with a known mean and standard deviation, we may convert the sample value of r to a deviation from expectation measured in z-scores which are normally distributed.

Let us denote the Z which corresponds to our sample value of r as Z_r, and the Z which corresponds to the hypothesized value of ρ (ρ_H) as Z_H.

The test value of z, the normal deviate in z-score form, is given by

$$z = \frac{Z_r - Z_H}{\sqrt{\frac{1}{N - 3}}} = (Z_r - Z_H)\sqrt{N - 3} \qquad (11\text{–}5.4)$$

Example 11–4. A random sample of 28 people shows a correlation between age and a measure of reactivity to novel stimuli of $-.35$. Can we reject the hypothesis of a zero correlation as a population value?

Here the hypothesized ρ is zero and the corresponding value of Z_H is found in Table G as zero. The sample value of r corresponds to a Z_r of $-.3654$. Our test value of z is found next, using (11–5.4):

$$\begin{aligned} z &= (-.365 - 0)\sqrt{28 - 3} = (-.365)(5) \\ &= -1.83 \end{aligned}$$

Use of Table C indicates that a random value of z will be between plus and minus 1.96 about 95 per cent of the time. The .05 level of confidence requires a value of z which is 1.96, or greater, in absolute value. Our test value of z does not fall in this region of rejection, so we do not have sufficient evidence to reject the hypothesis under test.

Example 11–5. Using the same data as given in Example 11–4, can we reject the suggestion that the true correlation in the population is .25?

In this case our value of Z_r remains at $-.365$, but the value of Z_H, which must correspond to ρ_H, is found in Table G as .2554. We then have a z-score for this test which is found by use of (11–5.4):

$$z = (-.365 - .255)\sqrt{25} = -3.10$$

This z-score is well into the region of rejection so that we may reject the hypothesis that the population correlation is .25. Here, the rejection may be made with confidence beyond the .01 level. There is only a very small probability that a random sample of 28 persons from a population with $\rho = .25$ will have an $r = -.35$.

The difference between two population correlations. On occasion one is interested in the degree of correlation between variables when two conditions or populations are used. For instance, men and women may have different degrees of correlation on two given variables. The only way to establish this difference is by inferential methods. We would probably select random samples of men and women and find the sample correlations in each group. The problem is then to find an appropriate test of the differences in sample r's which would allow us to reject the hypothesis that each population has the same correlation between the variables.

This type of test may also be accomplished by using the Z transformation of r rather than r directly. However, when two random values of Z are compared, the standard error must be computed for the difference between these values.

In general, if two random variables are selected independently of one another, the standard error of D, which is the difference between these variables, may be given in terms of the standard errors of each variable when taken separately. For example, if one were to select a random score, say X_1, from a normal distribution with $\sigma_1 = 5$ and a second random score, X_2, from a normal distribution with $\sigma_2 = 10$, then the standard deviation of the random sampling distribution of $X_1 - X_2$ (D) could be found from σ_1 and σ_2. The general formula for the standard error of D is

$$\sigma_D = \sqrt{\sigma_1{}^2 + \sigma_2{}^2} \tag{11–5.5}$$

where σ_1 and σ_2 are the standard errors of the two variables when sampled separately, and σ_D is the standard error of the difference between pairs of the two variables.

In the case of the correlation coefficient, we have considered the $RSD(Z)$ and found that its standard error is given in (11–5.3). If we now view the random sampling distribution of differences between two randomly selected Z's, the standard error of the distribution may be found from (11–5.5) and (11–5.3). Substitution yields

$$\sigma_{Z_1 - Z_2} = \sqrt{\frac{1}{N_1 - 3} + \frac{1}{N_2 - 3}} \tag{11–5.6}$$

In (11–5.6), N_1 refers to the sample size in one group, and N_2 to the sample size in another group. The statistic is the difference between Z_1, found from the correlation in group 1, and Z_2, found from the correlation in group 2.

The expected value of the difference between two such random values of Z is zero when we suppose that the two populations have the same ρ's. Moreover, the distribution of differences between the two Z's is normal. This permits us to use the normal distribution, in z-score form, to test the hypothesis of populations with equal correlations. The test may be set up as

$$z = \frac{Z_1 - Z_2}{\sqrt{\frac{1}{N_1 - 3} + \frac{1}{N_2 - 3}}} \tag{11–5.7}$$

Example 11–6. A random sample of 50 children in the sixth grade shows a correlation of .48 between reading speed and a standardized test of vocabulary. Another random sample of 75 children in the fourth grade shows the same two variables correlated at .30. Can one reject the hypothesis that these samples are drawn from populations with equal correlations between reading speed and the standardized vocabulary test performance?

Since the two samples are independent of one another, and both are random, (11–5.7) is appropriate to test the hypothesis of equal population correlations. We find the Z's which correspond to the sample r's. Table G shows that the first one of these, Z_1, is .523 while the second, Z_2, is .3095.

The N's are given in each case, so that the required z-score is found as

$$z = \frac{.523 - .310}{\sqrt{\frac{1}{47} + \frac{1}{72}}} = \frac{.213}{\sqrt{.0352}} = \frac{.213}{.188}$$
$$= 1.13$$

This value of z, on the normal curve, would not be in the region of rejection, using the .05 level of confidence. We do not have sufficient evidence to reject the hypothesis of equal population correlations.

It must be stressed that all the hypothesis-testing procedures given in this section require that the two population variables be normally distributed. Minor departures from this assumption will have very minor consequences on the tests given here. However, if the X and Y in the population are not close to being normally distributed, the tests given are of highly questionable validity. The effects of such departures on these tests do not seem to have been studied in a systematic fashion by statisticians.

11–6. Correlation and Causation. A correlation between two variables, no matter how high it may be, does not usually imply that either variable is the cause of the other. In other words, changes in one variable which are concomitant with changes in another variable, need not be caused by the latter variable.

Although this lack of causation between correlated variables is stressed in most statistics books, there *are* conditions under which a causal relationship between variables may be inferred from a significant correlation. Let us explore both cases, but keep in mind that causal relationships as shown by correlation are the exception rather than the rule.

In one sense, a true correlation in the real world probably does indicate a *connective* relationship between variables in the broadest sense of the term. However, in many cases, some factor not measured may be a common cause of both variables.

For instance, the age of weaning a child might be found to be correlated with a certain personality trait in a person's later life. A classic error would be committed if one were to assume that the correlation demonstrated that age of weaning caused differences in personality. A more likely state of affairs is that parental attitudes influence both age of weaning children and certain aspects of personality. There is no valid line of reasoning from such correlational data to indicate that the two measured variables are causally related, but there is no proof that they are not so related. At best (and sometimes no better case could be constructed) the correlation between two variables is *suggestive* of a causal relationship. Naturally, additional knowledge may either rule out such a relationship or support the notion that a cause and effect have been found.

The distinguishing feature of a correlational study which either permits cause and effect to be inferred or makes such an inference impossible is the distinction between *normative* and *manipulated* variables. As discussed in Chapter 6, when subjects are placed into treatment conditions on the basis of their existing characteristics, the differences between such treatments form a normative independent variable. When such treatment conditions are quantified and correlated with a dependent variable, the correlation is describing a relationship between two characteristics which existed prior to the investigation. There is no necessity that variation in the normative variable is causing variation in the other variable.

In contrast, when subjects are randomly assigned to treatment conditions which are scaled and measured, and these measurements show a significant correlation to some performance measure taken under these conditions, then one may validly argue that changes in the treatment variable have caused, at least indirectly, the changes in the measured variable.

As a means of illustration, assume that one were to find a significant positive correlation between the number of college courses taken by men and the annual income of these men. This should certainly not be interpreted to mean that education necessarily increases one's income, even on the average. It might simply be that the more intelligent men elect to take more education and that such men are better equipped, with or without education, to make high salaries in one field or another. Many college teachers feel that they would be in better paying jobs if they had never gone to college.

The only method of actually demonstrating a causal relationship conclusively would be to assign people to take a predetermined number of college courses and then measure income after some set period of time. This illustration indicates why normative variables are frequently studied rather than manipulated ones: it is often impossible to manipulate important features in the lives of humans. A large amount of animal research is done to overcome this limitation, even though the problem of generalizing results to humans is introduced.

One should not belittle the usefulness of a significant correlation even when cause and effect cannot be inferred. The possibility for prediction of an important variable like job performance from knowledge of an unimportant score on a standardized test may be exceedingly valuable, even though no one would suggest that the test performance causes job performance. The only real difficulty in working with mere correlation for prediction purposes is that changes in causal factors, which may be unknown, may wipe out the correlation being used for prediction. It is certainly better to know what causes what, than to depend upon a relationship which may vanish when unknown (and hence unforeseen) factors change.

In terms of research, a significant correlation between two variables has rarely demonstrated cause and effect relationships, but such a correlation has often suggested a line of research leading to knowledge about important manipulated independent variables.

11–7. Interpretations of r. We have seen that a correlation of 1, either plus or minus, indicates that one set of scores bears a perfect linear relationship to another set. We have also mentioned that a zero correlation means that knowledge of one score in a pair is of no use, on the average, for predicting its partner. However, what about the intermediate cases? For instance, is an r of .50 half as good as one of unity? If so, in what sense is the word half used? As stressed in an early chapter, the relationship of numbers to one another does not always reflect a similar relationship between whatever the numbers represent.

In the case of correlations, the usefulness of a correlation coefficient of a given magnitude is not directly proportional to its numeric distance from zero and unity. Rather, we shall find that the *square* of the correlation coefficient is a more meaningful number by which to judge its predictive value.

The unpredicted and predicted variance. In Section 11–3 the formula for r was derived by using the least-squares criterion. The variance of squared errors from prediction, in z-score terms, was given in (11–3.9). In that equation, β is set equal to r for the prediction equation (11–3.12) to apply. This gives the variance of squared errors in prediction, in z-score form, as

$$\sigma_{U_z}{}^2 = 1 - \left(\frac{\Sigma\, z_x z_y}{N}\right)^2 \qquad (11\text{–}7.1)$$

which is equal to

$$\sigma_{U_z}{}^2 = 1 - r^2 \qquad (11\text{–}7.2)$$

Let us examine the properties of $\sigma_{U_z}{}^2$. If r is zero, we have seen that the errors in prediction are equal to the z_y scores, since the *predicted* z_y is zero in each case. Then (11–7.2) is clearly unity as is the sum of any set of squared z-scores when divided by N. When r is one in absolute value, there are no errors in prediction and (11–7.2) gives the corresponding answer of zero.

The variance of the predictions, $\tilde{z}_y$'s, which we denote as $\sigma_{P_z}{}^2$ is equal to r^2. This fact is obtained by squaring and summing (11–3.12) and dividing by N:

$$\Sigma\,(\tilde{z}_y)^2 = \Sigma\,(rz_x)^2$$

$$= r^2\,\Sigma\,z_x{}^2 = r^2N$$

$$\sigma_{P_z}{}^2 = \frac{\Sigma\,(\tilde{z}_y)^2}{N} = \frac{r^2N}{N}$$

$$\sigma_{P_z}{}^2 = r^2 \qquad (11\text{–}7.3)$$

Let us now add (11–7.2) and (11–7.3) together:

$$\sigma_{U_z}{}^2 + \sigma_{P_z}{}^2 = 1 - r^2 + r^2 = 1 \qquad (11\text{–}7.4)$$

This equation shows that, *in terms of z-scores,* what we have called the *unpredicted variance* plus the *predicted variance,* as we will call $\sigma_{P_z}{}^2$, is equal to one. We may now speak of each of these entities as showing the *proportion* of the variance of the Y variable which is predictable or unpredictable by use of the correlation method.

If r is .80, we see that the variance of errors in prediction is $1 - .64 = .36$ compared to a total Y variance of 1, in z-score terms. The remaining proportion of this Y variance, .64, may be called the proportion of predicted variance.

This method of viewing the degree of usefulness of a correlation may be extended to the variance of Y as measured in raw score units as well as z-scores. Since z-scores have a constant divisor of s, we may convert the variance in z-scores to a variance in squared raw scores by multiplying by s^2. Denoting the variance of errors in prediction, in squared raw score units, as $\sigma_U{}^2$ and the variance of the predictions as $\sigma_P{}^2$, (11–7.2) and (11–7.3) become

$$\sigma_U{}^2 = s_y{}^2(1 - r^2) \qquad (11\text{–}7.5)$$

$$\sigma_P{}^2 = s_y{}^2r^2 \qquad (11\text{–}7.6)$$

The total variance of the Y measures (or the X measures if X is predicted from Y) can be broken into two additive parts, a predicted portion and an unpredicted portion:

$$s_y{}^2 = \sigma_U{}^2 + \sigma_P{}^2 \qquad (11\text{–}7.7)$$

Both predicted and unpredicted variances are functions of r^2. We may then scale the usefulness of r in terms of r^2 and make proportional comparisons of correlations of different magnitudes.

The standard error of estimate. A variance cannot be located along the same scale as raw-score units. This makes it difficult to understand the degree of error involved in predictions unless one converts back to raw-score units. This conversion is accomplished by changing a variance to a standard deviation.

The standard deviation of the errors in predicting Y from X is called the *standard error of estimate,* and is scaled in raw score units.

Taking the square root of both sides of (11–7.5), this standard error, σ_U, is:

$$\sigma_U = s_y\sqrt{1 - r^2} \tag{11–7.8}$$

Under certain very restrictive conditions, it is possible to use (11–7.8) as a means of giving the probability of errors in prediction of any specified magnitude. If one may assume that the variance of the Y scores is the same *at each value of* X (the assumption of *homoscedasticity*) and that the relationship between X and Y is truly linear, then our earlier assumption that Y is normally distributed permits one to use the normal curve for assessing error probabilities. For example, if one had a Y variable measured in feet and found that an X variable correlated .70 with Y, the standard error of estimate may be found from s_y. Assume this standard deviation to be 1.50 ft. Then σ_U is found as

$$\sigma_U = 1.50\sqrt{1 - .49} = 1.07$$

If the conditions described above are true, then the distribution of Y scores around a predicted Y value will be normal with a mean of $\tilde{Y}$ and a standard deviation of 1.07, σ_U. One would then be able to state that errors in prediction greater than (1.96)(1.07) or 2.10 would have a probability of occurrence of .05, as in any other normal-curve problem.

If the assumptions given do not hold, then (11–7.8) is only an average value, and predictions for some values of X may be far different in their degree of error than for other values of X.

Inspection of a scatter diagram is a good method of finding gross departures from all of the basic assumptions mentioned in this chapter.

In summary, it should be clearly recognized that the correlation coefficient is descriptive without assuming any sort of distribution or scale for X or Y. Moreover, the prediction equations and related predicted and unpredicted variance relationships are true without such assumptions. When one wishes to test hypotheses or make estimates about the probability of certain magnitudes of error for a given X, we need the more powerful conditions inherent in distributional assumptions about X and Y.

11–8. Correlations on Special Data. In this section, three special forms of Pearson's r will be demonstrated. Each of these is designed for data which is in special form as compared to the freely varying X and Y scores we have already considered. The special formulas given are simplifications of the basic equations for r. In fact, (11–3.11), (11–4.1), or (11–4.2) could each be used rather than the special formulas to be given here. However, these variant forms of getting the correlation are less work than the more basic equations.

Each of these correlations is given a special name. The name does not change the fact that, mathematically, Pearson's r is the type of correlation being obtained, the name merely calls attention to the special type of data on which the correlation is based.

A final measure of relationship will be introduced which is not based on Pearson's r. This measure, called Kendall's *tau* is closely related to probability theory and is an example of what are called *nonparametric* statistics. This term is used to describe statistics which may be studied without recourse to distributional assumptions about the population.

Point biserial correlation. The *point biserial correlation coefficient*, r_{pb}, is a variation of r for data which are continuous on one variable but dichotomous on the other. For example, if weight were to be correlated with the sex of the subjects, weight could be measured in continuous units, but the sex variable would have to be coded into two categories. By convention, one of the two categories on the dichotomized variable is assigned a score of 1, and the other category is assigned a score of 0. It is not important which is assigned the 1 and which the zero, although a reversal in this assignment would lead to a correlation with opposite sign. The use of any other two numbers would give the same result because r does not change when a variable is transformed in a linear fashion.

Let us use X for the continuous variable and Y for the two-valued variable. The derivation of the equation for r_{pb} from that of r is straightforward but will be omitted. One form of the result, which is easily applied, is

$$r_{pb} = \frac{(T_{xy}/T_y) - M_x}{s_x}\sqrt{\frac{p}{q}} \tag{11–8.1}$$

where: T_{xy} = the sum of cross products of X and Y
T_y = sum of Y
M_x = the mean of X
s_x = the standard deviation of X
p = the mean of Y
$q = 1 - p$

These terms will be discussed briefly prior to the giving of an example. The cross product term (T_{xy}/T_y) may also be expressed as the average X score for those persons who have a score of 1 on Y. The sum of the Y scores is simply equal to the number of persons who have a score of 1 on Y. The value p is the proportion of subjects who have a score of 1 on Y, i.e., T_y/N. These identifications permit the computation of r_{pb} to be done rather quickly as illustrated in the following example.

Example 11–7. An instructor is interested in finding the correlation between a single question on an objective test and the total points scored on the rest of the test. The following data give the tests scores for each student (X's) as well as whether or not the students got the single question correct or incorrect. A score of 1 is used to indicate that the single question was passed (correct) and a zero is used for an incorrect outcome.

Student	X	Y	Student	X	Y
1	42	0	11	18	1
2	57	1	12	29	0
3	40	1	13	45	1
4	38	1	14	35	1
5	33	1	15	41	1
6	37	1	16	47	1
7	36	0	17	30	0
8	41	1	18	36	0
9	30	0	19	18	0
10	32	0	20	27	0

Since the test scores have numerous possible values, we consider that X is at least approximately a continuous distribution. Clearly, the Y scores are dichotomized into pass and fail categories with the numbers 1 and 0 used for these Y scores. The measure of correlation which is descriptive of such data is r_{pb}. Let us work on the X measures first and obtain the mean and standard deviation. We find that

$$\Sigma X = 712, \quad \Sigma X^2 = 26{,}970$$

$$M_x = 35.60$$

$$s_x = \sqrt{\frac{\Sigma X^2}{N} - M^2} = \sqrt{81.14} = 9.01$$

The calculation of T_{xy} may be done by simply adding up the X scores which have corresponding Y's of 1. This total is 432.

The T_y is equal to the number of Y's which are 1; this is 11.

The proportion of Y's which are 1 is $\frac{11}{20}$ or .55. This value is p and q is then found as .45.

Substituting these values into (11–8.1) gives the r_{pb}:

$$r_{pb} = \frac{(\frac{432}{11}) - 35.60}{9.01}\sqrt{\frac{.55}{.45}} = \frac{3.673}{9.01}(1.105)$$
$$= .45$$

This is the same value as would have been obtained by use of (11–4.2) on these data, because the r_{pb} formula is derived from our basic definition of r.

A test of the hypothesis that a population correlation is zero may be made using r_{pb}, but the sampling distribution is somewhat different than for r. A reasonable test of the significance of r_{pb} may be made by use of the t distribution, see Table D. The resulting t has $(N - 2)$ degrees of freedom and may be found as:

$$t = r_{pb}\sqrt{\frac{N-2}{1-r_{pb}^2}} \qquad (11\text{–}8.2)$$

For example, the result in Example 11-7 would give a t of 2.14:

$$t = .45\sqrt{\frac{18}{1 - .2025}} = .45\sqrt{22.6} = 2.14$$

With 18 df, this value just exceeds the tabled value of 2.10 required for rejection at the .05 level of confidence.

The phi coefficient. When the type of restriction of data found in the correct application of r_{pb} is true for both X and Y variables, the correlation based on the data is called the *phi* coefficient, ϕ. This statistic is the same Pearson's r as discussed in this chapter, but the special name is again introduced to specify that the data being described are of a different sort than implied by the use of the symbol r.

The phi coefficient is based upon data in which both variables are dichotomous. For example, sex of subject and passing or failing an examination might be correlated. Usually one would prefer to use the test scores in this case and apply r_{pb}; however, in some cases limited information is available or the two variables have only two meaningful levels each.

In any case, when we have only two categories on each variable, there are only four possible pairs of X,Y scores. Using a 1 and 0 score for the two levels of each variable, the 4 possible pairs, giving the X first, are 00, 01, 10, and 11. This permits us to summarize all subjects (pairs) in a 2-by-2 frequency table. Table 11–2 illustrates such a frequency count of the 4 possible pairs of scores on X and Y.

TABLE 11–2

The Frequencies Resulting from Classifying 100 Persons on the Basis of Two Dichotomous Variables

Sex of Student (Y)	Results of an Examination (X)	
	Pass (1)	Fail (0)
Women (1)	39 a	27 b
Men (0)	18 c	16 d

In Table 11–2, the entries in the different cells, each of which is lettered, give the number of students with the different X, Y pairs. The 39 students in cell a are all women who passed the examination. The scoring is arbitrarily made; a 1 for passing, and a 1 for being a woman. These 39 people, if itemized separately, would each have an X of 1 and a Y of 1. The other entries may be interpreted in an analogous manner.

The ϕ resulting from such data may be most easily defined in terms of the letters within the cells of Table 11–2. In these terms, we have

$$\phi = \frac{ad - bc}{\sqrt{(a+b)(c+d)(a+c)(b+d)}} \tag{11–8.3}$$

where the letters stand for the frequencies found in the corresponding cells. It can be seen that the denominator of (11–8.3) contains factors which are the row and column totals from the table.

Example 11–8. Find the correlation, ϕ, for the data given in Table 11–2. Use of (11–8.3) permits the value of ϕ to be found by direct substitution from the cell frequencies.

$$\phi = \frac{(39)(16) - (27)(18)}{\sqrt{(66)(34)(57)(43)}} = \frac{138}{(8.124)(5.831)(7.550)(6.557)}$$
$$= \tfrac{138}{2340} = .059$$

Since the product of ad is larger than bc, the pairs of scores 1,1 and 0,0 predominate. The positive correlation means that there is some small tendency for women to pass the test a higher proportion of the time than men.

The reader may well realize that data in the form appropriate for the ϕ coefficient are also suitable for analysis by the chi square method. The similarity is more than perceptual. A contingency χ^2 test on this type of data tests the same hypothesis that is applicable to the ϕ. If one were to test the hypothesis that the population correlation were zero, the same question is asked as is the case in a contingency chi square.

If the *contingency* chi square is calculated *without* the correction for continuity (1 *df* is present), we find that:

$$\chi^2 = N\phi^2 \tag{11–8.4}$$

or that:

$$\phi = \sqrt{\frac{\chi^2}{N}} \tag{11–8.5}$$

Equation (11–8.5) presents an alternate method of finding ϕ for this type of problem; calculate the contingency chi square as shown in Section 9–7, without the correction for continuity, and then apply (11–8.5).

A test for the significance of ϕ, i.e., a test of the hypothesis that the population value is zero, may be most easily done by testing the contingency chi square calculated on the same data. However, this χ^2 for the hypothesis test should be done with the correction for continuity applied. No example of this will be given here; reference to Chapter 9 will permit such a test to be found in detail.

Spearman's rho. The important uses of r, in techniques beyond the scope of this text, require that interval scales of one or both X and Y are present. The meaningful application of the qualities of r often is handicapped by lack of such interval-scale requirements. In r_{pb} we have seen a case in which one variable, Y, may often consist of a nominal scale; however, if X measurements form an interval scale, r_{pb} may usually be so interpreted that the mathematical properties of correlation reflect meaningful corresponding properties of the measured characteristic.

In the case of ϕ, both X and Y may be considered as nominal scales. The interpretation of ϕ is best made in terms of probabilities of an element belonging to a certain class, given that the element belongs to some other class. The chapters on probability and frequency data are relevant to this type of interpretation.

Between the nominal scale and interval scale lies a scale with intermediate properties, the ordinal scale. When the numbers assigned to a characteristic imply order by virtue of the relative size of the numbers, but not distance, we have a situation which is neither nominal nor interval. The assignment of numbers is not arbitrary as in nominal data. But by assumption, the assignment of numbers to ordinal scales is not so done that equal differences in the numbers mean equal differences in the characteristic described by the scale.

The degree of relationship shown between two ordinal scales may also be described by use of our basic product moment correlation coefficient. The statistic, when applied to such ordinal scale data, is called *rho,* or Spearman's rho, ρ. The author of this book must apologize to the reader at this point. Rho in this context is a statistic based on sample data (although a population value is possible) which are ordinal in form. This use of rho should not be confused with the population value of r, which has the same symbol. The avoidance of this ambiguity would cause as much confusion as it would save. The statistic rho is usually called by that name, and a deviation here would violate, then, the common usage.

The data appropriate for the use of rho are rank-order numbers, hence, rho is often called the rank-order correlation. It should be kept clearly in mind that the formula to be given for ρ is valid only for rank-order data on both X and Y variables. Such data usually consists of the integers one through N, for N elements. A convention usually made in rank-ordering elements is that the highest or best element is given rank 1, i.e., first place. The element numbered N is in last place or is lowest with respect to the scaled attribute.

In the case of ties, one should note that elements are tied for two or more positions, not just one position. For example, if two golf players tie "for first place" they are actually tied for first and second place. Such ties are handled by giving each element a rank number which is the average of all ranks for which the group is tied. The next lowest position is given the number which would be used if no higher elements had not been tied. For example, the golfer who finished behind the two players tied for first and second would be given rank of 3.

The ranks are given separately on X and then on Y to a set of N elements. From the pairs of ranks we may now give the formula for ρ:

$$\rho = 1 - \frac{6 \Sigma D^2}{N(N^2 - 1)} \tag{11–8.6}$$

where the D's are differences for any element between its X and Y ranks. This equation is derived once again from the basic definition of r. The simplifications made depend upon certain algebraic properties of rank numbers. For example, the mean of N rank order numbers is always known from the value of N. This mean is equal to $(N + 1)/2$. If N is 4, the scores are 1, 2, 3, and 4, the mean of which is $\frac{5}{2}$ or 2.5. Similar algebraic expressions for the standard deviation of the first N integers permit (11–8.6) to be given in place of the formula for r.

Example 11–9. The order of finish in a baseball league in two successive years is given below. What is the correlation between these two sets of ranks?

Team	Year 1	Year 2
Q	1	3
R	2	2
S	3	1
T	4	7
U	5	5
V	6	8
W	7	6
X	8	4
Y	9	9
Z	10	10

Since we have rank-order data on each variable, (11–8.6) may be used to find the rank-order correlation between the variables, years. The only value in (11–8.6) which need be calculated, except for counting the pairs, is ΣD^2. This is obtained by finding the differences between each paired X and Y, squaring and adding. Since the D's are to be squared, we may omit the minus signs. In the order of teams given, we have the differences of: 2, 0, 2, 3, 0, 2, 1, 4, 0, and 0. Squaring these and adding the squares together, we obtain

$$\Sigma D^2 = 38$$

The N of 10 may be combined with this total of 38 in (11–8.6).

$$\rho = 1 - \frac{(6)(38)}{(10)(99)} = 1 - \frac{228}{990} = .77$$

A fairly high positive correlation holds between these two sets of ranks.

The value of ρ can be interpreted directly as any correlation in terms of the relationship between the rank numbers; however, it is difficult to imagine the meaning of the variance, for instance, of ranks in terms of the attributes represented by the numbers. Even the mean of ordinal-scale data is rather meaningless, since the operation of addition on ordinal data does not represent meaningful operations on ordinal characteristics in most instances.

Another difficulty in the use of ρ is that its random sampling distribution is peculiar: it does not converge very rapidly with increased N to a normal distribution. Moreover, when tied ranks are present, rho as found by (11–8.6) is not equal to r.

For these reasons, we will not discuss the use of rho for testing a population correlation.

Kendall's tau. The statistic τ, called Kendall's *tau* is a measure of relationship between pairs of rank-order numbers. The same data appropriate for Spearman's rho may be analyzed with τ. Although rho is a product moment correlation, tau is not. Tau is based on a probability analysis of the possible orders in which elements in a set of rankings may be placed, and the probability of two elements reversing their relative positions on two different rank-order scales.

This notion of one element's being higher than another on one scale, with the second higher than the first on a second scale, is called an **inversion**. In the statistic tau, the number of inversions between two rankings is compared to the number of possible inversions. This is quite different than what is done in rho. In the calculation of rho, the difference between two ranks for a single element is *squared*. This puts a good deal more weight, in a negative way, upon certain discrepancies than is true in the tau statistic.

Since order of elements is the only feature justifiably inferred from ordinal data, the tau statistic is a more legitimate measure of relationship for such data than rho.

The computation of tau will be illustrated in two ways. The definition of tau involves the number of inversions, which we may denote as n_{inv}, and the number of possible inversions. This latter figure comes from the number of possible pairs of elements which can be tested for inversions between the scales. If we have N elements, the number of possible pairs is equal to the number of combinations of N things taken two at a time. In Section 10–6 this value was denoted as ${}_NC_2$ and is equal to $N(N - 1)/2$.

The formula for tau may now be given as

$$\tau = 1 - \frac{2n_{inv}}{{}_NC_2} \qquad (11\text{–}8.7)$$

The first method of counting inversions, n_{inv}, is a graphic method but the second method may be used with the data in the same form as the first. Let us use both methods on the same data.

Suppose that an interior decorator shows 8 designs to each of two colleagues and asks them to rank-order the designs in their order of preference. The results might be summarized as:

	Design								
	a	b	c	d	e	f	g	h	
Judge 1	3	8	7	1	4	2	5	6	(X rankings)
Judge 2	4	7	8	3	5	1	2	6	(Y rankings)

In order to count the number of inversions between rankings, the *pairs* of ranks are so rearranged that one scale, the X's let us say, is in order. Take care that pairs of ranks are rearranged and not just the rankings on a single scale. After this has been done, lines are drawn between the identical rank numbers for all ranks. This procedure gives:

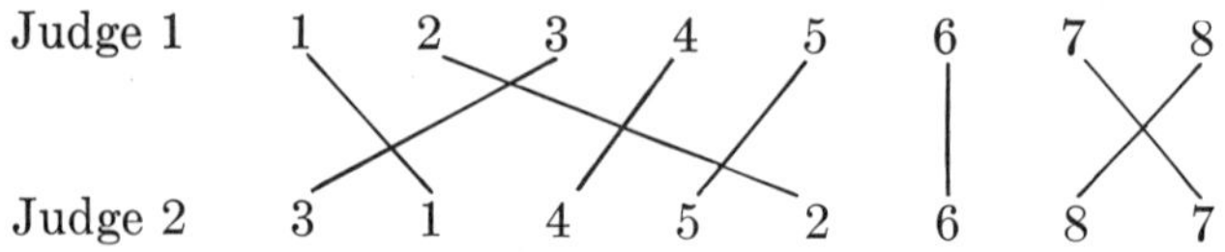

The number of inversions may now be counted by counting the number of intersections of the lines. In this case, we find that the number of inversions is 5.

Using the same display of the rankings, with one set in order, we may also count the inversions by examining the second set of ranks (judge 2). For *each* rank, count the number of ranks to the right of the one considered which are smaller. For the first rank of 3, we see that the ranks 1 and 2 are both to the right of 3; this gives a sum of 2 for the first rank. Continuing this process, we obtain a 0 for the rank 1, and a 1 for the rank 4, then the answers 1, 0, 0, and 1, in that order for the other ranks. The sum of these parts is the number of inversions, again n_{inv}, is 5.

The application of (11–8.7) will give the value of tau for these data:

$$\tau = 1 - \frac{2(5)}{{}_8C_2} = 1 - \frac{10}{(8)(7)/2} = 1 - \frac{10}{28}$$
$$= .64$$

The limits of tau are the same as for r, from plus one to minus one. But the same data will usually give different answers for tau than for rho, which is the counterpart of r for such data.

The interpretation of τ is made in terms of probability. If two randomly selected *elements* are considered, the comparison of their relative positions on X and Y rankings may yield two results, barring ties: an agreement in their relative positions, or an inversion. Let us denote the probability of an agreement as $P(A)$ and the probability of an inversion in their orders as $P(I)$. It can then be shown that:

$$P(A) - P(I) = \tau \qquad (11\text{–}8.8)$$

Since $P(A) + P(I)$ is one, assuming no ties, (11–8.8) may be solved for the probability that two random elements will be ranked in the same relative positions, $P(A)$:

$$P(A) = \frac{\tau + 1}{2} \qquad (11\text{–}8.9)$$

Using (11–8.9) to interpret our value of .64 for the data used for illustration, we find that $P(A)$ is 1.64/2 or .82. This value may be seen to be appropriate if we examine the fact that there were 5 inversions out of the 28 possible pairs. This means that $\frac{23}{28}$ is the proportion of agreements shown in these data; this is .82 as also given by (11–8.9).

The random sampling distribution of tau is approximately normal for an N of 10 or more, assuming no ties and a true population value of zero. One may test the hypothesis that the population tau is zero by use of the normally distributed value of z which is formed by finding:

$$z = \frac{\tau}{\sqrt{\dfrac{2(2N+5)}{9N(N-1)}}} \tag{11–8.10}$$

Example 11–10. Two groups of people are asked to vote for their favorite food from a list of 12 foods. Each group has 200 subjects but Group I consists of men and Group II of women. Assuming that the *first place* votes given below reflect an average *ranking* of the 12 foods, describe the degree of relationship between men and women on this set of foods.

	Foods listed											
	1	2	3	4	5	6	7	8	9	10	11	12
Men	28	9	35	19	17	12	42	10	3	14	6	5
Women	26	10	27	14	25	18	29	6	0	21	13	11

Although we assume that these totals reflect the general order of food preference, it would certainly be questionable to use the data as interval or ratio scale numbers for each group. If the data are then ordinal in nature, our first task is to rank the foods for each group. The statistics rho and tau require rank-order data, and rho may not be obtained from data not in such form, although tau can be with some difficulty. Inspection of the data from each group permits the following ranks to be assigned:

	Foods listed											
	1	2	3	4	5	6	7	8	9	10	11	12
Men	3	9	2	4	5	7	1	8	12	6	10	11
Women	3	10	2	7	4	6	1	11	12	5	8	9
Differences	0	1	0	3	1	1	0	3	0	1	2	2

The differences (unsigned) between pairs of ranks are also given so that rho may be used as one possible description of the relationship. We find from these differences, using (11–8.6):

$$\rho = 1 - \frac{6(30)}{12(143)} = 1 - \frac{180}{1{,}716}$$
$$= .895$$

which is a very high correlation.

Let us also calculate tau for these data. Placing the paired ranks so that the men's ranks are in order, we obtain:

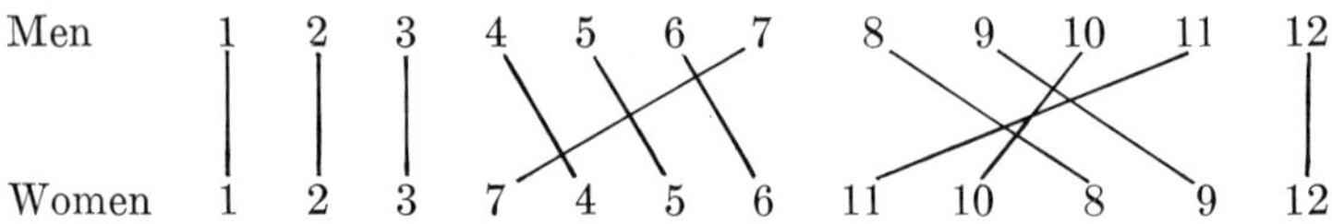

The number of intersections of the lines connecting the rank numbers is 8, which is n_{inv}. Equation (11–8.7) gives tau as:

$$\tau = 1 - \frac{2(8)}{{}_{12}C_2} = 1 - \frac{16}{\frac{132}{2}}$$
$$= .757$$

This is a lower value than rho but there is no easy method of comparing the values meaningfully.

The interpretation of this value of τ, by (11–8.9), is that about .88 is the probability of a randomly selected pair of foods showing the same relative preference for the men and women. We may well ask if this value of tau is likely to have been due to chance if the population value is actually zero. The z resulting from (11–8.10) provides a test of this hypothesis:

$$z = \frac{.757}{\sqrt{\frac{2(29)}{9(12)(11)}}} = \frac{.757}{\sqrt{.0488}} = \frac{.757}{.221}$$
$$= 3.43$$

This value of z on a normal distribution indicates that the correspondence between our two sets of ranks has only a remote probability of having occurred by chance. The hypothesis of a true zero relationship should be rejected.

11–9. Reliability and Validity. The concepts of *reliability* and *validity* are much more general than their inclusion within a discussion of correlation might imply. But a standard method of quantification of these concepts is by means of a correlation coefficient.

Roughly speaking, reliability is the degree to which one obtains the same result with a measuring device when the same entity is measured twice (or more). Validity is the degree to which a measuring device actually measures the characteristic of an entity which one is trying to measure.

Each of these general specifications must be explicitly defined in terms which allow the testing of an instrument and allow a quantitative answer for the degree of reliability and validity. A number of such definitions are found which are appropriate for different purposes related to the use of these concepts. Here, we will only examine a single definition of each concept in order to illustrate the use of correlation as a measure of reliability and validity and to show certain obvious relationships between the concepts.

Let us simply define the degree of reliability of a measuring device as the correlation between the results of measuring a representative set of entities on two different occasions. This correlation might take different forms, but we will limit our discussion to continuous measurement scales and use Pearson's r. Let us use the symbol r_r to indicate such a reliability coefficient.

By example, suppose one were interested in the reliability of an aptitude test of the pencil-and-paper type. The representative set of entities to be measured may be taken as a random sample of people from some population of interest. It is important to note that the test must actually be measuring the same amount of aptitude for each person on the two administrations. Such factors as changes in aptitude, motivation, and attitudes as well as the memory of the first administration must all be under control or assumed to be negligible in order to obtain an unbiased estimate of the reliability of the test. But ignoring these important experimental problems, our measure of reliability, statistically, would take the form of a correlation between the scores on the first administration of the test and the second. With N subjects, the N pairs of scores are correlated using the Pearson's r.

Until this measurement of reliability, we have considered a negative r to be just as good as a positive r of the same absolute value. However, it should be clear that a reliability coefficient will be positive, or randomly different from zero, unless something very strange is taking place. Certainly, a test would be considered useless if it could not be shown that the population correlation, ρ_r, were not greater than zero. In fact, as we shall note shortly, a rather high correlation of reliability is needed in order to make a single test useful in the measurement of some desired characteristic.

Let us assume that the aptitude test proves to be highly reliable. We may now ask the next crucial question. What is the test measuring? It may be presumed that a reliable test is measuring something, but one would certainly have an objective in mind for such a measuring device. Is the test measuring what one intends it to measure, and to what degree? The answer to this question requires that one have or obtain the true values of the quality to be measured for some representative set of entities. Such values are called the *criterion* values, or scores.

The definition of validity may be so made that the degree of validity of the instrument is equal to the correlation between the tests scores and the criterion scores for the entities to be considered. The estimate of the validity of our aptitude test would then come from the test scores of a random selection of people for whom the criterion scores were available. The definition and measuring of the criterion variable is a most difficult procedure in many cases. For example, most people would probably agree that no criterion for intelligence is available for finding the validity of an intelligence test.

However, once a criterion is established and measured, the correlation between a test set of scores and the paired criterion scores, denoted as r_v, may be obtained as the estimate of the validity of the test.

In the case of validity, two features might be mentioned. The validity coefficient may well be negative. For the purposes of measurement, the scale of units on the test need not be the same as on the criterion variable. A perfect negative r_v will permit one to obtain the desired criterion values from the test device. The second point is related to the first. Since a correlation is invariant under a linear transformation, the units of the test instrument may be consistently different than the criterion scale. For example, a spring scale which is set 10 lb. too high will still give a perfect positive r_v, within small errors of accuracy. Such a scale is a valid measuring device for weight, simply not weight in lb. It is the characteristic, ignoring units in the linear sense, that one is trying to measure.

The relationship between reliability and validity is most important. One may clearly see that a perfectly reliable test instrument may be measuring something completely different than one would wish. On the other hand, if a test instrument gives random results from one administration to the next, it can certainly not be used as a valid measuring device for any criterion variable. Without reliability, there can be no validity; but even with high reliability, the degree of validity may be zero.

Statistically, the relationship between reliability and validity may be stated in terms of a relationship between ρ_r and ρ_v for a given device. These population parameters are only estimated by any given sample used to study an instrument, but the population parameters are the average findings in a sample as well. It may be shown that:

$$|\rho_v| \leq \rho_r^2 \qquad (11\text{–}9.1)$$

The relationship shown in (11–9.1) may be stated: *The validity coefficient cannot exceed the square of the reliability coefficient.* Of course, this is a maximum figure for ρ_v. It is quite possible that the validity for an instrument is quite a bit less than the square of the reliability. Nonetheless, one may see that a moderately reliable test (say .70) has a *maximum* possible validity (.49 here) which is far too small for many practical purposes.

One may note the use of r^2 in this discussion. The prediction of a "true" score from a variable which is correlated as r with the true score accounts for only r^2 of the variance of the "truc" scores. A high reliability is needed for accurate prediction.

EXERCISES

1. (a) Use (11–3.11) to find the correlation between the X and Y pairs given below.

X	Y	X	Y
6	16	14	21
6	23	14	21
6	23	14	19
6	17	14	19
12	17	14	21
12	21	14	21
12	21	18	20
–	–	18	20

(b) Find each $\tilde{z}_y$, and each $\tilde{z}_y - z_y$. Now sum the squared errors in prediction and divide by N to obtain $\sigma_{U_z}{}^2$. Does your answer agree with (11–7.2)?

2. Find the correlation between X and Y for the following data, using any convenient formula.

X	Y
21.4	140
16.5	156
14.6	131
13.8	140
13.0	122
12.1	110
9.4	109
8.7	85

3. (a) For the r in Exercise 2, can you reject the hypothesis that $\rho = 0$?
   (b) Can you reject the hypothesis that the samples in exercises 1 and 2 were drawn from populations with equal correlations?
   (c) What assumptions are made for these tests?
4. If $r = -.65$, and $s_y = 101$, what would be σ_U^2 when predicting Y from X?
5. Calculate r_{pb} for the following data:

   Men's scores: 18, 24, 29, 30, 32, 33, and 35
   Women's scores: 10, 12, 13, 14, and 16
6. Find the value of ϕ and assess its significance:

	Pass	Fail
Young	62	28
Old	41	57

7. Find both rho and tau for the following ordinal data:

X	Y
27	121
16	150
14	180
15	175
38	62
72	20
6	200
18	170

Applications: Use information from all previous chapters where relevant.

8. Two groups of subjects are asked to rank-order 8 pictures of automobiles in terms of their appearance. Each person first ranks a set of 8 black-and-white pictures and then another set of 8 pictures of the same automobiles taken in color.

			Pictures							
Group	Subject	Ranking	A	B	C	D	E	F	G	H
I	1	1	1	7	4	3	5	6	8	2
		2	2	8	4	3	6	5	7	1
	2	1	3	1	4	6	5	8	2	7
		2	3	1	2	8	5	6	4	7
	3	1	1	2	8	5	3	6	4	7
		2	1	2	3	4	5	6	7	8
	4	1	2	3	7	5	6	4	1	8
		2	2	8	3	4	6	5	1	7

II	5	1	2	1	4	3	8	5	6	7
		2	2	1	3	4	6	5	7	8
	6	1	3	1	2	4	7	6	8	5
		2	3	1	2	4	7	6	8	5
	7	1	3	2	1	5	6	4	7	8
		2	3	1	2	5	4	6	7	8
	8	1	1	3	2	5	8	4	6	7
		2	1	3	2	5	6	4	7	8

The two groups differ in that Group II is told to disregard the color in the pictures before their second rankings. Given above are the two sets of rank orders for 4 subjects in each of the conditions. On the basis of these few subjects, can you conclude that the groups differ in the degree of consistency between rankings? (Hint: first develop a single measure of consistency or inconsistency for each subject.)

9. An experiment on verbal learning is conducted in which subjects are given various weights to lift while learning a list of paired associates. This learning task requires a person to learn a different response to each of a number of stimuli. The weights were used in an attempt to increase level of tension within subjects to variable degrees. It was planned that six different weights would be lifted in the various experimental treatments; however, four sets of these weights were used by different experimenters who worked together on the experiment, and it was later discovered that the weights were not as thought.

   The actual weight lifted (in pounds) for each subject is given below along with the number of trials needed to master the learning task. Did the lifted weights have a significant effect on learning?

Subject	Weight	Trials	Subject	Weight	Trials
1	1.2	27	13	4.0	22
2	3.9	19	14	14.0	15
3	2.0	20	15	2.0	40
4	0.8	31	16	10.0	15
5	6.7	15	17	7.0	19
6	14.2	13	18	13.2	8
7	9.5	17	19	1.0	21
8	0.9	22	20	10.4	12
9	9.9	10	21	3.8	19
10	7.2	9	22	7.6	18
11	13.7	11	23	4.1	26
12	2.2	20	24	1.9	32

10. An investigator needs a test that will predict at least 40 per cent of the variance of criterion scores. He develops a test which he hopes will do this. First, he administers the test twice to a group of 100 subjects in order to estimate its reliability. The $r_r = .70$; should he proceed with a validation of the test?

12

General Principles and Exceptions

In this chapter, we shall note several kinds of statistical considerations which have been omitted from our previous discussions. These problems range from general relationships among various statistical tests to conventions used in presenting data and the results of statistical tests. In addition, a brief discussion of *nonparametric* methods will be presented. Such methods form the basis of hypothesis testing when the more usual methods cannot be used because of violations of underlying assumptions.

12–1. Descriptive and Computational Considerations. **Data processing.** It may reasonably be claimed that all measurements of real things are approximations. The limitations of a measuring instrument allow only a certain number of digits to be given as an accurate indication of the "true" value for the entity measured. Of course, certain instruments are far better than others, but the fact remains that a precisely accurate measurement (ignoring frequency data) may not be made with any instrument. There are two basic problems which arise from such limitations of measurement. One of these is directly based upon the degree of accuracy of the basic measurement, and is called the significant-figure problem. The second problem, which we will consider first, has to do with the rounding of numbers.

There are two widely used methods of rounding numbers. Each of them may be illustrated by considering that one has one more digit than is desired for purposes of computation; but each is also relevant when a basic measurement is half way between two possible digits, one of which is to be chosen for the last digit in the measurement.

The first method in common usage is to *increase an odd number in the next to last position* if the final number is 5 or greater, and to *keep the next to last digit if it is an even number* when the last digit is 5 or less. Thus, 665.5 becomes 666. and 664.5 becomes 664. The other common method is to round up for all values ending in 5. In the examples above, this method would give 666. and 665.

Each method has problems inherent in its use. In the first case, the mean of the rounded scores is an unbiased estimate of the mean of the raw values, *assuming that the second to last position has an equal number of odd and even digits.* The second method, under the same assumption, will produce a biased mean, the degree of the bias being $\frac{1}{20}$ of the value of the next to last position.

However, the first method will yield a distribution which is not as representative of the shape of the raw scores as will the second. For example, if the numbers 00 through 99 are equally likely, the first method of rounding would give 9 tens, 11 twenties, 9 thirties, 11 forties, etc. while the second method would give 10 of each of the above rounded values. Moreover, this irregularity in the first method produces a variance of the rounded numbers which is not as close to the variance of the raw values as is the variance obtained from the second method.

Neither of these two methods can be recommended for all purposes. However, for most purposes involving statistical testing in the behavioral sciences, either method will be suitable unless the rounding produces only 1 or 2 digits in the resulting data.

The best method of rounding numbers would be to randomly round up or down when a number ends with a 5. Such a laborious procedure is not common in data processing. Another possible solution is to save the extra digit when a 5 occurs in the last position. This would produce some data which are more exact than other data, which again produces complications in the estimates of errors in measurement.

No matter which method of rounding is used, the most important problem in processing data relates to the number of *significant figures* in the answers resulting from such data. Although some text books include a statement that one can have no more significant (meaningful) figures in an answer than are present in the raw data, one finds that many statistical calculations can produce answers which are more exact than the raw data.

One times one is one, or is it? If the number 1. is multiplied by 1., the result is not necessarily equal to 1.0000 . . . . By convention, the number 1. is not as accurate as the number 1.0. The distinction here has to do with the number of significant figures. If a measuring device is of sufficient accuracy that one can distinguish between the tenths position in a reading, then the number above would be written as 1.0, indicating that the answer is closer to 1.0 than 0.9 or 1.1. However, the number 1. signifies that the measurement is closer to 1 than to 0 or 2.

Let us assume that our measurements are correct to the closest unit, i.e., the number just before the decimal point is the last significant figure. The decimal points can merely be moved in all results in order to convert to some other degree of accuracy. In this case, the number 1. may actually be any value between 0.5 and 1.5. Let us assume that the 11 discrete possibilities in this set are equally likely. This assumption will permit an *approximate* solution to the significant figures problem.

The multiplication of 1. by 1. may now be written as $(1 + e_1)(1 + e_2)$ where we assume that the two 1's are separate measures and that each e may take on any discrete value between $-.5$ and .5. We will also assume that all of the errors in measurement are uncorrelated with one another. We then find 121 possible permutations for the multiplication of 1. by 1., each of the 11 first values may be paired with each of the 11 second values. An examination of the outcomes of these 121 possible permutations, which are equally likely, shows that the mean outcome is 1.00 but the range is from .25 to 2.25. In the general case, this scheme yields a distribution of answers with a standard deviation which is dependent upon the values multiplied together.

Let us denote the first number as X and the second as Y, keeping in mind that the units position is the final significant digit of each. The mean of the possible results is simply XY; however, the standard deviation of the answer turns out to be

$$\sigma_{xy} = \sqrt{.1(X^2 + Y^2) + .01}$$

Although the possible answers are not normally distributed, a rough range of likely values may be obtained by adding and subtracting the standard deviation to the mean. For 1. times 1. we find the range so formed to be .542 to 1.458. This clearly indicates that only the first digit, 1, in the answer is meaningful. As another example, consider the multiplication of 56. by 78. The answer of 4,368. contains 4 digits, but only the first 2 of these (rounded off) are reliable. The standard deviation of possible values is about 64. so that our $1\sigma_{xy}$ range is from 4,304. to 4,431.

This kind of reasoning illustrates why the statement is often made that the number of significant figures in the answer is no more than the number found in the data. However, to advocate such a position for statistical computations would be to ignore all of the theory of statistics we have been studying. For instance, let us examine the degree of accuracy which may be claimed for the mean of N numbers.

We assume that the units position in the raw data contains the last significant digit. Let us denote a "true" value, carried to an additional significant figure, with the subscript T. A true score is then

$$X_T = X + e$$

where X is our raw score and e is an error which may vary from .5 to $-.5$. The average value of e is zero, if all values are equally likely, and the variance of the e's is found to be equal to .1 by squaring and adding each of the 11 possible values of e and converting to the variance. The σ_e is then equal to the square root of .1, about .316.

The true mean of the N numbers of a set may then be expressed as

$$M_T = \frac{\Sigma X_T}{N} = \frac{\Sigma X}{N} + \frac{\Sigma e}{N} = M + M_e$$

The mean error is zero, but we wish to find the variability of M_e, so that the number of significant figures in M may be established, statistically speaking.

If N is reasonably large, the central-limit theorem indicates that the distribution of M_e will be normal and that the standard deviation of this distribution is equal to

$$\sigma_{M_e} = \frac{\sigma_e}{\sqrt{N}} \qquad (12\text{–}1.1)$$

which is $(.316)/\sqrt{N}$.

We may now form a 95 per cent confidence interval for estimating the M_T:

$$M_T = M \pm 1.96\,\sigma_{M_e} = M \pm \frac{.62}{\sqrt{N}} \qquad (12\text{–}1.2)$$

To illustrate (12–1.2), assume that the mean of 10 scores is calculated as 14.5000. The application of (12–1.2) shows that the true value is probably covered by the range 14.500 $\pm$ .196, which is 14.696 to 14.304.

We find with 10 scores that the mean is accurate within the number of significant figures in the raw data but the next digit has only a slightly restricted range of possible values.

If one were to increase the number of scores to 100 or to 1,000, the 95 per cent confidence interval would shrink to 14.562 through 14.438, or to 14.5196 through 14.4804, respectively. Clearly, the number of significant digits in a sample mean depends upon the number of scores in the sample. It is quite possible to estimate a mean to 6 significant figures with data containing only 3 significant figures, if one wishes to obtain a sufficiently large sample.

Under the same set of assumptions, a more involved analysis of the sample variance, s^2, shows that the true variance, if another significant figure were available, would be equal to about $s^2 + .1$, on the average. This average figure indicates that an obtained sample variance is slightly too small, which occurs because a single value, such as 11., is used rather than a number of more exact values within the range 10.5 to 11.5. This is similar to finding a sum of squares between groups and ignoring the small contribution from the within-groups variation.

The accuracy of this average value may also be estimated. Approximately, for large N, the above assumptions lead to a standard deviation for the variance, around its mean of $s^2 + .1$, which is about

$$\sigma_{s^2} \approx \sqrt{\frac{4s^2}{10(N-1)}} \qquad (12\text{–}1.3)$$

In this case, the accuracy of the variance depends upon both the sample size and the observed variance. One should note that the variance of a set of numbers measures the dispersion of the numbers and not their absolute value, hence the number of significant figures in a variance is not really dependent upon the number of significant figures in the raw scores but upon the number of such figures available in deviation scores. For example, a set of 100 scores with 12 significant figures each (an unrealistic accuracy) having a sample variance of only 1.44 has a 95 per cent confidence interval of about $1.54 \pm .143$, by use of (12–1.3). Clearly, only 2 significant figures may be claimed at the most. However, if a sample of 100 scores of only 3 significant figures has a variance of 10,000, the true variance estimate of 10,000.1 forms the basis for a 95 per cent confidence interval going from 9,987.4 to 10,012.8. This shows about as many significant digits in the answer as in the raw scores.

Although the problem of significant digits is complex, the realistic application of statistics, especially involving hypothesis testing, permits one to use approximate rules of procedure. By carrying computations out 2 or 3 decimal places beyond the degree of accuracy in the raw data, an answer will be obtained which is realistic enough for use in the evaluation of a test statistic. The tables involved in this application rarely have more than 3 digits, hence, answers will be rounded anyway prior to entering such tables.

However, it must be clearly stated that raw data must have a sufficient number of significant figures to provide answers to the accuracy needed for the statistical tables. For example, attempting to perform an analysis of variance on small samples in which deviation scores from the grand mean have only one or two significant figures may well lead to results which are simply not valid. In such cases, one would have to calculate the range of "true" values in the answer and be certain that these did not overlap the critical values for a significance test.

Describing results. The presentation of statistical data requires one to compromise between conciseness and completeness. The selection of the statistics to be reported should be so done as to inform the reader and not to misinform him or to hide irregularities in the data.

Often the mean is used as the basic measure of central tendency; however, when skewed data are involved, it is necessary to report this fact. In such cases, the median and mode may also be given for descriptive purposes.

When an entire distribution is to be reported, it is common to use a *grouped frequency distribution*. Such a distribution presents the frequencies of scores within class intervals rather than for each possible score value. For descriptive purposes, one may select class intervals according to two criteria. First, an attempt is made to have between 10 and 20 classes; and second, the size of each interval should be a multiple of 1, 2, 3, 4, or 5. The size may be roughly calculated by dividing the range of scores by 15 and then adjusting the number and size of intervals until some reasonable compromise is reached.

Such distributions, as well as other descriptive statistics, may be presented graphically or by means of tables. In either case, the presentation should be made with titles and clearly labelled axes or rows and columns. A minor convention employed in graphing data may be mentioned here; the vertical axis is usually constructed so as to be about $\frac{3}{5}$ as high as the x-axis is long.

A more important feature of graphic presentation of data is that discrete variables usually are shown by means of bar graphs, whereas continous variables, with respect to the x-axis, are usually shown by use of a polygon. These methods emphasize the fact that values between the data points either cannot, or can, take place, depending upon the nature of the variable represented.

A common method of reporting scores is that of *percentiles* (or *centiles*). Such values simply give the relative position of a score value within some set of scores. They are quite similar to ranks when used to describe the position of a person's score within some set. For example, if a person *exceeds* 65 of 260 scores in some distribution, we can say that he falls at the 25th percentile, i.e., $(\frac{65}{260})$ times 100. The proportion of scores exceeded by a given value is multiplied by 100 to obtain a percentile.

A minor point, strictly speaking a person cannot attain a score at the 100th percentile within his own group, because he would have to exceed his own score at the same time.

Another common method of reporting data is by means of proportions. We have used this method frequently and little need be said about its descriptive use. However, statistics are often performed on proportions and many text books abound with special formulas for use with proportions. Such formulas are not given in our treatment of statistics for the simple reason that the same results are easily obtained by converting such data to the raw frequency form. If special formulas are sought for a given computation, the student quite directly may modify the formulas given in this text. For example, in Chapter 10 it was noted that the standard deviation of the binomial distribution is equal to $\sqrt{npq}$. This equation gives the standard deviation of the total number of one of the two possible events A and B. If one wished to find the standard deviation of the proportion of events of type A, the conversion to proportions is done quickly (since $P_A = f_A/n$):

$$\sigma_A = \sqrt{npq}$$

$$\sigma_{p_A} = \frac{\sigma_A}{n} = \frac{\sqrt{npq}}{n} = \sqrt{\frac{npq}{n^2}} = \sqrt{\frac{pq}{n}}$$

The use of proportions is facilitated by such special manipulations of basic equations, but one can simply work with the data in raw frequency form. Either method will result in the same statistical conclusions. The use of proportions for *reporting* data is sometimes more convenient than reporting data in raw frequency form.

In reading or writing a research report, certain conventions are of importance to the statistical content. Ordinarily, the recording of basic data and some preliminary manipulations on such data are reported in the *Method* or *Procedure* section of a research paper. The next section of such a report is typically called the *Results* section. Within a well-organized Results section, descriptive statistics, including tables and figures, are found first. These will include the basic measures of central tendency or other summarizing statistics which form the basis for hypothesis tests and estimation procedures. Then the details of methods used for hypothesis testing and the results of such testing are reported. The statistical interpretation of results is found in this section, but the experimental interpretation of the meaning of the results in the context of the experiment is usually reserved for the *Discussion* section.

This brief outline will aid the reading of experimental reports in so far as the statistical procedures are concerned.

Approximations. There are two main types of approximations used in statistical methods. The first of these is the sort in which an approximation is used for convenience in computation. For example, a rough estimate of a standard deviation of normally distributed data may be obtained by simply dividing the range of scores by 5 or 6, depending upon how large the sample is. Clearly, the approximation formed is rather gross, but the method is certainly convenient.

A very common method of finding means and standard deviations is illustrated by the use of grouped frequency distributions and coded scores. In such an approximation method, each class of scores is assigned a value equal to the middle of the interval of scores involved. All scores within such an interval are assumed to have this single midpoint value. Then the scale of midpoint values is reduced to a scale of integers by a linear transformation. For instance, if the intervals are 10 units wide, the division of each midpoint score by 10, and the possible addition of a constant, will produce a simple scale of numbers for computing a mean and standard deviation. Once the resulting computations on these coded scores are completed, the answers are transformed back to the original scale. However, the calculation of the standard deviation from such grouped data tends to be in error. If the scores are normally distributed, the assumption that each score in an interval has the midpoint value tends to produce a standard deviation which is too large. A correction for this tendency called Sheppard's correction is often applied. One should note that this is not the same type of problem relevant to (12–1.3), because the errors in grouping are not distributed rectangularly in a normal set of scores whereas errors of measurement were assumed to have a rectangular distribution.

The main point of this discussion is to convince the reader that approximation methods used for the sake of convenience should not be employed. The sacrifice of accuracy in such methods is no longer justified by the savings in time. Today, desk calculators are available to most persons. Moreover, high-speed digital computors are available to a large number of people working with statistical problems. In the opinion of the author, such devices make approximation methods, and certainly computational "tricks," obsolete. The residual usefulness of approximation formulas inherent to the analysis of incomplete data, which is sometimes the only sort available, is better discussed in a more advanced treatment of statistical methods.

The second type of approximation used in statistics involves the estimation of theoretical values of a statistic which are not directly known. Again, many of these approximations are used for convenience in lieu of more exact values which could be obtained. It is to be hoped that these uses will someday be replaced by the more exact methods available; however, in some instances, approximation formulas are used because nothing better is known.

For example, in certain types of mathematical theories of learning, statistical models are devised to predict performance of subjects in a certain kind of experimental situation. Some of these models are so complex that the *expected* values for the variance of performance at some point in the experiment have not been found. A comparison of the predictions of the theory and the data obtained would require some estimate of this parameter for proper evaluation. Approximation methods are often used to find minimum and maximum possible values for the parameter. Also, the statistical value may be estimated by simulation of the model on a high-speed computer. This is quite similar to the development of an empirical random sampling distribution of a statistic as illustrated in Chapter 4.

In general, it is better to have answers which are more precise than needed than to have the converse situation. It is a simple matter to round off a final F-ratio. Quite another tale is told if one finds the raw data rounded to a single significant digit.

12–2. Estimation of Parameters. Properties of good estimators. The only characteristic of a sample statistic used to estimate a population parameter which we have discussed thus far is that of unbiasedness. But other qualities are also desirable for such estimators. Four properties of estimators are usually considered by statisticians. A "good" estimator may be judged by whether or not (or to what degree) it is:

1. Unbiased
2. Consistent
3. Efficient
4. Sufficient

Before discussion of these properties it is well to state that a single estimator of a population value need not possess all four of these characteristics in order to be very useful.

As discussed in Chapter 5, an estimator, $\hat{S}_*$, of S_* is said to be *unbiased* if the mean of the $RSD(\hat{S}_*)$ is equal to S_*. In particular, we have found the sample mean to always be an unbiased estimator of μ. However, even the first score selected in a random sample will be an unbiased estimate of μ. Why use the mean if a single score will do as well? The answer is that the single score will not do as well, in other important respects.

An estimator is *consistent* if the values of $\hat{S}_*$ become closer to the value of S_* as the sample size increases. We have found this characteristic to be true for the mean; the standard error of the mean decreases as the sample size increases. Note that a single score from a sample will not get better, on the average, in estimating μ as the sample size increases. Such an estimator may be *unbiased,* but it is not *consistent.*

An *efficient* estimator is one which has a small sampling error, i.e., one which has a small variance (or standard deviation) for its random sampling distribution. The term small is relative, of course, and efficiency of an estimator is usually measured with respect to some other estimator for the same parameter. For example, the efficiency of $\hat{S}_*$ compared to $\hat{S}_*'$ may be given as:

$$\text{relative efficiency of } \hat{S}_* = \frac{\sigma_{\hat{S}_*'}^2}{\sigma_{\hat{S}_*}^2} \qquad (12\text{–}2.1)$$

where the variance of the respective random sampling distribution of each estimator is used. If $\hat{S}_*$ is more efficient than $\hat{S}_*'$, (12–2.1) will yield an answer greater than 1. The use of (12–2.1) is restricted to estimators which are both unbiased. Clearly, an estimator could have a very small standard error but be consistently far from the value of the parameter. An equation similar to (12–2.1) can be used for any estimator, biased or not, if the variance in the errors in prediction were used rather than the variance of the estimator.

The concept of *sufficient* estimators is somewhat more difficult to explain than the other characteristics. Basically, a *sufficient* estimator is one which uses all of the information in the sample which is of relevance to the parameter value. Clearly, a single score used as an estimate of the population mean fails to use all of the relevant information, the other scores, and is not a sufficient estimator. Another way of stating this concept is a kind of negative definition. An estimator is not sufficient if it can be improved by use of aspects of the sample not used in the estimator.

Certain features of the mean as an estimator of μ depend upon the population characteristics. However, in all cases the mean is unbiased and consistent. Its efficiency is usually high with respect to other estimators, but for some population distributions such as the rectangular distribution, other estimators, such as the median, are more efficient than the mean. The mean is usually considered to be a sufficient estimator as well.

Sometimes the term *best estimator* is employed. This term permits different definitions, as one might expect; however, a common definition of a best estimator, $\hat{S}_*$, of the parameter S_* is given as

$$\sigma_{\hat{S}_*-S_*}{}^2 \leq \sigma_{\hat{S}_*'-S_*}{}^2, \quad \text{for all } \hat{S}_*' \tag{12–2.2}$$

If (12–2.2), which compares the variances in errors of estimation, is true for any other estimator, $\hat{S}_*'$, then $\hat{S}_*$ is said to be the best estimator. It is possible, in special cases, that the best estimator for a given sample size is actually biased. However, a biased estimate that is usually quite close to the parameter value is often more useful than one which is unbiased but gives quite variable estimates.

Maximum-likelihood estimators. A certain method of developing an estimator for a given case defines a class of estimators called *maximum-likelihood* estimators. These estimators are found by examining possible values of the parameter in question and finding the value of S_* which makes the results of a given sample of scores more likely than would any other value of S_*. In our Question I procedures, we noted that the probability of some particular sample outcome could be found, if knowledge of the population were given. Each possible value for a population parameter would give a different probability for some given sample outcome. The maximum-likelihood technique determines which possible population value would make this probability greatest.

For example, assume that in sampling from a population with two types of elements, with replacement, we find that 3 of 4 elements are of type A. This is a binomial-type problem, and the probability of this particular outcome is (see 10–8.3)

$$P(3\ A\text{'s}) = 4\ p^3q \tag{12–2.3}$$

We wish to obtain the maximum-likelihood estimate of p. The value of q is always $1 - p$, so p is the only parameter for this distribution. One may see that different values of p give different probabilities for our given event when (12–2.3) is used. We will not show the general method of solution here as calculus is used. However, the three values for p, .70, .75, and .80, will be examined. Using these three possible values for p, one would find the probabilities of our event to be .4116, .4219, and .4096, respectively. Not only is our probability highest for .75 compared to the other two possible values of p, but one would find that any other value of p would lead to a smaller probability for 3 A's and 1 B than does $p = .75$. Therefore, .75 is the maximum-likelihood estimate of p.

Certain general features of maximum-likelihood estimators make them very common in estimation methods. First, a maximum-likelihood estimator may usually be developed. Second, such an estimator is usually a best estimator, (12–2.2), or close to it. Moreover, such an estimator is always consistent, efficient, and sufficient; however, it may be biased. A final characteristic is that the random sampling distribution of a maximum-likelihood estimator typically approaches a normal distribution as the sample size increases. This permits straightforward tests of hypotheses and formation of confidence intervals for estimating a parameter. Although the general method of finding a maximum-likelihood estimate is beyond the scope of this book, more advanced treatments of statistics usually emphasize this sort of estimator.

12–3. Relationships Between Statistics. In previous chapters we have dealt with four important sampling distributions pertinent to hypothesis testing in particular. These are the normal distribution (in z-score form), and those for chi square, t, and F. These distributions are closely related to one another.

A normally distributed z may be written as

$$z = \frac{X - \mu}{\sigma} \qquad (12\text{–}3.1)$$

Notice that the square of such a randomly selected z-score defines the chi square distribution, with 1 df:

$$z^2 = \chi^2 = \frac{(X - \mu)^2}{\sigma^2} \qquad (12\text{–}3.2)$$

The distribution of chi square is formed by adding together a certain number of randomly selected squared z-scores from the normal distribution.

Although these two distributions are different, they are similar to one another in that the population variance and standard deviation must be known rather than estimated. In the case of the normal deviate, a deviation score is used, but for chi square a squared deviation is employed.

A similar relationship holds between t and F, with 1 df in the numerator. The value of t is equal to a deviation score divided by an estimated standard deviation:

$$t = \frac{X - \mu}{\hat{\sigma}} \qquad (12\text{–}3.3)$$

Squaring the terms in (12–3.3) gives

$$t^2 = \frac{(X - \mu)^2}{\hat{\sigma}^2} = F\,(1, N - 1\ df) \qquad (12\text{–}3.4)$$

Although we have discussed the fact that a squared t is equal to F with 1 df in the numerator, note that $(X - \mu)^2$ is equal to a sort of variance because a single squared deviation divided by 1 is still the squared deviation. The denominator is clearly an estimated variance. These two statistics are different in that one employs a deviation score while the other involves a squared deviation score. The t and F statistics are similar in that each requires an estimated measure of dispersion as a denominator.

Now note the similarity of z and t. Each has a deviation score in the numerator, but the z-score has a known standard deviation in the denominator while the t has an estimated standard deviation. The same relationship holds between χ^2 and F. The numerators have squared, normally distributed deviates, while the denominator of χ^2 has a known variance and the denominator of F has an estimated variance.

Table 12–1 summarizes these similarities and differences. One may note that the normal distribution plays a basic role in each statistic, because the numerators of all four are based upon deviations which are presumed to be normally distributed. Also, the t and χ^2 distributions become normal with large samples.

TABLE 12–1

Relationships Between Statistics

Denominators	Numerators	
	Deviations	Squared deviations
Known dispersion	normal z	χ^2
Estimated dispersion	t	F

In cases where the population dispersion is known or assumed, either a z-score or chi square is typically used for hypothesis testing. For instance, in probability problems, such as the binomial type, the assumption of a population mean, p, is sufficient to specify the standard deviation of the binomial. As shown in Chapter 9, either chi square or a z-score may be used to test such an assumed population mean. However, when a population variance or standard deviation must be estimated, the distributions of t and F result.

In general, differences between two parameters from different populations are tested by use of a z or t distribution. We have seen that both chi square and F permit the simultaneous testing of more than two population parameters, although each could also be used in place of its counterpart, z or t.

The algebraic relationship between t and F disappears when the F has more than 1 *df* in the numerator. But the F distribution is still intimately related to chi square. If we form two independent values of chi square with v_1 and v_2 degrees of freedom, respectively, the F distribution is given by:

$$F(v_1, v_2) = \frac{\chi_1^2/v_1}{\chi_2^2/v_2} \qquad (12\text{–}3.5)$$

This equation actually forms the basis for the theoretical analysis of the distribution of F. The requirement that the two chi squares be *independent* of one another is at the heart of the assumption that the populations be normally distributed for analysis of variance.

Other relationships between statistics have been shown in our previous work. For example, in the last chapter we found that the phi coefficient was closely related to chi square. In more general terms, we have noted attempts to form normally distributed variables so as to permit simple hypothesis testing and interval estimation. The transformation of r to Fisher's Z was used with this goal in mind.

Many other interesting relationships between statistics might be given, but these few are offered in the hope that some degree of closure may be obtained within this introduction to statistics.

12–4. Nonparametric Statistics. A population without parameters is rather meaningless. The term *nonparametric statistics* does not strictly mean that no parameters are present in a population of scores, but is meant to stress the fact that few parameters, if any, need to be estimated in making a statistical test.

Usually, the use of nonparametric tests is required by abnormal types of distributions of scores. In other cases, the kind of elements in a population are such that nonparametric tests are the only type really appropriate. An important feature of such tests is that they test for different distributions of scores in two or more populations, *without assuming that any particular kind of distribution is present.* For example, the chi square statistic applied to frequency data is often considered to be nonparametric, because the scores in the population are simply elements of different kinds. The single population parameter is the proportion of elements of each type.

The development of nonparametric tests usually rests upon a probability analysis of ordinal relationships among numbers drawn from the population. This means that the variability of the scores in the population is actually ignored. For example, Kendall's tau is a good example of this method. The number of inversions between items on two rank orders is compared to the number of possible inversions; no estimate of population or sample dispersion is made. In such cases, the median is taken as the measure of central tendency rather than the mean, which is meaningless for ordinal data.

One may understand this nonparametric approach by considering two random samples of 10 scores each. If we use the median of the entire 20 scores and count the number of scores in each group above and below the median value, then a probability analysis may be used to test the assumption that the two sets are drawn from the same population. For example, suppose that one group had 8 of its 10 scores above the median while the second had only 2 scores above the median. The chance of a score being above the median is assumed to be .5 for each group. One could now calculate the probability of the outcome obtained, or one worse, and test the common population hypothesis. Notice that this type of analysis uses the probability of scores falling into different classes, but it ignores the variance of the scores themselves.

The advantages of nonparametric tests are that particular distributions of population scores need not be assumed, and the scales of numbers may well be ordinal rather than interval or ratio.

However, the indiscriminate use of such tests when parametric tests might be used is not wise. The disadvantages of these tests are many: they are not usually as sensitive to differences in population means as are their parametric counterparts; the hypothesis under test is not directed specifically to differences in means but only to differences in distributions between populations; and, since no parameters are assumed, parameter estimation is usually impossible with these procedures.

There are exceptions to these disadvantages, but, as a rule, parametric tests are more useful than nonparametric tests if the proper assumptions are true or nearly so.

Tchebycheff's inequality. A most useful property of numbers was proved by the Russian mathematician Tchebycheff. It will be presented here, because it concerns the proportion of scores found beyond a given z-score value, *no matter what the shape of the distribution might be.* For this reason, Tchebycheff's inequality, as the property is called, allows certain conclusions to be made without the need to specify the shape of a distribution. This is one of the essential properties of nonparametric tests.

If any set of scores is converted to z-scores (not necessarily normally distributed), then the proportion of z-scores which have absolute values greater than or equal to some number k is *always* less than or equal to $1/k^2$. Let us express such a proportion as a probability, assuming random selection. Then,

$$P(|z| \geq k) \leq \frac{1}{k^2} \qquad (12\text{–}4.1)$$

For example, if k is 2, we find that the proportion of scores in any distribution with z-scores of 2 or greater, in absolute value, must be less than or equal to $\frac{1}{4}$. This is far more than would be true for a normal distribution, but it is certainly useful to have some *maximum* figure available for the "tails" of all distributions.

Equation (12–4.1) can easily be converted to raw-score terms as well. When this is done, the result might be written as

$$P(|X - \mu| \geq m) \leq \frac{\sigma^2}{m^2} \qquad (12\text{–}4.2)$$

The expression in (12–4.2) is a restatement of (12–4.1), but it shows that the mean and variance of the set of scores must be known. Given this information, but none about the shape of the distribution, (12–4.2) is always true: a property of mathematics is involved, not really a statistical inference, except for the change from a proportion to a probability. For example, even if I.Q. scores were not normally distributed, knowledge that the mean is 100 and σ is 15 would allow one to conclude that the proportion of I.Q.'s 55 or less and 145 or more is less than .11!

$$P(|X - \mu| \geq 45) \leq \frac{\sigma^2}{m^2} = \frac{225}{2025} = .11$$

This inequality is not offered as a nonparametric test, but it does illustrate that certain conclusions may be drawn without assuming a particular distribution of scores.

Numerous nonparametric tests are available in the literature, but we will examine only one of these, the Mann-Whitney U test, which is both simple and characteristic of this class of nonparametric tests.

The Mann-Whitney U test. This test is designed to test the hypothesis that two *independent* random samples of scores were both drawn from the same population, or, one might say, from two identical populations. We must assume only an ordinal scale of measurement for the scores and nothing else!

The first step in this analysis is to rank all of the scores. If N_1 is the number of scores in group 1 and N_2 the number in group 2, the total number of $N_1 + N_2$ are ranked. The rank numbers will then run from 1 to $(N_1 + N_2)$. In this test it is essential that the *lowest score* be given the *rank* of *one*, and so forth, with the highest score given $(N_1 + N_2)$. Ties are dealt with as discussed in Chapter 11 (see Section 11–8). The ranks corresponding to the scores are used in the analysis, not the raw scores themselves. Next, select group 1 and add up the ranks within that group; we shall call this total T_1.

The statistic U is based on the value of T_1 and the N's involved:

$$U = N_1N_2 + \frac{N_1(N_1 + 1)}{2} - T_1 \qquad (12\text{–}4.3)$$

The average, or expected, value of U, if the hypothesis is correct, will be denoted as $E(U)$, the expected value of U, which is

$$E(U) = \frac{N_1N_2}{2} \qquad (12\text{–}4.4)$$

For samples in which each group has 10 or more scores (20 is better), the random sampling distribution of U is approximately normal. The standard error of U in such cases is based upon a probability analysis of all possible rank orders within each group. The solution is

$$\sigma_U = \sqrt{\frac{N_1N_2(N_1 + N_2 + 1)}{12}} \qquad (12\text{–}4.5)$$

We may now form a z-score which is approximately normally distributed if the groups are both drawn from the same population. The significance of z may be assessed by use of the normal-curve table. The .05 level of confidence for rejection requires, as usual, a z of 1.96:

$$z = \frac{U - E(U)}{\sigma_U} \qquad (12\text{–}4.6)$$

Example 12–1. An investigator asks 26 randomly selected students from a large class in sociology to estimate the average height, in *centimeters*, of Japanese men. His group contains 12 men and 14 women and their estimates are given below. Do the populations of men and women differ in this estimation task?

Men: 140, 148, 155, 175, 180, 220, 225, 230, 250, 387, 609, 700
Women: 72, 88, 99, 125, 140, 150, 185, 195, 210, 240, 248, 250, 280, 300

Although one might first think of a variance test for the means of the two groups, there are two reasons why the Mann-Whitney U might be preferred. First, the question asks about differences between the populations, not just mean differences, or differences in variances, which can also be tested with F. The second reason is the high degree of skewness in the sample data. Plotting each group of scores would reveal that the men show a decided positive skewness while the women show a negative skewness. This particular case is one in which the F-test may give erroneous results.

To perform the U test on these data, we must first rank all 26 scores, starting with rank 1 for the lowest score. This step will not be illustrated, but the results, broken into men and women, are found to be:

Men: 5.5, 7, 9, 10, 11, 15, 16, 17, 20.5, 24, 25, 26
Women: 1, 2, 3, 4, 5.5, 8, 12, 13, 14, 18, 19, 20.5, 22, 23

To find the value of U, let us use the men's ranks. The sum of these ranks, T_1, is equal to 186. Using (12–4.3) we find the value of U:

$$U = (12)(14) + \frac{(12)(13)}{2} - 186$$
$$= 60$$

The expected value of U is given by (12–4.4) and is (12)(14)/2 or 84. The standard error of U from (12–4.5) is 19.44:

$$\sigma_U = \sqrt{\frac{(12)(14)(27)}{12}} = \sqrt{378} = 19.44$$

Finally, the test value of z is formed by use of (12–4.6):

$$z = \frac{U - E(U)}{\sigma_U} = \frac{60 - 84}{19.44} = \frac{-24}{19.44}$$
$$= -1.23$$

This z-score does not fall into the .05 region of rejection so we do not have strong enough evidence by this test to reject the hypothesis.

It should be noted that the selection of the women's group for finding our test statistic would lead to the same z, with the opposite sign. Only T_1 and U need to be recomputed to illustrate this. The total of the ranks for the women is 165, so U is (using the appropriate N_1 from the women's group)

$$U = (12)(14) + \frac{(14)(15)}{2} - 165$$
$$= 108$$

and then

$$z = \frac{108 - 84}{19.44} = 1.23$$

If an analysis of variance had been performed on the raw scores, it is interesting that no mean difference could have been concluded but that the variance in the men's group is significantly greater than in the women's group. This result might merely be a reflection of departure from normal distributions, however, and not a valid indication of different population variances. In either method of analysis, the data are suggestive of differences, but certainly not significantly so.

The use of the Mann-Whitney U test for small groups requires tables of the distribution of U. These are given in a number of more advanced texts and should be used when the sample sizes fall below ten. Such texts will also provide a correction of the value of U for ties in the rankings. When ties are few, as in Example 12–1, the correction is minor and may be ignored for most purposes; however, when numerous tied ranks are found in the data, the correction should be employed. This correction will increase the absolute value of z, but usually only slightly. For example, this correction applied to the data in Example 12–1 produces no change in the z-score, carried to two decimal places.

12–5. Hypothesis Testing and Statistical Decisions. Types of error. We have noted that the rejection of some null hypothesis may be incorrect, i.e., the hypothesis may actually be true. This error is the type-I error, and its probability of occurrence is determined by the level of confidence which is used. If the .05 level is used, only 5 per cent of the times that H_0 is true will it be erroneously rejected.

The second type of error occurs when a null hypothesis is actually false but the statistical test does not permit its rejection. This is the type-II error, but the probability of its occurrence is not so easily stated. For example, if two populations have means which are nearly, but not quite, equal, the probability of a significant difference in two sample means will be just slightly greater than .05, assuming this level of confidence. However, if the two means are greatly different, the chances of a significant difference between any two random samples may be nearly perfect. In other words, without knowing the degree of difference in population means, the probability of the type-II error cannot be found.

If one wishes to be exact, another problem arises from even a two-group experiment in which the equality of population means is under test. Let us assume that such a study is conducted and that the first group shows a mean of 120 correct responses in a learning task while the second group averages only 85 correct responses. An analysis of variance may show that the null hypothesis may be rejected at the .01 level of confidence. Are we merely interested in stating that the two conditions used in the experiment do not produce the same average number of correct responses? Of course not; we wish to conclude that the first condition produces more correct responses, on the average, than the other condition. However, it is possible the second condition is actually better than the first, and that our significant F is based on a highly unlikely event. We would be in the situation of correctly rejecting the null hypothesis but concluding a population difference in the wrong direction.

Fortunately, the chances of such an error are remote if a true difference of any magnitude does exist. Let us simply consider the cases of rejecting H_0 or failing to reject.

The power of a statistical test. The probability of a type-I error is set at a given level by defining the region of rejection. This probability is called the **alpha level** (α) of the statistical procedure, and it remains constant under all appropriate applications of the test.

The probability of the type-II error does depend upon a number of factors. Rather than speaking in negative terms, let us consider the probability of rejecting H_0 when it is actually false. This is the opposite of a type-II error. Roughly speaking, the power of a statistical test is equal to the probability of rejecting H_0 when it is actually false.

In order to determine the power of a test, a number of factors must be kept in mind. First, the sample size will clearly have an influence on the probability of obtaining a "true" rejection of H_0. Also, the true extent of departure from H_0 in the populations will affect this probability. A final factor, one which will be combined with the degree of departure of H_0 conditions, is the population variance. We have seen that large variances require large mean differences for rejection to be very likely.

The power of a statistical test may be illustrated graphically. We shall restrict our attention to a simple test in which a population of normally distributed scores with known standard deviation is assumed to have a particular mean of μ_0. This assumed value for the mean constitutes the null hypothesis H_0 while the hypothesis H_1 is true if the population mean is anything other than μ_0.

We shall measure the possible values of the population mean in terms of the standard error of the mean, σ_M. This value takes into account both the standard deviation of the population scores and the sample size used to make the statistical test. A particular value for the true population mean would then be stated as

$$\mu = \mu_0 + k\sigma_M \tag{12–5.1}$$

To illustrate (12–5.1), say that the hypothesized mean is 100 and $\sigma = 10$. If we were to test this hypothesis with 25 random scores, a true population mean of 96 would be changed as follows

$$96 = 100 + k \frac{10}{\sqrt{25}} = 100 + k2$$

then

$$2k = -4$$
$$k = -2$$

We would then express the true population mean as

$$\mu = \mu_0 - 2\sigma_M$$

Any particular value of the population mean can be expressed in such a form. We now wish to study the probability of rejecting H_0 as a function of the different possible values for the population mean.

Under the specified conditions, a two-tailed z-score test is appropriate. We simply find the sample mean, M, deviate this from the expected value of μ_0, and divide by the standard error of the mean. This is the same kind of test illustrated in Chapter 5. We know that a test z which is greater than or equal to 1.96, in absolute value, will cause rejection of H_0, assuming the .05 level of confidence as usual.

Let us now find the probability of a sample value of z which is in the region of rejection when some particular H_1 is true. For example, assume that the population mean is actually $\mu_0 - 2\sigma_M$. We know that the sample means will be normally distributed around this population mean with a standard error of σ_M. If we test H_0, we wish to know how often the z will be in the region of rejection. Figure 12–1 illustrates the possible distribution of sample means, in terms of their z-score equivalents, and also shows the region of rejection. One may see that the probability of rejecting H_0 when this particular H_1 is true is equal to the proportion of sample means which would be obtained which yield z-scores in the region of rejection.

One may see in Fig. 12–1 that slightly more than half of the sample values of z which would be obtained from the assumed H_1 condition are in the region of rejection. More exactly, -1.96 is just .04 z-score units above the mean of the distribution shown. The normal-curve table indicates that this range includes .016 of a normal curve, and we then have .500 + .016 or .516 as the probability of rejecting H_0 in this illustration. It should be stated that some values of the sample mean would be 1.96 or above at the high end of the distribution. However, such a mean would be 3.96 z-scores from the mean of the means. The probability of this is too small to be shown in this example.

Let us summarize this result by designating β as the probability of failing to reject H_0. We have obtained $1 - \beta$, the probability of

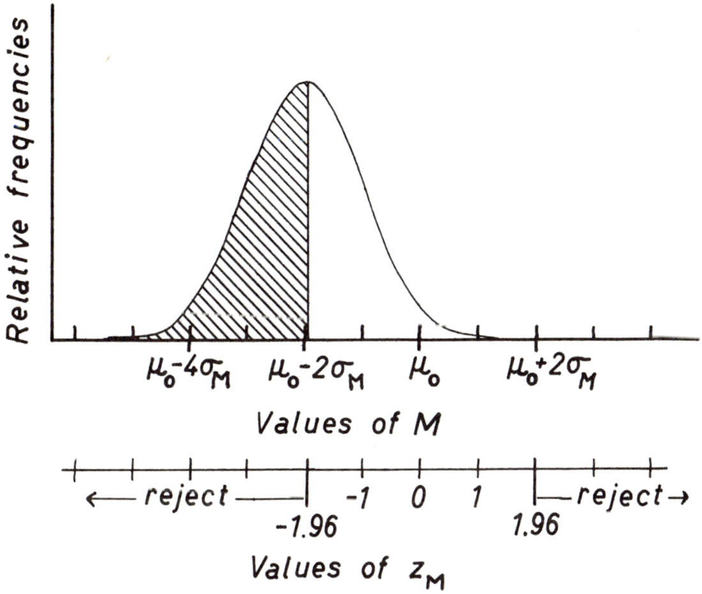

Fig. 12–1. An illustration of the sample means from a population with a true mean of $\mu_0 - 2\sigma_M$. The shaded area shows sample means which would cause rejection of the test that μ_0 is the population mean.

rejecting H_0, given that the population mean is actually $\mu_0 - 2\sigma_M$. We have a value of .516 for this probability.

If one were to consider other values for the true population mean, each would give a probability of rejecting H_0, found in the same way as just shown. A summary of all these probability values plotted as a function of the possible population means is called the *power curve* of a statistical test. For our z-score test being discussed, the power curve which is obtained is given in Fig. 12–2.

Examining Fig. 12–2, one can see that the probability of rejecting H_0 increases as the true population mean becomes more and more deviant from the hypothesized value. For instance, when μ is two standard errors from μ_0, we find the probability of rejecting H_0 is .516, as computed above. But when the deviation of the true mean is three standard errors, the probability of rejecting H_0 increases to about .85. For very extreme departures, the probability of rejecting H_0 is nearly unity. One should particularly note the value of the function shown when the true mean is equal to μ_0, i.e., when H_0 is actually true. The probability of rejecting H_0 is then .05, the alpha level used. Such a rejection is a type-I error and should naturally be equal to alpha.

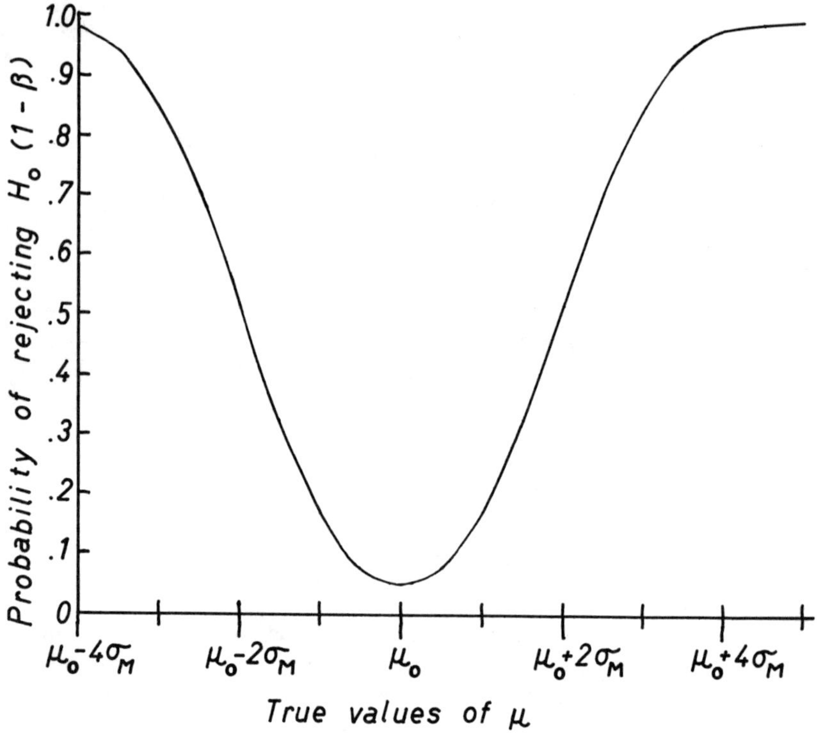

Fig. 12–2. The power curve for a test of H_0: $\mu = \mu_0$, with σ assumed to be known. Each point on the curve shows the probability of rejection for the corresponding true value of μ.

Any statistical test will have a power curve if we find an appropriate measure of the degree of departure of population values from the H_0 values. For tests which are designed to test the same type of hypothesis, a comparison may be made of their relative power (sensitivity) by comparing their power curves. If one of the two tests yields higher probabilities of rejecting H_0 for all values of H_1 than does the other test, then the first is said to be *uniformly more powerful* than the second. For instance, most of the parametric tests we have studied are uniformly more powerful than most of their nonparametric counterparts.

It should be noted that a power function such as shown in Fig. 12–2 does not directly show the influence of sample size. If one were to convert the scale of true population values to standard deviation units rather than standard error of the mean units, then the power curve would change as the sample size changed. With a larger sample, the probability of rejecting a false H_0 would be higher at all points than would be true for a lesser sample size.

Statistical decisions. There is nothing magical about the .05 level of confidence. The ultimate reason for using a statistical test is to convince people, including one's self, that certain results are not due to chance. The critical person requires that estimates of chance be precise rather than subjective guesses based on some unknown sort of reasoning or perceptions of sets of data. However, once a statement about the likelihood of chance results has been made, the decision which each person must make for himself is whether or not the evidence presented is strong enough to conclude that something other than chance is actually involved. If a well controlled experiment with a die which, after examination, is clearly balanced and precisely formed—if such a study were to show that the number 3 has occurred much more often than chance would suggest, even at the .01 level of confidence, a person might well be unconvinced that the results were anything except chance. It is well to keep in mind that in a world where events take place continually, the laws of chance themselves would predict that some very remarkable oddities are bound to take place.

On the other hand, a probability of .06 related to a statistical test may not warrant an asterisk in a scientific journal, but this "non-significant" result must be interpreted in its own right, ignoring the conventions, with due caution. Such a value, .06, means that only 6 per cent of the time a null hypothesis is true would such deviant results take place. For many purposes, such a result is convincing to most people.

There is a system called decision theory in which the probability level needed for rejecting a hypothesis is varied from case to case. This method requires, however, that one be able to place some values on the consequences of making errors in our decisions. When this is possible, as it often is in business or military settings, then the arbitrary .05 level of confidence becomes just another number. In science, the consequences of errors involve so many different and often subjective factors that no one has seriously suggested such a decision-making analysis in the general case.

There remains a recourse for the unhappy scientist who has obtained suggestive results from an experiment but has missed the traditional level of significance; any valid scientific finding is repeatable and another study may be conducted. A second attempt would usually be made with more subjects or observations than the first; this will produce a greater chance of significance at any arbitrary level of confidence.

It is well to remember that statistical testing is a valuable tool; but, a trivial difference from a null hypothesis will produce significant results if large enough samples are taken. Today, the behavioral sciences can probably do without knowledge which requires thousands of subjects in each condition for its demonstration.

EXERCISES

1. Find the 95 per cent confidence interval for the measurement of the true sample mean in each of the following cases. For each, express the final answer as a number in which the last figure is within one unit of the limits of the confidence interval:
   (a) $M = 58.75120$, based on 100 scores, each of which is accurate ($\pm .5$) to the units position.
   (b) $M = 127.43650$, based on 500 scores accurate to the tenths position (use a linear transformation).
   (c) $M = .215860$, based on 36 scores accurate to the hundredths position.
2. A sample variance of 144.000 is based on 100 scores accurate to the units position. What is the approximate 95 per cent confidence interval for the variance which would have been obtained had another significant figure been available for each score?
3. The standard error of the median is $1.253\sigma_M$ for large samples taken from a normal population:
   (a) What is the relative efficiency of the mean compared to the median for estimating μ?
   (b) Twenty-five random samples of 10 scores each were taken from a rectangular distribution with known mean. The variance in errors between M's and μ was 49.14 while the variance in errors between the medians and μ was 79.37. What is your best estimate of the relative efficiency of the mean as an estimate of μ in this case?
4. An observer finds that 67 cars out of 500 have trailor hitches. Assuming his sample to be reasonably random, what is the 95 per cent confidence interval for estimating the population *proportion?*
5. An investigator has reason to believe that a certain slum area has a population with a mean I.Q. of 90. He estimates that σ is about 10. If he tests H_0: $\mu = 100$ (with a z-score), what is his probability of rejecting H_0 with a sample size of 25 ($\alpha = .05$)?

Applications: Use information from all previous chapters where relevant.

6. A short arithmetic test is given to two randomly selected groups of children. One group consists of third-graders while the second consists of fourth-graders. Their scores are given below (correct answers). Is there any difference between the populations represented?

   Third graders: 0, 2, 1, 1, 3, 5, and 0
   Fourth graders: 4, 6, 8, 9, 9, 10, 11, and 11

7. An experimenter measures the degree of departure from a theoretical prediction for the data of each subject in two independent conditions. These departures yield a 1 *df* χ^2 for each person. The sum of the 15 χ^2 values in

the experimental group is 6.453, which is not significantly different from prediction. The corresponding total for the 15 control subjects is 25.405, which is significant as expected by the investigator.

However, a nonsignificant test never proves that the null hypothesis is exactly true. It is *not* possible, just because the one group differs significantly from some H_0 while the other does not, to conclude that the two groups differ significantly from one another.

How can the investigator prove that departures from prediction are significantly greater in the control group than in the experimental group?

8. In drawing samples of size 4 from a population of scores with $\mu = 200$ and $\sigma = 16$, the σ_M will be $\sigma/\sqrt{N}$ as always. However, if the population is not normal, the sample means may not be normally distributed. Using (12–4.1) or (12.–4.2), calculate the maximum probability of obtaining a sample mean which is more than 24 points from μ, for any population distribution.

9. Since the χ^2 formula for frequency data (two groups) is based on the normal approximation to the binomial, it can be shown that

$$\chi^2(1\ df) = \frac{(X - \mu)^2}{\sigma^2} = \frac{(f_{o,1} - f_{e,1})^2}{f_{e,1}} + \frac{(f_{o,2} - f_{e,2})^2}{f_{e,2}}$$

where $X = f_{o,1}$, $\mu = f_{e,1} = Np$, and $\sigma^2 = Npq$. For an exercise in algebra, start with the frequency formula and show that it is equal to $(X - \mu)^2/\sigma^2$. The following are needed: $p + q = 1, f_{e,2} = Nq = N - f_{e,1}, f_{o,2} = N - X$.

Appendix

TABLE A

Squares, Square Roots, and Reciprocals

N	N^2	$\sqrt{N}$	$\sqrt{10N}$	1000/N	N	N^2	$\sqrt{N}$	$\sqrt{10N}$	1000/N
					30	900	5.4772	17.321	33.333
1	1	1.0000	3.1623	1000.0	31	961	5.5678	17.607	32.258
2	4	1.4142	4.4721	500.00	32	1 024	5.6569	17.889	31.250
3	9	1.7321	5.4772	333.33	33	1 089	5.7446	18.166	30.303
4	16	2.0000	6.3246	250.00	34	1 156	5.8310	18.439	29.412
5	25	2.2361	7.0711	200.00	35	1 225	5.9161	18.708	28.571
6	36	2.4495	7.7460	166.67	36	1 296	6.0000	18.974	27.778
7	49	2.6458	8.3666	142.86	37	1 369	6.0828	19.235	27.027
8	64	2.8284	8.9443	125.00	38	1 444	6.1644	19.494	26.316
9	81	3.0000	9.4868	111.11	39	1 521	6.2450	19.748	25.641
10	100	3.1623	10.000	100.00	40	1 600	6.3246	20.000	25.000
11	121	3.3166	10.488	90.909	41	1 681	6.4031	20.248	24.390
12	144	3.4641	10.954	83.333	42	1 764	6.4807	20.494	23.810
13	169	3.6056	11.402	76.923	43	1 849	6.5574	20.736	23.256
14	196	3.7417	11.832	71.429	44	1 936	6.6333	20.976	22.727
15	225	3.8730	12.247	66.667	45	2 025	6.7082	21.213	22.222
16	256	4.0000	12.649	62.500	46	2 116	6.7823	21.448	21.739
17	289	4.1231	13.038	58.824	47	2 209	6.8557	21.679	21.277
18	324	4.2426	13.416	55.556	48	2 304	6.9282	21.909	20.833
19	361	4.3589	13.784	52.632	49	2 401	7.0000	22.136	20.408
20	400	4.4721	14.142	50.000	50	2 500	7.0711	22.361	20.000
21	441	4.5826	14.491	47.619	51	2 601	7.1414	22.583	19.608
22	484	4.6904	14.832	45.455	52	2 704	7.2111	22.804	19.231
23	529	4.7958	15.166	43.478	53	2 809	7.2801	23.022	18.868
24	576	4.8990	15.492	41.667	54	2 916	7.3485	23.238	18.519
25	625	5.0000	15.811	40.000	55	3 025	7.4162	23.452	18.182
26	676	5.0990	16.125	38.462	56	3 136	7.4833	23.664	17.857
27	729	5.1962	16.432	37.037	57	3 249	7.5498	23.875	17.544
28	784	5.2915	16.733	35.714	58	3 364	7.6158	24.083	17.241
29	841	5.3852	17.029	34.483	59	3 481	7.6811	24.290	16.949

N	N^2	$\sqrt{N}$	$\sqrt{10N}$	1000/N	N	N^2	$\sqrt{N}$	$\sqrt{10N}$	1000/N
60	3 600	7.7460	24.495	16.667	110	12 100	10.488	33.166	9.0909
61	3 721	7.8103	24.698	16.393	111	12 321	10.536	33.317	9.0090
62	3 844	7.8740	24.900	16.129	112	12 544	10.583	33.466	8.9286
63	3 969	7.9373	25.100	15.873	113	12 769	10.630	33.615	8.8496
64	4 096	8.0000	25.298	15.625	114	12 996	10.677	33.764	8.7719
65	4 225	8.0623	25.495	15.385	115	13 225	10.724	33.912	8.6957
66	4 356	8.1240	25.690	15.152	116	13 456	10.770	34.059	8.6207
67	4 489	8.1854	25.884	14.925	117	13 689	10.817	34.205	8.5470
68	4 624	8.2462	26.077	14.706	118	13 924	10.863	34.351	8.4746
69	4 761	8.3066	26.268	14.493	119	14 161	10.909	34.496	8.4034
70	4 900	8.3666	26.458	14.286	120	14 400	10.954	34.641	8.3333
71	5 041	8.4262	26.646	14.085	121	14 641	11.000	34.785	8.2645
72	5 184	8.4853	26.833	13.889	122	14 884	11.045	34.929	8.1967
73	5 329	8.5440	27.019	13.699	123	15 129	11.091	35.071	8.1301
74	5 476	8.6023	27.203	13.514	124	15 376	11.136	35.214	8.0645
75	5 625	8.6603	27.386	13.333	125	15 625	11.180	35.355	8.0000
76	5 776	8.7178	27.568	13.158	126	15 876	11.225	35.496	7.9365
77	5 929	8.7750	27.749	12.987	127	16 129	11.269	35.637	7.8740
78	6 084	8.8318	27.928	12.821	128	16 384	11.314	35.777	7.8125
79	6 241	8.8882	28.107	12.658	129	16 641	11.358	35.917	7.7519
80	6 400	8.9443	28.284	12.500	130	16 900	11.402	36.056	7.6923
81	6 561	9.0000	28.461	12.346	131	17 161	11.446	36.194	7.6336
82	6 724	9.0554	28.636	12.195	132	17 424	11.489	36.332	7.5758
83	6 889	9.1104	28.810	12.048	133	17 689	11.533	36.469	7.5188
84	7 056	9.1652	28.983	11.905	134	17 956	11.576	36.606	7.4627
85	7 225	9.2195	29.155	11.765	135	18 225	11.619	36.742	7.4074
86	7 396	9.2736	29.326	11.628	136	18 496	11.662	36.878	7.3529
87	7 569	9.3274	29.496	11.494	137	18 769	11.705	37.014	7.2993
88	7 744	9.3808	29.665	11.364	138	19 044	11.747	37.148	7.2464
89	7 921	9.4340	29.833	11.236	139	19 321	11.790	37.283	7.1942
90	8 100	9.4868	30.000	11.111	140	19 600	11.832	37.417	7.1429
91	8 281	9.5394	30.166	10.989	141	19 881	11.874	37.550	7.0922
92	8 464	9.5917	30.332	10.870	142	20 164	11.916	37.683	7.0423
93	8 649	9.6437	30.496	10.753	143	20 449	11.958	37.815	6.9930
94	8 836	9.6954	30.659	10.638	144	20 736	12.000	37.947	6.9444
95	9 025	9.7468	30.822	10.526	145	21 025	12.042	38.079	6.8966
96	9 216	9.7980	30.984	10.417	146	21 316	12.083	38.210	6.8493
97	9 409	9.8489	31.145	10.309	147	21 609	12.124	38.341	6.8027
98	9 604	9.8995	31.305	10.204	148	21 904	12.166	38.471	6.7568
99	9 801	9.9499	31.464	10.101	149	22 201	12.207	38.601	6.7114
100	10 000	10.000	31.623	10.000	150	22 500	12.247	38.730	6.6667
101	10 201	10.050	31.781	9.9010	151	22 801	12.288	38.859	6.6225
102	10 404	10.100	31.937	9.8039	152	23 104	12.329	38.987	6.5789
103	10 609	10.149	32.094	9.7087	153	23 409	12.369	39.115	6.5359
104	10 816	10.198	32.249	9.6154	154	23 716	12.410	39.243	6.4935
105	11 025	10.247	32.404	9.5238	155	24 025	12.450	39.370	6.4516
106	11 236	10.296	32.558	9.4340	156	24 336	12.490	39.497	6.4103
107	11 449	10.344	32.711	9.3458	157	24 649	12.530	39.623	6.3694
108	11 664	10.392	32.863	9.2593	158	24 964	12.570	39.749	6.3291
109	11 881	10.440	33.015	9.1743	159	25 281	12.610	39.875	6.2893

N	N^2	$\sqrt{N}$	$\sqrt{10N}$	1000/N	N	N^2	$\sqrt{N}$	$\sqrt{10N}$	1000/N
160	25 600	12.649	40.000	6.2500	210	44 100	14.491	45.826	4 7619
161	25 921	12.689	40.125	6.2112	211	44 521	14.526	45.935	4.7393
162	26 244	12.728	40.249	6.1728	212	44 944	14.560	46.043	4.7170
163	26 569	12.767	40.373	6.1350	213	45 369	14.595	46.152	4.0948
164	26 896	12.806	40.497	6.0976	214	45 796	14.629	46.260	4.6729
165	27 225	12.845	40.620	6.0606	215	46 225	14.663	46.368	4.6512
166	27 556	12.884	40.743	6.0241	216	46 656	14.697	46.476	4.6296
167	27 889	12.923	40.866	5.9880	217	47 089	14.731	46.583	4.6083
168	28 224	12.961	40.988	5.9524	218	47 524	14.765	46.690	4.5872
169	28 561	13.000	41.110	5.9172	219	47 961	14.799	46.797	4.5662
170	28 900	13.038	41.231	5.8824	220	48 400	14.832	46.904	4.5455
171	29 241	13.077	41.352	5.8480	221	48 841	14.866	47.011	4.5249
172	29 584	13.115	41.473	5.8140	222	49 284	14.900	47.117	4.5045
173	29 929	13.153	41.593	5.7803	223	49 729	14.933	47.223	4.4843
174	30 276	13.191	41.713	5.7471	224	50 176	14.967	47.329	4.4643
175	30 625	13.229	41.833	5.7143	225	50 625	15.000	47.434	4.4444
176	30 976	13.267	41.952	5.6818	226	51 076	15.033	47.539	4.4248
177	31 329	13.304	42.071	5.6497	227	51 529	15.067	47.645	4.4053
178	31 684	13.342	42.190	5.6180	228	51 984	15.100	47.749	4.3860
179	32 041	13.379	42.308	5.5866	229	52 441	15.133	47.854	4.3668
180	32 400	13.416	42.426	5.5556	230	52 900	15.166	47.958	4.3478
181	32 761	13.454	42.544	5.5249	231	53 361	15.199	48.062	4.3290
182	33 124	13.491	42.661	5.4945	232	53 824	15.232	48.166	4.3103
183	33 489	13.528	42.779	5.4645	233	54 289	15.264	48.270	4.2918
184	33 856	13.565	42.895	5.4348	234	54 756	15.297	48.374	4.2735
185	34 225	13.601	43.012	5.4054	235	55 225	15.330	48.477	4.2553
186	34 596	13.638	43.128	5.3763	236	55 696	15.362	48.580	4.2373
187	34 969	13.675	43.244	5.3476	237	56 169	15.395	48.683	4.2194
188	35 344	13.711	43.359	5.3191	238	56 644	15.427	48.785	4.2017
189	35 721	13.748	43.474	5.2910	239	57 121	15.460	48.888	4.1841
190	36 100	13.784	43.589	5.2632	240	57 600	15.492	48.990	4.1667
191	36 481	13.820	43.704	5.2356	241	58 081	15.524	49.092	4.1494
192	36 864	13.856	43.818	5.2083	242	58 564	15.556	49.194	4.1322
193	37 249	13.892	43.932	5.1813	243	59 049	15.588	49.295	4.1152
194	37 636	13.928	44.045	5.1546	244	59 536	15.621	49.396	4.0984
195	38 025	13.964	44.159	5.1282	245	60 025	15.652	49.497	4.0816
196	38 416	14.000	44.272	5.1020	246	60 516	15.684	49.598	4.0650
197	38 809	14.036	44.385	5.0761	247	61 009	15.716	49.699	4.0486
198	39 204	14.071	44.497	5.0505	248	61 504	15.748	49.800	4.0323
199	39 601	14.107	44.609	5.0251	249	62 001	15.780	49.900	4.0161
200	40 000	14.142	44.721	5.0000	250	62 500	15.811	50.000	4.0000
201	40 401	14.177	44.833	4.9751	251	63 001	15.843	50.100	3.9841
202	40 804	14.213	44.944	4.9505	252	63 504	15.875	50.200	3.9683
203	41 209	14.248	45.056	4.9261	253	64 009	15.906	50.299	3.9526
204	41 616	14.283	45.166	4.9020	254	64 516	15.937	50.398	3.9370
205	42 025	14.318	45.277	4.8780	255	65 025	15.969	50.498	3.9216
206	42 436	14.353	45.387	4.8544	256	65 536	16.000	50.596	3.9063
207	42 849	14.387	45.497	4.8309	257	66 049	16.031	50.695	3.8911
208	43 264	14.422	45.607	4.8077	258	66 564	16.062	50.794	3.8760
209	43 681	14.457	45.717	4.7847	259	67 081	16.093	50.892	3.8610

N	N^2	$\sqrt{N}$	$\sqrt{10N}$	1000/N	N	N^2	$\sqrt{N}$	$\sqrt{10N}$	1000/N
260	67 600	16.125	50.990	3.8462	310	96 100	17.607	55.678	3.2258
261	68 121	16.155	51.088	3.8314	311	96 721	17.635	55.767	3.2154
262	68 644	16.186	51.186	3.8168	312	97 344	17.664	55.857	3.2051
263	69 169	16.217	51.284	3.8023	313	97 969	17.692	55.946	3.1949
264	69 696	16.248	51.381	3.7879	314	98 596	17.720	56.036	3.1847
265	70 225	16.279	51.478	3.7736	315	99 225	17.748	56.125	3.1746
266	70 756	16.310	51.575	3.7594	316	99 856	17.776	56.214	3.1646
267	71 289	16.340	51.672	3.7453	317	100 489	17.804	56.303	3.1546
268	71 824	16.371	51.769	3.7313	318	101 124	17.833	56.391	3.1447
269	72 361	16.401	51.865	3.7175	319	101 761	17.861	56.480	3.1348
270	72 900	16.432	51.962	3.7037	320	102 400	17.889	56.569	3.1250
271	73 441	16.462	52.058	3.6900	321	103 041	17.916	56.657	3.1153
272	73 984	16.492	52.154	3.6765	322	103 684	17.944	56.745	3.1056
273	74 529	16.523	52.249	3.6630	323	104 329	17.972	56.833	3.0960
274	75 076	16.553	52.345	3.6496	324	104 976	18.000	56.921	3.0864
275	75 625	16.583	52.440	3.6364	325	105 625	18.028	57.009	3.0769
276	76 176	16.613	52.536	3.6232	326	106 276	18.055	57.096	3.0675
277	76 729	16.643	52.631	3.6101	327	106 929	18.083	57.184	3.0581
278	77 284	16.673	52.726	3.5971	328	107 584	18.111	57.271	3.0488
279	77 841	16.703	52.820	3.5842	329	108 241	18.138	57.359	3.0395
280	78 400	16.733	52.915	3.5714	330	108 900	18.166	57.446	3.0303
281	78 961	16.763	53.009	3.5587	331	109 561	18.193	57.533	3.0211
282	79 524	16.793	53.104	3.5461	332	110 224	18.221	57.619	3.0120
283	80 089	16.823	53.198	3.5336	333	110 889	18.248	57.706	3.0030
284	80 656	16.852	53.292	3.5211	334	111 556	18.276	57.793	2.9940
285	81 225	16.882	53.385	3.5088	335	112 225	18.303	57.879	2.9851
286	81 796	16.912	53.479	3.4965	336	112 896	18.330	57.966	2.9762
287	82 369	16.941	53.572	3.4843	337	113 569	18.358	58.052	2.9674
288	82 944	16.971	53.666	3.4722	338	114 244	18.385	58.138	2.9586
289	83 521	17.000	53.759	3.4602	339	114 921	18.412	58.224	2.9499
290	84 100	17.029	53.852	3.4483	340	115 600	18.439	58.310	2.9412
291	84 681	17.059	53.944	3.4364	341	116 281	18.466	58.395	2.9326
292	85 264	17.088	54.037	3.4247	342	116 964	18.493	58.481	2.9240
293	85 849	17.117	54.129	3.4130	343	117 649	18.520	58.566	2.9155
294	86 436	17.146	54.222	3.4014	344	118 336	18.547	58.652	2.9070
295	87 025	17.176	54.314	3.3898	345	119 025	18.574	58.737	2.8986
296	87 616	17.205	54.406	3.3784	346	119 716	18.601	58.822	2.8902
297	88 209	17.234	54.498	3.3670	347	120 409	18.628	58.907	2.8818
298	88 804	17.263	54.589	3.3557	348	121 104	18.655	58.992	2.8736
299	88 401	17.292	54.681	3.3445	349	121 801	18.682	59.076	2.8653
300	90 000	17.321	54.772	3.3333	350	122 500	18.708	59.161	2.8571
301	90 601	17.349	54.863	3.3223	351	123 201	18.735	59.245	2.8490
302	91 204	17.378	54.955	3.3113	352	123 904	18.762	59.330	2.8409
303	91 809	17.407	55.045	3.3003	353	124 609	18.788	59.414	2.8329
304	92 416	17.436	55.136	3.2895	354	125 316	18.815	59.498	2.8249
305	93 025	17.464	55.227	3.2787	355	126 025	18.841	59.582	2.8169
306	93 636	17.493	55.317	3.2680	356	126 736	18.868	59.666	2.8090
307	94 249	17.521	55.408	3.2573	357	127 449	18.894	59.749	2.8011
308	94 864	17.550	55.498	3.2468	358	128 164	18.921	59.833	2.7933
309	95 481	17.578	55.588	3.2362	359	128 881	18.947	59.917	2.7855

N	N^2	$\sqrt{N}$	$\sqrt{10N}$	1000/N	N	N^2	$\sqrt{N}$	$\sqrt{10N}$	1000/N
360	129 600	18.974	60.000	2.7778	410	168 100	20.248	64.031	2.4390
361	130 321	19.000	60.083	2.7701	411	168 921	20.273	64.109	2.4331
362	131 044	19.026	60.166	2.7624	412	169 744	20.298	64.187	2.4272
363	131 769	19.053	60.249	2.7548	413	170 569	20.322	64.265	2.4213
364	132 496	19.079	60.332	2.7473	414	171 396	20.347	64.343	2.4155
365	133 225	19.105	60.415	2.7397	415	172 225	20.372	64.420	2.4096
366	133 956	19.131	60.498	2.7322	416	173 056	20.396	64.498	2.4038
367	134 689	19.157	60.581	2.7248	417	173 889	20.421	64.576	2.3981
368	135 424	19.183	60.663	2.7174	418	174 724	20.445	64.653	2.3923
369	136 161	19.209	60.745	2.7100	419	175 561	20.469	64.730	2.3866
370	136 900	19.235	60.828	2.7027	420	176 400	20.494	64.807	2.3810
371	137 641	19.261	60.910	2.6954	421	177 241	20.518	64.885	2.3753
372	138 384	19.287	60.992	2.6882	422	178 084	20.543	64.962	2.3697
373	139 129	19.313	61.074	2.6810	423	178 929	20.567	65.038	2.3641
374	139 876	19.339	61.156	2.6738	424	179 776	20.591	65.115	2.3585
375	140 625	19.365	61.237	2.6667	425	180 625	20.616	65.192	2.3529
376	141 376	19.391	61.319	2.6596	426	181 476	20.640	65.269	2.3474
377	142 129	19.416	61.400	2.6525	427	182 329	20.664	65.345	2.3419
378	142 884	19.442	61.482	2.6455	428	183 184	20.688	65.422	2.3364
379	143 641	19.468	61.563	2.6385	429	184 041	20.712	65.498	2.3310
380	144 400	19.494	61.644	2.6316	430	184 900	20.736	65.574	2.3256
381	145 161	19.519	61.725	2.6247	431	185 761	20.761	65.651	2.3202
382	145 924	19.545	61.806	2.6178	432	186 624	20.785	65.727	2.3148
383	146 689	19.570	61.887	2.6110	433	187 489	20.809	65.803	2.3095
384	147 456	19.596	61.968	2.6042	434	188 356	20.833	65.879	2.3041
385	148 225	19.621	62.048	2.5974	435	189 225	20.857	65.955	2.2989
386	148 996	19.647	62.129	2.5907	436	190 096	20.881	66.030	2.2936
387	149 769	19.672	62.209	2.5840	437	190 969	20.905	66.106	2.2883
388	150 544	19.698	62.290	2.5773	438	191 844	20.928	66.182	2.2831
389	151 321	19.723	62.370	2.5707	439	192 721	20.952	66.257	2.2779
390	152 100	19.748	62.450	2.5641	440	193 600	20.976	66.333	2.2727
391	152 881	19.774	62.530	2.5575	441	194 481	21.000	66.408	2.2676
392	153 664	19.799	62.610	2.5510	442	195 364	21.024	66.483	2.2624
393	154 449	19.824	62.690	2.5445	443	196 249	21.048	66.558	2.2573
394	155 236	19.849	62.769	2.5381	444	197 136	21.071	66.633	2.2523
395	156 025	19.875	62.849	2.5316	445	198 025	21.095	66.708	2.2472
396	156 816	19.900	62.929	2.5253	446	198 916	21.119	66.783	2.2422
397	157 609	19.925	63.008	2.5189	447	199 809	21.142	66.858	2.2371
398	158 404	19.950	63.087	2.5126	448	200 704	21.166	66.933	2.2321
399	159 201	19.975	63.166	2.5063	449	201 601	21.190	67.007	2.2272
400	160 000	20.000	63.246	2.5000	450	202 500	21.213	67.082	2.2222
401	160 801	20.025	63.325	2.4938	451	203 401	21.237	67.157	2.2173
402	161 604	20.050	63.403	2.4876	452	204 304	21.260	67.231	2.2124
403	162 409	20.075	63.482	2.4814	453	205 209	21.284	67.305	2.2075
404	163 216	20.100	63.561	2.4752	454	206 116	21.307	67.380	2.2026
405	164 025	20.125	63.640	2.4691	455	207 025	21.331	67.454	2.1978
406	164 836	20.149	63.718	2.4631	456	207 936	21.354	67.528	2.1930
407	165 649	20.174	63.797	2.4570	457	208 849	21.378	67.602	2.1882
408	166 464	20.199	63.875	2.4510	458	209 764	21.401	67.676	2.1834
409	167 281	20.224	63.953	2.4450	459	210 681	21.424	67.750	2.1786

N	N^2	$\sqrt{N}$	$\sqrt{10N}$	1000/N
460	211 600	21.448	67.823	2.1739
461	212 521	21.471	67.897	2.1692
462	213 444	21.494	67.971	2.1645
463	214 369	21.517	68.044	2.1508
464	215 296	21.541	68.118	2.1552
465	216 225	21.564	68.191	2.1505
466	217 156	21.587	68.264	2.1459
467	218 089	21.610	68.337	2.1413
468	219 024	21.633	68.411	2.1368
469	219 961	21.656	68.484	2.1322
470	220 900	21.679	68.557	2.1277
471	221 841	21.703	68.629	2.1231
472	222 784	21.726	68.702	2.1186
473	223 729	21.749	68.775	2.1142
474	224 676	21.772	68.848	2.1097
475	225 625	21.794	68.920	2.1053
476	226 576	21.817	68.993	2.1008
477	227 529	21.840	69.065	2.0964
478	228 484	21.863	69.138	2.0921
479	229 441	21.886	69.210	2.0877
480	230 400	21.909	69.282	2.0833
481	231 361	21.932	69.354	2.0790
482	232 324	21.955	69.426	2.0747
483	233 289	21.977	69.498	2.0704
484	234 256	22.000	69.570	2.0661
485	235 225	22.023	69.642	2.0619
486	236 196	22.045	69.714	2.0576
487	237 169	22.068	69.785	2.0534
488	238 144	22.091	69.857	2.0492
489	239 121	22.113	69.929	2.0450
490	240 100	22.136	70.000	2.0408
491	241 081	22.159	70.071	2.0367
492	242 064	22.181	70.143	2.0325
493	243 049	22.204	70.214	2.0284
494	244 036	22.226	70.285	2.0243
495	245 025	22.249	70.356	2.0202
496	246 016	22.271	70.427	2.0161
497	247 009	22.294	70.498	2.0121
498	248 004	22.316	70.569	2.0080
499	249 001	22.338	70.640	2.0040
500	250 000	22.361	70.711	2.0000
501	251 001	22.383	70.781	1.9960
502	252 004	22.405	70.852	1.9920
503	253 009	22.428	70.922	1.9881
504	254 016	22.450	70.993	1.9841
505	255 025	22.472	71.063	1.9802
506	256 036	22.494	71.134	1.9763
507	257 049	22.517	71.204	1.9724
508	258 064	22.539	71.274	1.9685
509	259 081	22.561	71.344	1.9646
510	260 100	22.583	71.414	1.9608
511	261 121	22.605	71.484	1.9569
512	262 144	22.627	71.554	1.9531
513	263 169	22.650	71.624	1.9493
514	264 196	22.672	71.694	1.9455
515	265 225	22.694	71.764	1.9417
516	266 256	22.716	71.833	1.9380
517	267 289	22.738	71.903	1.9342
518	268 324	22.760	71.972	1.9305
519	269 361	22.782	72.042	1.9268
520	270 400	22.804	72.111	1.9231
521	271 441	22.825	72.180	1.9194
522	272 484	22.847	72.250	1.9157
523	273 529	22.869	72.319	1.9120
524	274 576	22.891	72.388	1.9084
525	275 625	22.913	72.457	1.9048
526	276 676	22.935	72.526	1.9011
527	277 729	22.956	72.595	1.8975
528	278 784	22.978	72.664	1.8939
529	279 841	23.000	72.732	1.8904
530	280 900	23.022	72.801	1.8868
531	281 961	23.043	72.870	1.8832
532	283 024	23.065	72.938	1.8797
533	284 089	23.087	73.007	1.8762
534	285 156	23.108	73.075	1.8727
535	286 225	23.130	73.144	1.8692
536	287 296	23.152	73.212	1.8657
537	288 369	23.173	73.280	1.8622
538	289 444	23.195	73.348	1.8587
539	290 521	23.216	73.417	1.8553
540	291 600	23.238	73.485	1.8519
541	292 681	23.259	73.553	1.8484
542	293 764	23.281	73.621	1.8450
543	294 849	23.302	73.689	1.8416
544	295 936	23.324	73.756	1.8382
545	297 025	23.345	73.824	1.8349
546	298 116	23.367	73.892	1.8315
547	299 209	23.388	73.959	1.8282
548	300 304	23.409	74.027	1.8248
549	301 401	23.431	74.095	1.8215
550	302 500	23.452	74.162	1.8182
551	303 601	23.473	74.229	1.8149
552	304 704	23.495	74.297	1.8116
553	305 809	23.516	74.364	1.8083
554	306 916	23.537	74.431	1.8051
555	308 025	23.558	74.498	1.8018
556	309 136	23.580	74.565	1.7986
557	310 249	23.601	74.632	1.7953
558	311 364	23.622	74.699	1.7921
559	312 481	23.643	74.766	1.7889

N	N^2	$\sqrt{N}$	$\sqrt{10N}$	1000/N
560	313 600	23.664	74.833	1.7857
561	314 721	23.685	74.900	1.7825
562	315 844	23.707	74.967	1.7794
563	316 969	23.728	75.033	1.7762
564	318 096	23.749	75.100	1.7731
565	319 225	23.770	75.166	1.7699
566	320 356	23.791	75.233	1.7668
567	321 489	23.812	75.299	1.7637
568	322 624	23.833	75.366	1.7606
569	323 761	23.854	75.432	1.7575
570	324 900	23.875	75.498	1.7544
571	326 041	23.896	75.565	1.7513
572	327 184	23.917	75.631	1.7483
573	328 329	23.937	75.697	1.7452
574	329 476	23.958	75.763	1.7422
575	330 625	23.979	75.829	1.7391
576	331 776	24.000	75.895	1.7361
577	332 929	24.021	75.961	1.7331
578	334 084	24.042	76.026	1.7301
579	335 241	24.062	76.092	1.7271
580	336 400	24.083	76.158	1.7241
581	337 561	24.104	76.223	1.7212
582	338 724	24.125	76.289	1.7182
583	339 889	24.145	76.354	1.7153
584	341 056	24.166	76.420	1.7123
585	342 225	24.187	76.485	1.7094
586	343 396	24.207	76.551	1.7065
587	344 569	24.228	76.616	1.7036
588	345 744	24.249	76.681	1.7007
589	346 921	24.269	76.746	1.6978
590	348 100	24.290	76.811	1.6949
591	349 281	24.310	76.877	1.6920
592	350 464	24.331	76.942	1.6892
593	351 649	24.352	77.006	1.6863
594	352 836	24.372	77.071	1.6835
595	354 025	24.393	77.136	1.6807
596	355 216	24.413	77.201	1.6779
597	356 409	24.434	77.266	1.6750
598	357 604	24.454	77.330	1.6722
599	358 801	24.474	77.395	1.6694
600	360 000	24.495	77.460	1.6667
601	361 201	24.515	77.524	1.6639
602	362 404	24.536	77.589	1.6611
603	363 609	24.556	77.653	1.6584
604	364 816	24.576	77.717	1.6556
605	366 025	24.597	77.782	1.6529
606	367 236	24.617	77.846	1.6502
607	368 449	24.637	77.910	1.6474
608	369 664	24.658	77.974	1.6447
609	370 881	24.678	78.038	1.6420
610	372 100	24.698	78.103	1.6393
611	373 321	24.718	78.166	1.6367
612	374 544	24.739	78.230	1.6340
613	375 769	24.759	78.294	1.6313
614	376 996	24.779	78.358	1.6287
615	378 225	24.799	78.422	1.6260
616	379 456	24.819	78.486	1 6234
617	380 689	24.839	78.549	1.6207
618	381 924	24.860	78.613	1.6181
619	383 161	24.880	78.677	1.6155
620	384 400	24.900	78.740	1.6129
621	385 641	24.920	78.804	1.6103
622	386 884	24.940	78.867	1.6077
623	388 129	24.960	78.930	1.6051
624	389 376	24.980	78.994	1.6026
625	390 625	25.000	79.057	1.6000
626	391 876	25.020	79.120	1.5974
627	393 129	25.040	79.183	1.5949
628	394 384	25.060	79.246	1.5924
629	395 641	25.080	79.310	1.5898
630	396 900	25.100	79.373	1.5873
631	398 161	25.120	79.436	1.5848
632	399 424	25.140	79.498	1.5823
633	400 689	25.159	79.561	1.5798
634	401 956	25.179	79.624	1.5773
635	403 225	25.199	79.687	1.5748
636	404 496	25.219	79.750	1.5723
637	405 769	25.239	79.812	1.5699
638	407 044	25.259	79.875	1.5674
639	408 321	25.278	79.937	1.5649
640	409 600	25.298	80.000	1.5625
641	410 881	25.318	80.062	1.5601
642	412 164	25.338	80.125	1.5576
643	413 449	25.357	80.187	1.5552
644	414 736	25.377	80.250	1.5528
645	416 025	25.397	80.312	1.5504
646	417 316	25.417	80.374	1.5480
647	418 609	25.436	80.436	1.5456
648	419 904	25.456	80.498	1.5432
649	421 201	25.475	80.561	1.5408
650	422 500	25.495	80.623	1.5385
651	423 801	25.515	80.685	1.5361
652	425 104	25.534	80.747	1.5337
653	426 409	25.554	80.808	1.5314
654	427 716	25.573	80.870	1.5291
655	429 025	25.593	80.932	1.5267
656	430 336	25.613	80.994	1.5244
657	431 649	25.632	81.056	1 5221
658	432 964	25.652	81.117	1.5198
659	434 281	25.671	81.179	1.5175

N	N^2	$\sqrt{N}$	$\sqrt{10N}$	1000/N
660	435 600	25.690	81.240	1.5152
661	436 921	25.710	81.302	1.5129
662	438 244	25.729	81.363	1.5106
663	439 569	25.749	81.425	1.5083
664	440 896	25.768	81.486	1.5060
665	442 225	25.788	81.548	1.5038
666	443 556	25.807	81.609	1.5015
667	444 889	25.826	81.670	1.4993
668	446 224	25.846	81.731	1.4970
669	447 561	25.865	81.792	1.4948
670	448 900	25.884	81.854	1.4925
671	450 241	25.904	81.915	1.4903
672	451 584	25.923	81.976	1.4881
673	452 929	25.942	82.037	1.4859
674	454 276	25.962	82.098	1.4837
675	455 625	25.981	82.158	1.4815
676	456 976	26.000	82.219	1.4793
677	458 329	26.019	82.280	1.4771
678	459 684	26.038	82.341	1.4749
679	461 041	26.058	82.401	1.4728
680	462 400	26.077	82.462	1.4706
681	463 761	26.096	82.523	1.4684
682	465 124	26.115	82.583	1.4663
683	466 489	26.134	82.644	1.4641
684	467 856	26.153	82.704	1.4620
685	469 225	26.173	82.765	1.4599
686	470 596	26.192	82.825	1.4577
687	471 969	26.211	82.885	1.4556
688	473 344	26.230	82.946	1.4535
689	474 721	26.249	83.006	1.4514
690	476 100	26.268	83.066	1.4493
691	477 481	26.287	83.126	1.4472
692	478 864	26.306	83.187	1.4451
693	480 249	26.325	83.247	1.4430
694	481 636	26.344	83.307	1.4409
695	483 025	26.363	83.367	1.4388
696	484 416	26.382	83.427	1.4368
697	485 809	26.401	83.487	1.4347
698	487 204	26.420	83.546	1.4327
699	488 601	26.439	83.606	1.4306
700	490 000	26.458	83.666	1.4286
701	491 401	26.476	83.726	1.4265
702	492 804	26.495	83.785	1.4245
703	494 209	26.514	83.845	1.4225
704	495 616	26.533	83.905	1.4205
705	497 025	26.552	83.964	1.4184
706	498 436	26.571	84.024	1.4164
707	499 849	26.589	84.083	1.4144
708	501 264	26.608	84.143	1.4124
709	502 681	26.627	84.202	1.4104
710	504 100	26.646	84.262	1.4085
711	505 521	26.665	84.321	1.4065
712	506 944	26.683	84.380	1.4045
713	508 369	26.702	84.439	1.4025
714	509 796	26.721	84.499	1.4006
715	511 225	26.739	84.558	1.3986
716	512 656	26.758	84.617	1.3966
717	514 089	26.777	84.676	1.3947
718	515 524	26.796	84.735	1.3928
719	516 961	26.814	84.794	1.3908
720	518 400	26.833	84.853	1.3889
721	519 841	26.851	84.912	1.3870
722	521 284	26.870	84.971	1.3850
723	522 729	26.889	85.029	1.3831
724	524 176	26.907	85.088	1.3812
725	525 625	26.926	85.147	1.3793
726	527 076	26.944	85.206	1.3774
727	528 529	26.963	85.264	1.3755
728	529 984	26.981	85.323	1.3736
729	531 441	27.000	85.382	1.3717
730	532 900	27.019	85.440	1.3699
731	534 361	27.037	85.499	1.3680
732	535 824	27.056	85.557	1.3661
733	537 289	27.074	85.615	1.3643
734	538 756	27.092	85.674	1.3624
735	540 225	27.111	85.732	1.3605
736	541 696	27.129	85.790	1.3587
737	543 169	27.148	85.849	1.3569
738	544 644	27.166	85.907	1.3550
739	546 121	27.185	85.965	1.3532
740	547 600	27.203	86.023	1.3514
741	549 081	27.221	86.081	1.3495
742	550 564	27.240	86.139	1.3477
743	552 049	27.258	86.197	1.3459
744	553 536	27.276	86.255	1.3441
745	555 025	27.295	86.313	1.3423
746	556 516	27.313	86.371	1.3405
747	558 009	27.331	86.429	1.3387
748	559 504	27.350	86.487	1.3359
749	561 001	27.368	86.545	1.3351
750	562 500	27.386	86.603	1.3333
751	564 001	27.404	86.660	1.3316
752	565 504	27.423	86.718	1.3298
753	567 009	27.441	86.776	1.3280
754	568 516	27.459	86.833	1.3263
755	570 025	27.477	86.891	1.3245
756	571 536	27.495	86.948	1.3228
757	573 049	27.514	87.006	1.3210
758	574 564	27.532	87.063	1.3193
759	576 081	27.550	87.121	1.3175

N	N^2	$\sqrt{N}$	$\sqrt{10N}$	1000/N
760	577 600	27.568	87.178	1.3158
761	579 121	27.586	87.235	1.3141
762	580 644	27.604	87.293	1.3123
763	582 169	27.622	87.350	1.3106
764	583 696	27.641	87.407	1.3089
765	585 225	27.659	87.464	1.3072
766	586 756	27.677	87.521	1.3055
767	588 289	27.695	87.579	1.3038
768	589 824	27.713	87.636	1.3021
769	591 361	27.731	87.693	1.3004
770	592 900	27.749	87.750	1.2987
771	594 441	27.767	87.807	1.2970
772	595 984	27.785	87.864	1.2953
773	597 529	27.803	87.920	1.2937
774	599 076	27.821	87.977	1.2920
775	600 625	27.839	88.034	1.2903
776	602 176	27.857	88.091	1.2887
777	603 729	27.875	88.148	1.2870
778	605 284	27.893	88.204	1.2853
779	606 841	27.911	88.261	1.2837
780	608 400	27.928	88.318	1.2821
781	609 961	27.946	88.374	1.2804
782	611 524	27.964	88.431	1.2788
783	613 089	27.982	88.487	1.2771
784	614 656	28.000	88.544	1.2755
785	616 225	28.018	88.600	1.2739
786	617 796	28.036	88.657	1.2723
787	619 369	28.054	88.713	1.2706
788	620 944	28.071	88.769	1.2690
789	622 521	28.089	88.826	1.2674
790	624 100	28.107	88.882	1.2658
791	625 681	28.125	88.938	1.2642
792	627 264	28.142	88.994	1.2626
793	628 849	28.160	89.051	1.2610
794	630 436	28.178	89.107	1.2594
795	632 025	28.196	89.163	1.2579
796	633 616	28.213	89.219	1.2563
797	635 209	28.231	89.275	1.2547
798	636 804	28.249	89.331	1.2531
799	638 401	28.267	89.387	1.2516
800	640 000	28.284	89.443	1.2500
801	641 601	28.302	89.499	1.2484
802	643 204	28.320	89.554	1.2469
803	644 809	28.337	89.610	1.2453
804	646 416	28.355	89.666	1.2438
805	648 025	28.373	89.722	1.2422
806	649 636	28.390	89.778	1.2407
807	651 249	28.408	89.833	1.2392
808	652 864	28.425	89.889	1.2376
809	654 481	28.443	89.944	1.2361

N	N^2	$\sqrt{N}$	$\sqrt{10N}$	1000/N
810	656 100	28.461	90.000	1.2346
811	657 721	28.478	90.056	1.2330
812	659 344	28.496	90.111	1.2315
813	660 969	28.513	90.167	1.2300
814	662 596	28.531	90.222	1.2285
815	664 225	28.548	90.277	1.2270
816	665 856	28.566	90.333	1.2255
817	667 489	28.583	90.388	1.2240
818	669 124	28.601	90.443	1.2225
819	670 761	28.618	90.499	1.2210
820	672 400	28.636	90.554	1.2195
821	674 041	28.653	90.609	1.2180
822	675 684	28.671	90.664	1.2165
823	677 329	28.688	90.719	1.2151
824	678 976	28.705	90.774	1.2136
825	680 625	28.723	90.830	1.2121
826	682 276	28.740	90.885	1.2107
827	683 929	28.758	90.940	1.2092
828	685 584	28.775	90.995	1.2077
829	687 241	28.792	91.049	1.2063
830	688 900	28.810	91.104	1.2048
831	690 561	28.827	91.159	1.2034
832	692 224	28.844	91.214	1.2019
833	693 889	28.862	91.269	1.2005
834	695 556	28.879	91.324	1.1990
835	697 225	28.896	91.378	1.1976
836	698 896	28.914	91.433	1.1962
837	700 569	28.931	91.488	1.1947
838	702 244	28.948	91.542	1.1933
839	703 921	28.966	91.597	1.1919
840	705 600	28.983	91.652	1.1905
841	707 281	29.000	91.706	1.1891
842	708 964	29.017	91.761	1.1876
843	710 649	29.034	91.815	1.1862
844	712 336	29.052	91.869	1.1848
845	714 025	29.069	91.924	1.1834
846	715 716	29.086	91.978	1.1820
847	717 409	29.103	92.033	1.1806
848	719 104	29.120	92.087	1.1792
849	720 801	29.138	92.141	1.1779
850	722 500	29.155	92.195	1.1765
851	724 201	29.172	92.250	1.1751
852	725 904	29.189	92.304	1.1737
853	727 609	29.206	92.358	1.1723
854	729 316	29.223	92.412	1.1710
855	731 025	29.240	92.466	1.1696
856	732 736	29.257	92.520	1.1682
857	734 449	29.275	92.574	1.1669
858	736 164	29.292	92.628	1.1655
859	737 881	29.309	92.682	1.1641

N	N^2	$\sqrt{N}$	$\sqrt{10N}$	1000/N
860	739 600	29.326	92.736	1.1628
861	741 321	29.343	92.790	1.1614
862	743 044	29.360	92.844	1.1601
863	744 769	29.377	92.898	1.1587
864	746 496	29.394	92.952	1.1574
865	748 225	29.411	93.005	1.1561
866	749 956	29.428	93.059	1.1547
867	751 689	29.445	93.113	1.1534
868	753 424	29.462	93.167	1.1521
869	755 161	29.479	93.220	1.1507
870	756 900	29.496	93.274	1.1494
871	758 641	29.513	93.327	1.1481
872	760 384	29.530	93.381	1.1468
873	762 129	29.547	93.434	1.1455
874	763 876	29.563	93.488	1.1442
875	765 625	29.580	93.541	1.1429
876	767 376	29.597	93.595	1.1416
877	769 129	29.614	93.648	1.1403
878	770 884	29.631	93.702	1.1390
879	772 641	29.648	93.755	1.1377
880	774 400	29.665	93.808	1.1364
881	776 161	29.682	93.862	1.1351
882	777 924	29.698	93.915	1.1338
883	779 689	29.715	93.968	1.1325
884	781 456	29.732	94.021	1.1312
885	783 225	29.749	94.074	1.1299
886	784 996	29.766	94.128	1.1287
887	786 769	29.783	94.181	1.1274
888	788 544	29.799	94.234	1.1261
889	790 321	29.816	94.287	1.1249
890	792 100	29.833	94.340	1.1236
891	793 881	29.850	94.393	1.1223
892	795 664	29.866	94.446	1.1211
893	797 449	29.883	94.499	1.1198
894	799 236	29.900	94.552	1.1186
895	801 025	29.917	94.604	1.1173
896	802 816	29.933	94.657	1.1161
897	804 609	29.950	94.710	1.1148
898	806 404	29.967	94.763	1.1136
899	808 201	29.983	94.816	1.1123
900	810 000	30.000	94.868	1.1111
901	811 801	30.017	94.921	1.1099
902	813 604	30.033	94.974	1.1086
903	815 409	30.050	94.026	1.1074
904	817 216	30.067	95.079	1.1062
905	819 025	30.083	95.131	1.1050
906	820 836	30.100	95.184	1.1038
907	822 649	30.116	95.237	1.1025
908	824 464	30.133	95.289	1.1013
909	826 281	30.150	95.341	1.1001
910	828 100	30.166	95.394	1.0989
911	829 921	30.183	95.446	1.0977
912	831 744	30.199	95.499	1.0965
913	833 569	30.216	95.551	1.0953
914	835 396	30.232	95.603	1.0941
915	837 225	30.249	95.656	1.0929
916	839 056	30.265	95.708	1.0917
917	840 889	30.282	95.760	1.0905
918	842 724	30.299	95.812	1.0893
919	844 561	30.315	95.864	1.0881
920	846 400	30.332	95.917	1.0870
921	848 241	30.348	95.969	1.0858
922	850 084	30.364	96.021	1.0846
923	851 929	30.381	96.073	1.0834
924	853 776	30.397	96.125	1.0823
925	855 625	30.414	96.177	1.0811
926	857 476	30.430	96.229	1.0799
927	859 329	30.447	96.281	1.0787
928	861 184	30.463	96.333	1.0776
929	863 041	30.480	96.385	1.0764
930	864 900	30.496	96.437	1.0753
931	866 761	30.512	96.488	1.0741
932	868 624	30.529	96.540	1.0730
933	870 489	30.545	96.592	1.0718
934	872 356	30.561	96.644	1.0707
935	874 225	30.578	96.695	1.0695
936	876 096	30.594	96.747	1.0684
937	877 969	30.610	96.799	1.0672
938	879 844	30.627	96.850	1.0661
939	881 721	30.643	96.902	1.0650
940	883 600	30.659	96.954	1.0638
941	885 481	30.676	96.005	1.0627
942	887 364	30.692	97.057	1.0616
943	889 249	30.708	97.108	1.0604
944	891 136	30.725	97.160	1.0593
945	893 025	30.741	97.211	1.0582
946	894 916	30.757	97.263	1.0571
947	896 809	30.773	97.314	1.0560
948	898 704	30.790	97.365	1.0549
949	900 601	30.806	97.417	1.0537
950	902 500	30.822	97.468	1.0526
951	904 401	30.838	97.519	1.0515
952	906 304	30.855	97.570	1.0504
953	908 209	30.871	97.622	1.0493
954	910 116	30.887	97.673	1.0482
955	912 025	30.903	97.724	1.0471
956	913 936	30.919	97.775	1.0460
957	915 849	30.935	97.826	1.0449
958	917 764	30.952	97.877	1.0438
959	919 681	30.968	97.929	1.0428

N	N^2	$\sqrt{N}$	$\sqrt{10N}$	1000/N	N	N^2	$\sqrt{N}$	$\sqrt{10N}$	1000/N
960	921 600	30.984	97.980	1.0417	980	960 400	31.305	98.995	1.0204
961	923 521	31.000	98.031	1.0406	981	962 361	31.321	99.045	1.0194
962	925 444	31.016	98.082	1.0395	982	964 324	31.337	99.096	1.0183
963	927 369	31.032	98.133	1.0384	983	966 289	31.353	99.146	1.0173
964	929 296	31.048	98.184	1.0373	984	968 256	31.369	99.197	1.0163
965	931 225	31.064	98.234	1.0363	985	970 225	31.385	99.247	1.0152
966	933 156	31.081	98.285	1.0352	986	972 196	31.401	99.298	1.0142
967	935 089	31.097	98.336	1.0341	987	974 169	31.417	99.348	1.0132
968	937 024	31.113	98.387	1.0331	988	976 144	31.432	99.398	1.0121
969	938 961	31.129	98.438	1.0320	989	978 121	31.448	99.448	1.0111
970	940 900	31.145	98.489	1.0309	990	980 100	31.464	99.499	1.0101
971	942 841	31.161	98.539	1.0299	991	982 081	31.480	99.549	1.0091
972	944 784	31.177	98.590	1.0288	992	984 064	31.496	99.599	1.0081
973	946 729	31.193	98.641	1.0277	993	986 049	31.512	99.649	1.0070
974	948 676	31.209	98.691	1.0267	994	988 036	31.528	99.700	1.0060
975	950 625	31.225	98.742	1.0256	995	990 025	31.544	99.750	1.0050
976	952 576	31.241	98.793	1.0246	996	992 016	31.559	99.800	1.0040
977	954 529	31.257	98.843	1.0235	997	994 009	31.575	99.850	1.0030
978	956 484	31.273	98.894	1.0225	998	996 004	31.591	99.900	1.0020
979	958 441	31.289	98.944	1.0215	999	998 001	31.607	99.950	1.0010

TABLE B

Sample Page from a Table of Random Numbers

15051	20470	19502	94699	95147	52871	99244	81393	78603	18726	94585
80381	48101	83920	04613	73955	19133	47150	61227	77218	40008	39924
29141	38085	91842	26565	05879	37284	88568	04208	21778	88357	76929
02799	39990	37655	36482	85097	56476	64346	79595	46675	00011	80407
51449	15917	32184	30629	30039	54019	47393	92614	26171	54992	72852
76779	83463	45656	68585	26352	78347	85647	27430	17373	20066	67412
75037	58696	50677	16209	69869	91985	45048	90111	03200	91720	10861
80001	99118	65191	88610	09578	14524	52501	51600	35350	39129	26566
05951	46640	95459	93114	90594	65581	38851	90685	77273	47120	57455
90633	70543	79018	72234	97123	07775	81845	36963	47133	59148	08987
38229	10455	27657	46640	95432	89691	49108	13813	60785	20257	92119
62329	19314	70381	58123	76821	88854	41104	81363	74737	20057	66279
28895	06337	96385	12571	00587	54691	34113	79459	29122	35685	82328
99233	79993	98015	22931	36997	51506	23191	70634	90665	74780	25535
72962	90968	13746	52181	10252	01444	65148	83075	95655	18447	58543
30978	75133	71142	56299	41461	27462	21509	53597	92909	64228	64337
78431	96519	29832	27230	91607	96299	01486	70603	86745	69072	89215
87685	90276	24472	35857	04516	61442	04947	17062	79949	92301	85755
41292	05650	07721	74968	49823	06097	65384	13471	16741	38580	55783
74961	48948	93204	02225	65947	86155	92883	60828	25752	00887	93346
20524	26508	98482	83133	03056	73164	17090	83597	62981	03479	27674
48905	87667	88024	34009	66034	97297	30184	72681	54775	44862	66156
13091	67731	16171	64950	57453	90317	29841	28386	40789	40763	37304
91101	30939	70108	22803	20540	28553	62205	03396	16962	67091	33721
28950	13441	12831	34133	82147	75860	64859	45733	78480	02916	55075
83623	66257	26118	48192	95718	26596	09789	41735	62748	73427	51063
66041	98252	53496	79886	84261	48585	46357	59019	92356	92910	64381
84032	19103	43208	52747	83311	26088	44268	89451	18053	07711	73693
85297	82266	91186	41899	83906	02772	36536	92115	61708	39207	36600
00380	27946	84016	17028	75548	24679	62460	36282	59286	26773	32610
85637	26070	41910	85351	89180	83191	10585	44379	03813	70753	06103
66204	19246	61672	34583	40148	58006	61675	34957	88345	75429	09303
78968	65742	59666	75909	71172	60099	31680	65659	48938	91987	45245
15531	82449	14791	86949	95316	74696	14707	76191	07617	61487	10746
65185	87849	11440	54729	38956	04241	25987	31290	15340	57833	39372
57876	44989	82478	18650	84748	11364	44843	63690	94975	30740	44405
07175	04444	52206	13455	14596	61851	57660	17096	84262	48705	61816
53243	47290	79328	12199	52625	67612	00854	89146	78783	41971	93232
05807	28070	99925	69219	08145	29650	03841	74388	74935	45504	48901
87123	17763	70420	63098	18621	81032	32002	07240	12892	41951	90667
74944	46757	10548	39641	92643	29851	29659	04966	19579	04620	74935
45495	47809	46325	54837	52894	02202	63037	10641	51640	40440	95640
16435	99024	53015	17831	79118	85150	63234	36120	38403	32932	27216
89829	66815	98100	33893	51091	69692	69194	05007	24855	85183	67567
95116	48878	84252	47385	91588	93734	70672	95625	14604	62807	81032

32079	17158	92293	84732	09350	85056	51202	83984	12853	36918	41356
13813	60816	24174	97389	42057	04280	31068	86748	69366	27099	74657
09692	29202	45937	04809	99294	87840	10273	04205	21437	44368	02399
88340	74820	30686	37415	05520	90950	11506	63240	36862	34151	84354
60600	96299	01555	79567	43036	30575	23111	60294	56902	19245	61578
22499	81362	74679	12527	94957	28396	42057	04344	39255	42794	99374
98224	49792	02159	57436	88177	53783	16906	59861	01026	11251	30265
83081	96404	15034	18274	36340	66759	90782	89789	61693	37349	96958
86473	34010	66247	24836	82807	60972	44338	98505	86079	83073	95317
74864	36393	73618	75695	43626	06665	38728	74794	27315	02623	17332
14747	81270	62808	81175	50502	93641	58573	34813	69817	85389	94083
15615	93266	10290	06317	93762	74233	54931	65064	72248	98913	38713
72930	86879	86317	13848	65328	06274	88267	65361	10488	31864	89397
11089	09449	97893	07153	01683	96066	71418	91796	20605	37003	52306
26397	84130	31739	73240	26840	41346	12564	99712	41823	74152	44582
30030	52855	97262	25708	95313	74258	58200	86671	59490	53166	37303
91011	19307	69526	47722	35081	04319	36097	35419	48046	76896	98451
79170	91886	32217	34974	90562	61390	98279	56861	13994	84162	35885
08122	26680	20618	38656	65579	38644	63985	33023	38867	92724	40305
78229	70431	64570	08429	66330	35533	62644	60025	22196	42243	28222
19570	03508	31445	35391	44355	00759	76838	91051	24483	37256	84930
34914	82870	69117	95043	39506	75243	85274	79312	10164	90144	07471
42683	85039	48925	90263	22916	35092	05753	21021	90583	64098	47605
19924	49068	08646	94272	39966	34567	38135	98357	66955	16092	54778
45276	19508	95463	93599	53257	49057	07314	22437	73340	39846	19093
41941	89380	08931	31059	85574	18041	06181	76284	19611	08801	14277
20653	43140	43952	48713	62861	88059	38596	57794	34420	19083	40649
22641	99650	33736	30839	57184	55685	62355	22707	08167	32474	68091
62606	55146	92775	46905	29702	10490	32160	27574	35972	19380	79014
71684	26208	59811	94533	73636	78013	42588	72726	60642	01802	11329
40417	92728	40803	42467	57239	62730	71057	45246	15639	96208	01417
61720	40876	51896	73549	66760	91022	20798	61931	68047	56962	27092
73854	06073	62393	27678	49420	54146	63824	12268	61546	18385	50577
03388	15960	37720	44786	56374	51201	83895	01399	59377	38624	61467
08174	33360	82394	07804	85683	32097	19490	93088	87312	42177	19701

TABLE C

Area Under the Normal Curve

$z = \frac{x}{\sigma}$	0.00	0.01	0.02	0.03	0.04	0.05	0.06	0.07	0.08	0.09
0.0	.0000	.0040	.0080	.0120	.0160	.0199	.0239	.0279	.0319	.0359
0.1	.0398	.0438	.0478	.0517	.0557	.0596	.0636	.0675	.0714	.0753
0.2	.0793	.0832	.0871	.0910	.0948	.0987	.1026	.1064	.1103	.1141
0.3	.1179	.1217	.1255	.1293	.1331	.1368	.1406	.1443	.1480	.1517
0.4	.1554	.1591	.1628	.1664	.1700	.1736	.1772	.1808	.1844	.1879
0.5	.1915	.1950	.1985	.2019	.2054	.2088	.2123	.2157	.2190	.2224
0.6	.2257	.2291	.2324	.2357	.2389	.2422	.2454	.2486	.2517	.2549
0.7	.2580	.2611	.2642	.2673	.2704	.2734	.2764	.2794	.2823	.2852
0.8	.2881	.2910	.2939	.2967	.2995	.3023	.3051	.3078	.3106	.3133
0.9	.3159	.3186	.3212	.3238	.3264	.3289	.3315	.3340	.3365	.3389
1.0	.3413	.3438	.3461	.3485	.3508	.3531	.3554	.3577	.3599	.3621
1.1	.3643	.3665	.3686	.3708	.3729	.3749	.3770	.3790	.3810	.3830
1.2	.3849	.3869	.3888	.3907	.3925	.3944	.3962	.3980	.3997	.4015
1.3	.4032	.4049	.4066	.4082	.4099	.4115	.4131	.4147	.4162	.4177
1.4	.4192	.4207	.4222	.4236	.4251	.4265	.4279	.4292	.4306	.4319
1.5	.4332	.4345	.4357	.4370	.4382	.4394	.4406	.4418	.4429	.4441
1.6	.4452	.4463	.4474	.4484	.4495	.4505	.4515	.4525	.4535	.4545
1.7	.4554	.4564	.4573	.4582	.4591	.4599	.4608	.4616	.4625	.4633
1.8	.4641	.4649	.4656	.4664	.4671	.4678	.4686	.4693	.4699	.4706
1.9	.4713	.4719	.4726	.4732	.4738	.4744	.4750	.4756	.4761	.4767
2.0	.4772	.4778	.4783	.4788	.4793	.4798	.4803	.4808	.4812	.4817
2.1	.4821	.4826	.4830	.4834	.4838	.4842	.4846	.4850	.4854	.4857
2.2	.4861	.4864	.4868	.4871	.4875	.4878	.4881	.4884	.4887	.4890
2.3	.4893	.4896	.4898	.4901	.4904	.4906	.4909	.4911	.4913	.4916
2.4	.4918	.4920	.4922	.4925	.4927	.4929	.4931	.4932	.4934	.4936
2.5	.4938	.4940	.4941	.4943	.4945	.4946	.4948	.4949	.4951	.4952
2.6	.4953	.4955	.4956	.4957	.4959	.4960	.4961	.4962	.4963	.4964
2.7	.4965	.4966	.4967	.4968	.4969	.4970	.4971	.4972	.4973	.4974
2.8	.4974	.4975	.4976	.4977	.4977	.4978	.4979	.4979	.4980	.4981
2.9	.4981	.4982	.4982	.4983	.4984	.4984	.4985	.4985	.4986	.4986
3.0	.4987	.4987	.4987	.4988	.4988	.4989	.4989	.4989	.4990	.4990
3.1	.4990	.4991	.4991	.4991	.4992	.4992	.4992	.4992	.4993	.4993
3.2	.4993	.4993	.4994	.4994	.4994	.4994	.4994	.4995	.4995	.4995
3.3	.4995	.4995	.4995	.4996	.4996	.4996	.4996	.4996	.4996	.4997
3.4	.4997	.4997	.4997	.4997	.4997	.4997	.4997	.4997	.4997	.4998
3.5	.4998	.4998	.4998	.4998	.4998	.4998	.4998	.4998	.4998	.4998
3.6	.4998	.4998	.4999	.4999	.4999	.4999	.4999	.4999	.4999	.4999

TABLE D*

Student's *t* Distribution

Degrees of Freedom	Level of Significance												
	.9	.8	.7	.6	.5	.4	.3	.2	.1	.05	.02	.01	.001
1	.158	.325	.510	.727	1.000	1.376	1.963	3.078	6.314	12.706	31.821	63.657	636.619
2	.142	.289	.445	.617	.816	1.061	1.386	1.886	2.910	4.303	6.965	9.925	31.598
3	.137	.277	.424	.584	.765	.978	1.250	1.638	2.353	3.182	4.541	5.841	12.941
4	.134	.271	.414	.569	.741	.941	1.190	1.533	2.132	2.776	3.747	4.604	8.610
5	.132	.267	.408	.559	.727	.920	1.156	1.476	2.015	2.571	3.365	4.032	6.859
6	.131	.265	.404	.553	.718	.906	1.134	1.440	1.943	2.447	3.143	3.707	5.959
7	.130	.263	.402	.549	.711	.896	1.119	1.415	1.895	2.365	2.998	3.499	5.405
8	.130	.262	.399	.546	.706	.889	1.108	1.397	1.860	2.306	2.896	3.355	5.041
9	.129	.261	.398	.543	.703	.883	1.100	1.383	1.833	2.262	2.821	3.250	4.781
10	.129	.260	.397	.542	.700	.879	1.093	1.372	1.812	2.228	2.764	3.169	4.587
11	.129	.260	.396	.540	.697	.876	1.088	1.363	1.796	2.201	2.718	3.106	4.437
12	.128	.259	.395	.539	.695	.873	1.083	1.356	1.782	2.179	2.681	3.055	4.318
13	.128	.259	.394	.538	.694	.870	1.079	1.350	1.771	2.160	2.650	3.012	4.221
14	.128	.258	.393	.537	.692	.868	1.076	1.345	1.761	2.145	2.624	2.977	4.140
15	.128	.258	.393	.536	.691	.866	1.074	1.341	1.753	2.131	2.602	2.947	4.073
16	.128	.258	.392	.535	.690	.865	1.071	1.337	1.746	2.120	2.583	2.921	4.015
17	.128	.257	.392	.534	.689	.863	1.069	1.333	1.740	2.110	2.567	2.898	3.965
18	.127	.257	.392	.534	.688	.862	1.067	1.330	1.734	2.101	2.552	2.878	3.922
19	.127	.257	.391	.533	.688	.861	1.066	1.328	1.729	2.093	2.539	2.861	3.883
20	.127	.257	.391	.533	.687	.860	1.064	1.325	1.725	2.086	2.528	2.845	3.850
21	.127	.257	.391	.532	.686	.859	1.063	1.323	1.721	2.080	2.518	2.831	3.819
22	.127	.256	.390	.532	.686	.858	1.061	1.321	1.717	2.074	2.508	2.819	3.792
23	.127	.256	.390	.532	.685	.858	1.060	1.319	1.714	2.069	2.500	2.807	3.767
24	.127	.256	.390	.531	.685	.857	1.059	1.318	1.711	2.064	2.492	2.797	3.745
25	.127	.256	.390	.531	.684	.856	1.058	1.316	1.708	2.060	2.485	2.787	3.725
26	.127	.256	.390	.531	.684	.856	1.058	1.315	1.706	2.056	2.479	2.779	3.707
27	.127	.256	.389	.531	.684	.855	1.057	1.314	1.703	2.052	2.473	2.771	3.690
28	.127	.256	.389	.530	.683	.855	1.056	1.313	1.701	2.048	2.467	2.763	3.674
29	.127	.256	.389	.530	.683	.854	1.055	1.311	1.699	2.045	2.462	2.756	3.659
30	.127	.256	.389	.530	.683	.854	1.055	1.310	1.697	2.042	2.457	2.750	3.646
40	.126	.255	.388	.529	.681	.851	1.050	1.303	1.684	2.021	2.423	2.704	3.551
60	.126	254	.387	.527	.679	.848	1.046	1.296	1.671	2.000	2.390	2.660	3.460
120	.126	.254	.386	.526	.677	.845	1.041	1.289	1.658	1.980	2.358	2.617	3.373
∞	.126	.253	.385	.524	.674	.842	1.036	1.282	1.645	1.960	2.326	2.576	3.291

* Table D is taken from Table III of Fisher, R. A., and Yates, F.: *Statistical Tables for Biological, Agricultural, and Medical Research,* published by Oliver & Boyd Ltd., Edinburgh, by permission of the authors and publishers.

TABLE E*

DF$_2$ = Degrees of Freedom for Denominator	DF$_1$ = Degrees of Freedom for Numerator											
DF$_2$	1	2	3	4	5	6	7	8	9	10	11	12
1	161	200	216	225	230	234	237	239	241	242	243	244
	4,052	**4,999**	**5,403**	**5,625**	**5,764**	**5,859**	**5,928**	**5,981**	**6,022**	**6,056**	**6,082**	**6,106**
2	18.51	19.00	19.16	19.25	19.30	19.33	19.36	19.37	19.38	19.39	19.40	19.41
	98.49	**99.01**	**99.17**	**99.25**	**99.30**	**99.33**	**99.34**	**99.36**	**99.38**	**99.40**	**99.41**	**99.42**
3	10.13	9.55	9.28	9.12	9.01	8.94	8.88	8.84	8.81	8.78	8.76	8.74
	34.12	**30.81**	**29.46**	**28.71**	**28.24**	**27.91**	**27.67**	**27.49**	**27.34**	**27.23**	**27.13**	**27.05**
4	7.71	6.94	6.59	6.39	6.26	6.16	6.09	6.04	6.00	5.96	5.93	5.91
	21.20	**18.00**	**16.69**	**15.98**	**15.52**	**15.21**	**14.98**	**14.80**	**14.66**	**14.54**	**14.45**	**14.37**
5	6.61	5.79	5.41	5.19	5.05	4.95	4.88	4.82	4.78	4.74	4.70	4.68
	16.26	**13.27**	**12.06**	**11.39**	**10.97**	**10.67**	**10.45**	**10.27**	**10.15**	**10.05**	**9.96**	**9.89**
6	5.99	5.14	4.76	4.53	4.39	4.28	4.21	4.15	4.10	4.06	4.03	4.00
	13.74	**10.92**	**9.78**	**9.15**	**8.75**	**8.47**	**8.26**	**8.10**	**7.98**	**7.87**	**7.79**	**7.72**
7	5.59	4.74	4.35	4.12	3.97	3.87	3.79	3.73	3.68	3.63	3.60	3.57
	12.25	**9.55**	**8.45**	**7.85**	**7.46**	**7.19**	**7.00**	**6.84**	**6.71**	**6.62**	**6.54**	**6.47**
8	5.32	4.46	4.07	3.84	3.69	3.58	3.50	3.44	3.39	3.34	3.31	3.28
	11.26	**8.65**	**7.59**	**7.01**	**6.63**	**6.37**	**6.19**	**6.03**	**5.91**	**5.82**	**5.74**	**5.67**
9	5.12	4.26	3.86	3.63	3.48	3.37	3.29	3.23	3.18	3.13	3.10	3.07
	10.56	**8.02**	**6.99**	**6.42**	**6.06**	**5.80**	**5.62**	**5.47**	**5.35**	**5.26**	**5.18**	**5.11**
10	4.96	4.10	3.71	3.48	3.33	3.22	3.14	3.07	3.02	2.97	2.94	2.91
	10.04	**7.56**	**6.55**	**5.99**	**5.64**	**5.39**	**5.21**	**5.06**	**4.95**	**4.85**	**4.78**	**4.71**
11	4.84	3.98	3.59	3.36	3.20	3.09	3.01	2.95	2.90	2.86	2.82	2.79
	9.65	**7.20**	**6.22**	**5.67**	**5.32**	**5.07**	**4.88**	**4.74**	**4.63**	**4.54**	**4.46**	**4.40**
12	4.75	3.88	3.49	3.26	3.11	3.00	2.92	2.85	2.80	2.76	2.72	2.69
	9.33	**6.93**	**5.95**	**5.41**	**5.06**	**4.82**	**4.65**	**4.50**	**4.39**	**4.30**	**4.22**	**4.16**
13	4.67	3.80	3.41	3.18	3.02	2.92	2.84	2.77	2.72	2.67	2.63	2.60
	9.07	**6.70**	**5.74**	**5.20**	**4.86**	**4.62**	**4.44**	**4.30**	**4.19**	**4.10**	**4.02**	**3.96**
14	4.60	3.74	3.34	3.11	2.96	2.85	2.77	2.70	2.65	2.60	2.56	2.53
	8.86	**6.51**	**5.56**	**5.03**	**4.69**	**4.46**	**4.28**	**4.14**	**4.03**	**3.94**	**3.86**	**3.80**
15	4.54	3.68	3.29	3.06	2.90	2.79	2.70	2.64	2.59	2.55	2.51	2.48
	8.68	**6.36**	**5.42**	**4.89**	**4.56**	**4.32**	**4.14**	**4.00**	**3.89**	**3.80**	**3.73**	**3.67**
16	4.49	3.63	3.24	3.01	2.85	2.74	2.66	2.59	2.54	2.49	2.45	2.42
	8.53	**6.23**	**5.29**	**4.77**	**4.44**	**4.20**	**4.03**	**3.89**	**3.78**	**3.69**	**3.61**	**3.55**
17	4.45	3.59	3.20	2.96	2.81	2.70	2.62	2.55	2.50	2.45	2.41	2.38
	8.40	**6.11**	**5.18**	**4.67**	**4.34**	**4.10**	**3.93**	**3.79**	**3.68**	**3.59**	**3.52**	**3.45**
18	4.41	3.55	3.16	2.93	2.77	2.66	2.58	2.51	2.46	2.41	2.37	2.34
	8.28	**6.01**	**5.09**	**4.58**	**4.25**	**4.01**	**3.85**	**3.71**	**3.60**	**3.51**	**3.44**	**3.37**
19	4.38	3.52	3.13	2.90	2.74	2.63	2.55	2.48	2.43	2.38	2.34	2.31
	8.18	**5.93**	**5.01**	**4.50**	**4.17**	**3.94**	**3.77**	**3.63**	**3.52**	**3.43**	**3.36**	**3.30**
20	4.35	3.49	3.10	2.87	2.71	2.60	2.52	2.45	2.40	2.35	2.31	2.28
	8.10	**5.85**	**4.94**	**4.43**	**4.10**	**3.87**	**3.71**	**3.56**	**3.45**	**3.37**	**3.30**	**3.23**
21	4.32	3.47	3.07	2.84	2.68	2.57	2.49	2.42	2.37	2.32	2.28	2.25
	8.02	**5.78**	**4.87**	**4.37**	**4.04**	**3.81**	**3.65**	**3.51**	**3.40**	**3.31**	**3.24**	**3.17**
22	4.30	3.44	3.05	2.82	2.66	2.55	2.47	2.40	2.35	2.30	2.26	2.23
	7.94	**5.72**	**4.82**	**4.31**	**3.99**	**3.76**	**3.59**	**3.45**	**3.35**	**3.26**	**3.18**	**3.12**
23	4.28	3.42	3.03	2.80	2.64	2.53	2.45	2.38	2.32	2.28	2.24	2.20
	7.88	**5.66**	**4.76**	**4.26**	**3.94**	**3.71**	**3.54**	**3.41**	**3.30**	**3.21**	**3.14**	**3.07**
24	4.26	3.40	3.01	2.78	2.62	2.51	2.43	2.36	2.30	2.26	2.22	2.18
	7.82	**5.61**	**4.72**	**4.22**	**3.90**	**3.67**	**3.50**	**3.36**	**3.25**	**3.17**	**3.09**	**3.03**
25	4.24	3.38	2.99	2.76	2.60	2.49	2.41	2.34	2.28	2.24	2.20	2.16
	7.77	**5.57**	**4.68**	**4.18**	**3.86**	**3.63**	**3.46**	**3.32**	**3.21**	**3.13**	**3.05**	**2.99**
26	4.22	3.37	2.98	2.74	2.59	2.47	2.39	2.32	2.27	2.22	2.18	2.15
	7.72	**5.53**	**4.64**	**4.14**	**3.82**	**3.59**	**3.42**	**3.29**	**3.17**	**3.09**	**3.02**	**2.96**

F Distribution†

DF$_1$ = Degrees of Freedom for Numerator												
14	16	20	24	30	40	50	75	100	200	500	∞	DF$_2$
245 **6,142**	246 **6,169**	248 **6,208**	249 **6,234**	250 **6,258**	251 **6,286**	252 **6,302**	253 **6,323**	253 **6,334**	254 **6,352**	254 **6,361**	254 **6,366**	1
19.42 **99.43**	19.43 **99.44**	19.44 **99.45**	19.45 **99.46**	19.46 **99.47**	19.47 **99.48**	19.47 **99.48**	19.48 **99.49**	19.49 **99.49**	19.49 **99.49**	19.50 **99.50**	19.50 **99.50**	2
8.71 **26.92**	8.69 **26.83**	8.66 **26.69**	8.64 **26.60**	8.62 **26.50**	8.60 **26.41**	8.58 **26.35**	8.57 **26.27**	8.56 **26.23**	8.54 **26.18**	8.54 **26.14**	8.53 **26.12**	3
5.87 **14.24**	5.84 **14.15**	5.80 **14.02**	5.77 **13.93**	5.74 **13.83**	5.71 **13.74**	5.70 **13.69**	5.68 **13.61**	5.66 **13.57**	5.65 **13.52**	5.64 **13.48**	5.63 **13.46**	4
4.64 **9.77**	4.60 **9.68**	4.56 **9.55**	4.53 **9.47**	4.50 **9.38**	4.46 **9.29**	4.44 **9.24**	4.42 **9.17**	4.40 **9.13**	4.38 **9.07**	4.37 **9.04**	4.36 **9.02**	5
3.96 **7.60**	3.92 **7.52**	3.87 **7.39**	3.84 **7.31**	3.81 **7.23**	3.77 **7.14**	3.75 **7.09**	3.72 **7.02**	3.71 **6.99**	3.69 **6.94**	3.68 **6.90**	3.67 **6.88**	6
3.52 **6.35**	3.49 **6.27**	3.44 **6.15**	3.41 **6.07**	3.38 **5.98**	3.34 **5.90**	3.32 **5.85**	3.29 **5.78**	3.28 **5.75**	3.25 **5.70**	3.24 **5.67**	3.23 **5.65**	7
3.23 **5.56**	3.20 **5.48**	3.15 **5.36**	3.12 **5.28**	3.08 **5.20**	3.05 **5.11**	3.03 **5.06**	3.00 **5.00**	2.98 **4.96**	2.96 **4.91**	2.94 **4.88**	2.93 **4.86**	8
3.02 **5.00**	2.98 **4.92**	2.93 **4.80**	2.90 **4.73**	2.86 **4.64**	2.82 **4.56**	2.80 **4.51**	2.77 **4.45**	2.76 **4.41**	2.73 **4.36**	2.72 **4.33**	2.71 **4.31**	9
2.86 **4.60**	2.82 **4.52**	2.77 **4.41**	2.74 **4.33**	2.70 **4.25**	2.67 **4.17**	2.64 **4.12**	2.61 **4.05**	2.59 **4.01**	2.56 **3.96**	2.55 **3.93**	2.54 **3.91**	10
2.74 **4.29**	2.70 **4.21**	2.65 **4.10**	2.61 **4.02**	2.57 **3.94**	2.53 **3.86**	2.50 **3.80**	2.47 **3.74**	2.45 **3.70**	2.42 **3.66**	2.41 **3.62**	2.40 **3.60**	11
2.64 **4.05**	2.60 **3.98**	2.54 **3.86**	2.50 **3.78**	2.46 **3.70**	2.42 **3.61**	2.40 **3.56**	2.36 **3.49**	2.35 **3.46**	2.32 **3.41**	2.31 **3.38**	2.30 **3.36**	12
2.55 **3.85**	2.51 **3.78**	2.46 **3.67**	2.42 **3.59**	2.38 **3.51**	2.34 **3.42**	2.32 **3.37**	2.28 **3.30**	2.26 **3.27**	2.24 **3.21**	2.22 **3.18**	2.21 **3.16**	13
2.48 **3.70**	2.44 **3.62**	2.39 **3.51**	2.35 **3.43**	2.31 **3.34**	2.27 **3.26**	2.24 **3.21**	2.21 **3.14**	2.19 **3.11**	2.16 **3.06**	2.14 **3.02**	2.13 **3.00**	14
2.43 **3.56**	2.39 **3.48**	2.33 **3.36**	2.29 **3.29**	2.25 **3.20**	2.21 **3.12**	2.18 **3.07**	2.15 **3.00**	2.12 **2.97**	2.10 **2.92**	2.08 **2.89**	2.07 **2.87**	15
2.37 **3.45**	2.33 **3.37**	2.28 **3.25**	2.24 **3.18**	2.20 **3.10**	2.16 **3.01**	2.13 **2.96**	2.09 **2.89**	2.07 **2.86**	2.04 **2.80**	2.02 **2.77**	2.01 **2.75**	16
2.33 **3.35**	2.29 **3.27**	2.23 **3.16**	2.19 **3.08**	2.15 **3.00**	2.11 **2.92**	2.08 **2.86**	2.04 **2.79**	2.02 **2.76**	1.99 **2.70**	1.97 **2.67**	1.96 **2.65**	17
2.29 **3.27**	2.25 **3.19**	2.19 **3.07**	2.15 **3.00**	2.11 **2.91**	2.07 **2.83**	2.04 **2.78**	2.00 **2.71**	1.98 **2.68**	1.95 **2.62**	1.93 **2.59**	1.92 **2.57**	18
2.26 **3.19**	2.21 **3.12**	2.15 **3.00**	2.11 **2.92**	2.07 **2.84**	2.02 **2.76**	2.00 **2.70**	1.96 **2.63**	1.94 **2.60**	1.91 **2.54**	1.90 **2.51**	1.88 **2.49**	19
2.23 **3.13**	2.18 **3.05**	2.12 **2.94**	2.08 **2.86**	2.04 **2.77**	1.99 **2.69**	1.96 **2.63**	1.92 **2.56**	1.90 **2.53**	1.87 **2.47**	1.85 **2.44**	1.84 **2.42**	20
2.20 **3.07**	2.15 **2.99**	2.09 **2.88**	2.05 **2.80**	2.00 **2.72**	1.96 **2.63**	1.93 **2.58**	1.89 **2.51**	1.87 **2.47**	1.84 **2.42**	1.82 **2.38**	1.81 **2.36**	21
2.18 **3.02**	2.13 **2.94**	2.07 **2.83**	2.03 **2.75**	1.98 **2.67**	1.93 **2.58**	1.91 **2.53**	1.87 **2.46**	1.84 **2.42**	1.81 **2.37**	1.80 **2.33**	1.78 **2.31**	22
2.14 **2.97**	2.10 **2.89**	2.04 **2.78**	2.00 **2.70**	1.96 **2.62**	1.91 **2.53**	1.88 **2.48**	1.84 **2.41**	1.82 **2.37**	1.79 **2.32**	1.77 **2.28**	1.76 **2.26**	23
2.13 **2.93**	2.09 **2.85**	2.02 **2.74**	1.98 **2.66**	1.94 **2.58**	1.89 **2.49**	1.86 **2.44**	1.82 **2.36**	1.80 **2.33**	1.76 **2.27**	1.74 **2.23**	1.73 **2.21**	24
2.11 **2.89**	2.06 **2.81**	2.00 **2.70**	1.96 **2.62**	1.92 **2.54**	1.87 **2.45**	1.84 **2.40**	1.80 **2.32**	1.77 **2.29**	1.74 **2.23**	1.72 **2.19**	1.71 **2.17**	25
2.10 **2.86**	2.05 **2.77**	1.99 **2.66**	1.95 **2.58**	1.90 **2.50**	1.85 **2.41**	1.82 **2.36**	1.78 **2.28**	1.76 **2.25**	1.72 **2.19**	1.70 **2.15**	1.69 **2.13**	26

DF$_2$ = Degrees of Freedom for Denominator

† Right tail of the distribution for $P = 0.05$ (light-face type), 0.01 (bold-face type).

DF_2 = Degrees of Freedom for Denominator

	DF_1 = Degrees of Freedom for Numerator											
DF_2	1	2	3	4	5	6	7	8	9	10	11	12
27	4.21 **7.68**	3.35 **5.49**	2.96 **4.60**	2.73 **4.11**	2.57 **3.79**	2.46 **3.56**	2.37 **3.39**	2.30 **3.26**	2.25 **3.14**	2.20 **3.06**	2.16 **2.98**	2.13 **2.93**
28	4.20 **7.64**	3.34 **5.45**	2.95 **4.57**	2.71 **4.07**	2.56 **3.76**	2.44 **3.53**	2.36 **3.36**	2.29 **3.23**	2.24 **3.11**	2.19 **3.03**	2.15 **2.95**	2.12 **2.90**
29	4.18 **7.60**	3.33 **5.42**	2.93 **4.54**	2.70 **4.04**	2.54 **3.73**	2.43 **3.50**	2.35 **3.33**	2.28 **3.20**	2.22 **3.08**	2.18 **3.00**	2.14 **2.92**	2.10 **2.87**
30	4.17 **7.56**	3.32 **5.39**	2.92 **4.51**	2.69 **4.02**	2.53 **3.70**	2.42 **3.47**	2.34 **3.30**	2.27 **3.17**	2.21 **3.06**	2.16 **2.98**	2.12 **2.90**	2.09 **2.84**
32	4.15 **7.50**	3.30 **5.34**	2.90 **4.46**	2.67 **3.97**	2.51 **3.66**	2.40 **3.42**	2.32 **3.25**	2.25 **3.12**	2.19 **3.01**	2.14 **2.94**	2.10 **2.86**	2.07 **2.80**
34	4.13 **7.44**	3.28 **5.29**	2.88 **4.42**	2.65 **3.93**	2.49 **3.61**	2.38 **3.38**	2.30 **3.21**	2.23 **3.08**	2.17 **2.97**	2.12 **2.89**	2.08 **2.82**	2.05 **2.76**
36	4.11 **7.39**	3.26 **5.25**	2.86 **4.38**	2.63 **3.89**	2.48 **3.58**	2.36 **3.35**	2.28 **3.18**	2.21 **3.04**	2.15 **2.94**	2.10 **2.86**	2.06 **2.78**	2.03 **2.72**
38	4.10 **7.35**	3.25 **5.21**	2.85 **4.34**	2.62 **3.86**	2.46 **3.54**	2.35 **3.32**	2.26 **3.15**	2.19 **3.02**	2.14 **2.91**	2.09 **2.82**	2.05 **2.75**	2.02 **2.69**
40	4.08 **7.31**	3.23 **5.18**	2.84 **4.31**	2.61 **3.83**	2.45 **3.51**	2.34 **3.29**	2.25 **3.12**	2.18 **2.99**	2.12 **2.88**	2.07 **2.80**	2.04 **2.73**	2.00 **2.66**
42	4.07 **7.27**	3.22 **5.15**	2.83 **4.29**	2.59 **3.80**	2.44 **3.49**	2.32 **3.26**	2.24 **3.10**	2.17 **2.96**	2.11 **2.86**	2.06 **2.77**	2.02 **2.70**	1.99 **2.64**
44	4.06 **7.24**	3.21 **5.12**	2.82 **4.26**	2.58 **3.78**	2.43 **3.46**	2.31 **3.24**	2.23 **3.07**	2.16 **2.94**	2.10 **2.84**	2.05 **2.75**	2.01 **2.68**	1.98 **2.62**
46	4.05 **7.21**	3.20 **5.10**	2.81 **4.24**	2.57 **3.76**	2.42 **3.44**	2.30 **3.22**	2.22 **3.05**	2.14 **2.92**	2.09 **2.82**	2.04 **2.73**	2.00 **2.66**	1.97 **2.60**
48	4.04 **7.19**	3.19 **5.08**	2.80 **4.22**	2.56 **3.74**	2.41 **3.42**	2.30 **3.20**	2.21 **3.04**	2.14 **2.90**	2.08 **2.80**	2.03 **2.71**	1.99 **2.64**	1.96 **2.58**
50	4.03 **7.17**	3.18 **5.06**	2.79 **4.20**	2.56 **3.72**	2.40 **3.41**	2.29 **3.18**	2.20 **3.02**	2.13 **2.88**	2.07 **2.78**	2.02 **2.70**	1.98 **2.62**	1.95 **2.56**
55	4.02 **7.12**	3.17 **5.01**	2.78 **4.16**	2.54 **3.68**	2.38 **3.37**	2.27 **3.15**	2.18 **2.98**	2.11 **2.85**	2.05 **2.75**	2.00 **2.66**	1.97 **2.59**	1.93 **2.53**
60	4.00 **7.08**	3.15 **4.98**	2.76 **4.13**	2.52 **3.65**	2.37 **3.34**	2.25 **3.12**	2.17 **2.95**	2.10 **2.82**	2.04 **2.72**	1.99 **2.63**	1.95 **2.56**	1.92 **2.50**
65	3.99 **7.04**	3.14 **4.95**	2.75 **4.10**	2.51 **3.62**	2.36 **3.31**	2.24 **3.09**	2.15 **2.93**	2.08 **2.79**	2.02 **2.70**	1.98 **2.61**	1.94 **2.54**	1.90 **2.47**
70	3.98 **7.01**	3.13 **4.92**	2.74 **4.08**	2.50 **3.60**	2.35 **3.29**	2.23 **3.07**	2.14 **2.91**	2.07 **2.77**	2.01 **2.67**	1.97 **2.59**	1.93 **2.51**	1.89 **2.45**
80	3.96 **6.96**	3.11 **4.88**	2.72 **4.04**	2.48 **3.56**	2.33 **3.25**	2.21 **3.04**	2.12 **2.87**	2.05 **2.74**	1.99 **2.64**	1.95 **2.55**	1.91 **2.48**	1.88 **2.41**
100	3.94 **6.90**	3.09 **4.82**	2.70 **3.98**	2.46 **3.51**	2.30 **3.20**	2.19 **2.99**	2.10 **2.82**	2.03 **2.69**	1.97 **2.59**	1.92 **2.51**	1.88 **2.43**	1.85 **2.36**
125	3.92 **6.84**	3.07 **4.78**	2.68 **3.94**	2.44 **3.47**	2.29 **3.17**	2.17 **2.95**	2.08 **2.79**	2.01 **2.65**	1.95 **2.56**	1.90 **2.47**	1.86 **2.40**	1.83 **2.33**
150	3.91 **6.81**	3.06 **4.75**	2.67 **3.91**	2.43 **3.44**	2.27 **3.14**	2.16 **2.92**	2.07 **2.76**	2.00 **2.62**	1.94 **2.53**	1.89 **2.44**	1.85 **2.37**	1.82 **2.30**
200	3.89 **6.76**	3.04 **4.71**	2.65 **3.88**	2.41 **3.41**	2.26 **3.11**	2.14 **2.90**	2.05 **2.73**	1.98 **2.60**	1.92 **2.50**	1.87 **2.41**	1.83 **2.34**	1.80 **2.28**
400	3.86 **6.70**	3.02 **4.66**	2.62 **3.83**	2.39 **3.36**	2.23 **3.06**	2.12 **2.85**	2.03 **2.69**	1.96 **2.55**	1.90 **2.46**	1.85 **2.37**	1.81 **2.29**	1.78 **2.23**
1,000	3.85 **6.66**	3.00 **4.62**	2.61 **3.80**	2.38 **3.34**	2.22 **3.04**	2.10 **2.82**	2.02 **2.66**	1.95 **2.53**	1.89 **2.43**	1.84 **2.34**	1.80 **2.26**	1.76 **2.20**
∞	3.84 **6.64**	2.99 **4.60**	2.60 **3.78**	2.37 **3.32**	2.21 **3.02**	2.09 **2.80**	2.01 **2.64**	1.94 **2.51**	1.88 **2.41**	1.83 **2.32**	1.79 **2.24**	1.75 **2.18**

DF$_1$ = Degrees of Freedom for Numerator													DF$_2$ = Degrees of Freedom for Denominator
14	16	20	24	30	40	50	75	100	200	500	∞	DF$_2$	
2.08 **2.83**	2.03 **2.74**	1.97 **2.63**	1.93 **2.55**	1.88 **2.47**	1.84 **2.38**	1.80 **2.33**	1.76 **2.25**	1.74 **2.21**	1.71 **2.16**	1.68 **2.12**	1.67 **2.10**	27	
2.06 **2.80**	2.02 **2.71**	1.96 **2.60**	1.91 **2.52**	1.87 **2.44**	1.81 **2.35**	1.78 **2.30**	1.75 **2.22**	1.72 **2.18**	1.69 **2.13**	1.67 **2.09**	1.65 **2.06**	28	
2.05 **2.77**	2.00 **2.68**	1.94 **2.57**	1.90 **2.49**	1.85 **2.41**	1.80 **2.32**	1.77 **2.27**	1.73 **2.19**	1.71 **2.15**	1.68 **2.10**	1.65 **2.06**	1.64 **2.03**	29	
2.04 **2.74**	1.99 **2.66**	1.93 **2.55**	1.89 **2.47**	1.84 **2.38**	1.79 **2.29**	1.76 **2.24**	1.72 **2.16**	1.69 **2.13**	1.66 **2.07**	1.64 **2.03**	1.62 **2.01**	30	
2.02 **2.70**	1.97 **2.62**	1.91 **2.51**	1.86 **2.42**	1.82 **2.34**	1.76 **2.25**	1.74 **2.20**	1.69 **2.12**	1.67 **2.08**	1.64 **2.02**	1.61 **1.98**	1.59 **1.96**	32	
2.00 **2.66**	1.95 **2.58**	1.89 **2.47**	1.84 **2.38**	1.80 **2.30**	1.74 **2.21**	1.71 **2.15**	1.67 **2.08**	1.64 **2.04**	1.61 **1.98**	1.59 **1.94**	1.57 **1.91**	34	
1.98 **2.62**	1.93 **2.54**	1.87 **2.43**	1.82 **2.35**	1.78 **2.26**	1.72 **2.17**	1.69 **2.12**	1.65 **2.04**	1.62 **2.00**	1.59 **1.94**	1.56 **1.90**	1.55 **1.87**	36	
1.96 **2.59**	1.92 **2.51**	1.85 **2.40**	1.80 **2.32**	1.76 **2.22**	1.71 **2.14**	1.67 **2.08**	1.63 **2.00**	1.60 **1.97**	1.57 **1.90**	1.54 **1.86**	1.53 **1.84**	38	
1.95 **2.56**	1.90 **2.49**	1.84 **2.37**	1.79 **2.29**	1.74 **2.20**	1.69 **2.11**	1.66 **2.05**	1.61 **1.97**	1.59 **1.94**	1.55 **1.88**	1.53 **1.84**	1.51 **1.81**	40	
1.94 **2.54**	1.89 **2.46**	1.82 **2.35**	1.78 **2.26**	1.73 **2.17**	1.68 **2.08**	1.64 **2.02**	1.60 **1.94**	1.57 **1.91**	1.54 **1.85**	1.51 **1.80**	1.49 **1.78**	42	
1.92 **2.52**	1.88 **2.44**	1.81 **2.32**	1.76 **2.24**	1.72 **2.15**	1.66 **2.06**	1.63 **2.00**	1.58 **1.92**	1.56 **1.88**	1.52 **1.82**	1.50 **1.78**	1.48 **1.75**	44	
1.91 **2.50**	1.87 **2.42**	1.80 **2.30**	1.75 **2.22**	1.71 **2.13**	1.65 **2.04**	1.62 **1.98**	1.57 **1.90**	1.54 **1.86**	1.51 **1.80**	1.48 **1.76**	1.46 **1.72**	46	
1.90 **2.48**	1.86 **2.40**	1.79 **2.28**	1.74 **2.20**	1.70 **2.11**	1.64 **2.02**	1.61 **1.96**	1.56 **1.88**	1.53 **1.84**	1.50 **1.78**	1.47 **1.73**	1.45 **1.70**	48	
1.90 **2.46**	1.85 **2.39**	1.78 **2.26**	1.74 **2.18**	1.69 **2.10**	1.63 **2.00**	1.60 **1.94**	1.55 **1.86**	1.52 **1.82**	1.48 **1.76**	1.46 **1.71**	1.44 **1.68**	50	
1.88 **2.43**	1.83 **2.35**	1.76 **2.23**	1.72 **2.15**	1.67 **2.06**	1.61 **1.96**	1.58 **1.90**	1.52 **1.82**	1.50 **1.78**	1.46 **1.71**	1.43 **1.66**	1.41 **1.64**	55	
1.86 **2.40**	1.81 **2.32**	1.75 **2.20**	1.70 **2.12**	1.65 **2.03**	1.59 **1.93**	1.56 **1.87**	1.50 **1.79**	1.48 **1.74**	1.44 **1.68**	1.41 **1.63**	1.39 **1.60**	60	
1.85 **2.37**	1.80 **2.30**	1.73 **2.18**	1.68 **2.09**	1.63 **2.00**	1.57 **1.90**	1.54 **1.84**	1.49 **1.76**	1.46 **1.71**	1.42 **1.64**	1.39 **1.60**	1.37 **1.56**	65	
1.84 **2.35**	1.79 **2.28**	1.72 **2.15**	1.67 **2.07**	1.62 **1.98**	1.56 **1.88**	1.53 **1.82**	1.47 **1.74**	1.45 **1.69**	1.40 **1.62**	1.37 **1.56**	1.35 **1.53**	70	
1.82 **2.32**	1.77 **2.24**	1.70 **2.11**	1.65 **2.03**	1.60 **1.94**	1.54 **1.84**	1.51 **1.78**	1.45 **1.70**	1.42 **1.65**	1.38 **1.57**	1.35 **1.52**	1.32 **1.49**	80	
1.79 **2.26**	1.75 **2.19**	1.68 **2.06**	1.63 **1.98**	1.57 **1.89**	1.51 **1.79**	1.48 **1.73**	1.42 **1.64**	1.39 **1.59**	1.34 **1.51**	1.30 **1.46**	1.28 **1.43**	100	
1.77 **2.23**	1.72 **2.15**	1.65 **2.03**	1.60 **1.94**	1.55 **1.85**	1.49 **1.75**	1.45 **1.68**	1.39 **1.59**	1.36 **1.54**	1.31 **1.46**	1.27 **1.40**	1.25 **1.37**	125	
1.76 **2.20**	1.71 **2.12**	1.64 **2.00**	1.59 **1.91**	1.54 **1.83**	1.47 **1.72**	1.44 **1.66**	1.37 **1.56**	1.34 **1.51**	1.29 **1.43**	1.25 **1.37**	1.22 **1.33**	150	
1.74 **2.17**	1.69 **2.09**	1.62 **1.97**	1.57 **1.88**	1.52 **1.79**	1.45 **1.69**	1.42 **1.62**	1.35 **1.53**	1.32 **1.48**	1.26 **1.39**	1.22 **1.33**	1.19 **1.28**	200	
1.72 **2.12**	1.67 **2.04**	1.60 **1.92**	1.54 **1.84**	1.49 **1.74**	1.42 **1.64**	1.38 **1.57**	1.32 **1.47**	1.28 **1.42**	1.22 **1.32**	1.16 **1.24**	1.13 **1.19**	400	
1.70 **2.09**	1.65 **2.01**	1.58 **1.89**	1.53 **1.81**	1.47 **1.71**	1.41 **1.61**	1.36 **1.54**	1.30 **1.44**	1.26 **1.38**	1.19 **1.28**	1.13 **1.19**	1.08 **1.11**	1,000	
1.69 **2.07**	1.64 **1.99**	1.57 **1.87**	1.52 **1.79**	1.46 **1.69**	1.40 **1.59**	1.35 **1.52**	1.28 **1.41**	1.24 **1.36**	1.17 **1.25**	1.11 **1.15**	1.00 **1.00**	∞	

TABLE F*

Chi Square Distribution

Degrees of Freedom† DF	Level of Significance								
	.99	.98	.95	.90	.50	.10	.05	.02	.01
1	.000157	.000628	.00393	.0158	.455	2.706	3.841	5.412	6.635
2	.0201	.0404	.103	.211	1.386	4.605	5.991	7.824	9.210
3	.115	.185	.352	.584	2.366	6.251	7.815	9.837	11.345
4	.297	.429	.711	1.064	3.357	7.779	9.488	11.668	13.277
5	.554	.752	1.145	1.610	4.351	9.236	11.070	13.388	15.086
6	.872	1.134	1.635	2.204	5.348	10.645	12.592	15.033	16.812
7	1.239	1.564	2.167	2.833	6.346	12.017	14.067	16.622	18.475
8	1.646	2.032	2.733	3.490	7.344	13.362	15.507	18.168	20.090
9	2.088	2.532	3.325	4.168	8.343	14.684	16.919	19.679	21.666
10	2.558	3.059	3.940	4.865	9.342	15.987	18.307	21.161	23.209
11	3.053	3.609	4.575	5.578	10.341	17.275	19.675	22.618	24.725
12	3.571	4.178	5.226	6.304	11.340	18.549	21.026	24.054	26.217
13	4.107	4.765	5.892	7.042	12.340	19.812	22.362	25.472	27.688
14	4.660	5.368	6.571	7.790	13.339	21.064	23.685	26.873	29.141
15	5.229	5.985	7.261	8.547	14.339	22.307	24.996	28.259	30.578
16	5.812	6.614	7.962	9.312	15.338	23.542	26.296	29.633	32.000
17	6.408	7.255	8.672	10.085	16.338	24.769	27.587	30.995	33.409
18	7.015	7.906	9.390	10.865	17.338	25.989	28.869	32.346	34.805
19	7.633	8.567	10.117	11.651	18.338	27.204	30.144	33.687	36.191
20	8.260	9.237	10.851	12.443	19.337	28.412	31.410	35.020	37.566
21	8.897	9.915	11.591	13.240	20.337	29.615	32.671	36.343	38.932
22	9.542	10.600	12.338	14.041	21.337	30.813	33.924	37.659	40.289
23	10.196	11.293	13.091	14.848	22.337	32.007	35.172	38.968	41.638
24	10.856	11.992	13.848	15.659	23.337	33.196	36.415	40.270	42.980
25	11.524	12.697	14.611	16.473	24.337	34.382	37.652	41.566	44.314
26	12.198	13.409	15.379	17.292	25.336	35.563	38.885	42.856	45.642
27	12.879	14.125	16.151	18.114	26.336	36.741	40.113	44.140	46.963
28	13.565	14.847	16.928	18.939	27.336	37.916	41.337	45.419	48.278
29	14.256	15.574	17.708	19.768	28.336	39.087	42.557	46.693	49.588
30	14.953	16.306	18.493	20.599	29.336	40.256	43.773	47.962	50.892

* Table F is taken from Table IV of Fisher, R. A., and Yates, F.: *Statistical Tables for Biological, Agricultural, and Medical Research*, published by Oliver & Boyd Ltd., Edinburgh; by permission of the authors and publishers.

TABLE G*

Transformation of *r* to *Z*

r	.000	.001	.002	.003	.004	.005	.006	.007	.008	.009
.00	.0000	.0010	.0020	.0030	.0040	.0050	.0060	.0070	.0080	.0090
.01	.0100	.0110	.0120	.0130	.0140	.0150	.0160	.0170	.0180	.0190
.02	.0200	.0210	.0220	.0230	.0240	.0250	.0260	.0270	.0280	.0290
.03	.0300	.0310	.0320	.0330	.0340	.0350	.0360	.0370	.0380	.0390
.04	.0400	.0410	.0420	.0430	.0440	.0450	.0460	.0470	.0480	.0490
.05	.0500	.0510	.0520	.0530	.0541	.0551	.0561	.0571	.0581	.0591
.06	.0601	.0611	.0621	.0631	.0641	.0651	.0661	.0671	.0681	.0691
.07	.0701	.0711	.0721	.0731	.0741	.0751	.0761	.0772	.0782	.0792
.08	.0802	.0812	.0822	.0832	.0842	.0852	.0862	.0872	.0882	.0892
.09	.0902	.0913	.0923	.0933	.0943	.0953	.0963	.0973	.0983	.0993
.10	.1003	.1013	.1024	.1034	.1044	.1054	.1064	.1074	.1084	.1094
.11	.1104	.1115	.1125	.1135	.1145	.1155	.1165	.1175	.1186	.1196
.12	.1206	.1216	.1226	.1236	.1246	.1257	.1267	.1277	.1287	.1297
.13	.1307	.1318	.1328	.1338	.1348	.1358	.1368	.1379	.1389	.1399
.14	.1409	.1419	.1430	.1440	.1450	.1460	.1471	.1481	.1491	.1501
.15	.1511	.1522	.1532	.1542	.1552	.1563	.1573	.1583	.1593	.1604
.16	.1614	.1624	.1634	.1645	.1655	.1665	.1676	.1686	.1696	.1706
.17	.1717	.1727	.1737	.1748	.1758	.1768	.1779	.1789	.1799	.1809
.18	.1820	.1830	.1841	.1851	.1861	.1872	.1882	.1892	.1903	.1913
.19	.1923	.1934	.1944	.1955	.1965	.1975	.1986	.1996	.2007	.2017
.20	.2027	.2038	.2048	.2059	.2069	.2079	.2090	.2100	.2111	.2121
.21	.2132	.2142	.2153	.2163	.2174	.2184	.2195	.2205	.2216	.2226
.22	.2237	.2247	2258	.2268	.2279	.2289	.2300	.2310	.2321	.2331
.23	.2342	.2352	.2363	.2374	.2384	.2395	.2405	2416	.2427	.2437
.24	.2448	.2458	.2469	2480	.2490	.2501	.2512	.2522	.2533	.2543
.25	.2554	.2565	.2575	.2586	.2597	.2608	.2618	.2629	.2640	.2650
.26	.2661	.2672	.2683	.2693	.2704	.2715	.2726	.2736	.2747	.2758
.27	.2769	.2779	.2790	.2801	.2812	.2823	.2833	.2844	.2855	.2866
.28	.2877	.2888	.2899	.2909	.2920	.2931	.2942	.2953	.2964	.2975
.29	.2986	.2997	.3008	.3018	.3029	.3040	.3051	.3062	.3073	.3084
.30	.3095	.3106	.3117	.3128	.3139	.3150	.3161	.3172	.3183	.3194
.31	.3205	.3217	.3228	.3239	.3250	.3261	.3272	.3283	.3294	.3305
.32	.3316	.3328	.3339	.3350	.3361	.3372	.3383	.3395	.3406	.3417
.33	.3428	.3440	.3451	.3462	.3473	.3484	.3496	.3507	.3518	.3530
.34	.3541	.3552	.3564	.3575	.3586	.3598	.3609	.3620	.3632	.3643
.35	.3654	.3666	.3677	.3689	.3700	.3712	.3723	.3734	.3746	.3757
.36	.3769	.3780	.3792	.3803	.3815	.3826	.3838	.3850	.3861	.3873
.37	.3884	.3896	.3907	.3919	.3931	.3942	.3954	.3966	.3977	.3989
.38	.4001	.4012	.4024	.4036	.4047	.4059	.4071	.4083	.4094	.4106
.39	.4118	.4130	.4142	.4153	.4165	.4177	.4189	.4201	.4213	.4225
.40	.4236	.4248	.4260	.4272	.4284	.4296	.4308	.4320	.4332	.4344
.41	.4356	.4368	.4380	.4392	.4404	.4416	.4428	.4441	.4453	.4465
.42	.4477	.4489	.4501	.4513	.4526	.4538	.4550	.4562	.4574	.4587
.43	.4599	.4611	.4624	.4636	.4648	.4660	.4673	.4685	.4698	.4710
.44	.4722	.4735	.4747	.4760	.4772	.4784	.4797	.4809	.4822	.4834

* Table G is taken from Table Z of P.H. DuBois: *An Introduction to Psychological Statistics.* New York: Harper & Row, Publishers, Incorporated, 1965, with the permission of author and publisher.

r	.000	.001	.002	.003	.004	.005	.006	.007	.008	.009
.45	.4847	.4860	.4872	.4885	.4897	.4910	.4922	.4935	.4948	.4960
.46	.4973	.4986	.4999	.5011	.5024	.5037	.5049	.5062	.5075	.5088
.47	.5101	.5114	.5126	.5139	.5152	.5165	.5178	.5191	.5204	.5217
.48	.5230	.5243	.5256	.5269	.5282	.5295	.5308	.5321	.5334	.5347
.49	.5361	.5374	.5387	.5400	.5413	.5427	.5440	.5453	.5466	.5480
.50	.5493	.5506	.5520	.5533	.5547	.5560	.5573	.5587	.5600	.5614
.51	.5627	.5641	.5654	.5668	.5682	.5695	.5709	.5722	.5736	.5750
.52	.5763	.5777	.5791	.5805	.5818	.5832	.5846	.5860	.5874	.5888
.53	.5901	.5915	.5929	.5943	.5957	.5971	.5985	.5999	.6013	.6027
.54	.6042	.6056	.6070	.6084	.6098	.6112	.6127	.6141	.6155	.6169
.55	.6184	.6198	.6213	.6227	.6241	.6256	.6270	.6285	.6299	.6314
.56	.6328	.6343	.6358	.6372	.6387	.6401	.6416	.6431	.6446	.6460
.57	.6475	.6490	.6505	.6520	.6535	.6550	.6565	.6580	.6595	.6610
.58	.6625	.6640	.6655	.6670	.6685	.6700	.6716	.6731	.6746	.6761
.59	.6777	.6792	.6807	.6823	.6838	.6854	.6869	.6885	.6900	.6916
.60	.6931	.6947	.6963	.6978	.6994	.7010	.7026	.7042	.7057	.7073
.61	.7089	.7105	.7121	.7137	.7153	.7169	.7185	.7201	.7218	.7234
.62	.7250	.7266	.7283	.7299	.7315	.7332	.7348	.7365	.7381	.7398
.63	.7414	.7431	.7447	.7464	.7481	.7498	.7514	.7531	.7548	.7565
.64	.7582	.7599	.7616	.7633	.7650	.7667	.7684	.7701	.7718	.7736
.65	.7753	.7770	.7788	.7805	.7823	.7840	.7858	.7875	.7893	.7910
.66	.7928	.7946	.7964	.7981	.7999	.8017	.8035	.8053	.8071	.8089
.67	.8107	.8126	.8144	.8162	.8180	.8199	.8217	.8236	.8254	.8273
.68	.8291	.8310	.8328	.8347	.8366	.8385	.8404	.8423	.8441	.8460
.69	.8480	.8499	.8518	.8537	.8556	.8576	.8595	.8614	.8634	.8653
.70	.8673	.8693	.8712	.8732	.8752	.8772	.8792	.8812	.8832	.8852
.71	.8872	.8892	.8912	.8933	.8953	.8973	.8994	.9014	.9035	.9056
.72	.9076	.9097	.9118	.9139	.9160	.9181	.9202	.9223	.9245	.9266
.73	.9287	.9309	.9330	.9352	.9373	.9395	.9417	.9439	.9461	.9483
.74	.9505	.9527	.9549	.9571	.9594	.9616	.9639	.9661	.9684	.9707
.75	.9730	.9752	.9775	.9798	.9822	.9845	.9868	.9892	.9915	.9939
.76	.9962	.9986	1.0010	1.0034	1.0058	1.0082	1.0106	1.0130	1.0154	1.0179
.77	1.0203	1.0228	1.0253	1.0277	1.0302	1.0327	1.0352	1.0378	1.0403	1.0428
.78	1.0454	1.0479	1.0505	1.0531	1.0557	1.0583	1.0609	1.0635	1.0661	1.0688
.79	1.0714	1.0741	1.0768	1.0795	1.0822	1.0849	1.0876	1.0903	1.0931	1.0958
.80	1.0986	1.1014	1.1042	1.1070	1.1098	1.1127	1.1155	1.1184	1.1212	1.1241
.81	1.1270	1.1299	1.1329	1.1358	1.1388	1.1417	1.1447	1.1477	1.1507	1.1538
.82	1.1568	1.1599	1.1630	1.1660	1.1692	1.1723	1.1754	1.1786	1.1817	1.1849
.83	1.1881	1.1914	1.1946	1.1979	1.2011	1.2044	1.2077	1.2111	1.2144	1.2178
.84	1.2212	1.2246	1.2280	1.2315	1.2349	1.2384	1.2419	1.2454	1.2490	1.2526
.85	1.2562	1.2598	1.2634	1.2671	1.2707	1.2745	1.2782	1.2819	1.2857	1.2895
.86	1.2933	1.2972	1.3011	1.3050	1.3089	1.3129	1.3169	1.3209	1.3249	1.3290
.87	1.3331	1.3372	1.3414	1.3456	1.3498	1.3540	1.3583	1.3626	1.3670	1.3714
.88	1.3758	1.3802	1.3847	1.3892	1.3938	1.3984	1.4030	1.4077	1.4124	1.4171
.89	1.4219	1.4268	1.4316	1.4365	1.4415	1.4465	1.4516	1.4567	1.4618	1.4670
.90	1.4722	1.4775	1.4828	1.4882	1.4937	1.4992	1.5047	1.5103	1.5160	1.5217
.91	1.5275	1.5334	1.5393	1.5453	1.5513	1.5574	1.5636	1.5698	1.5762	1.5826
.92	1.5890	1.5956	1.6022	1.6089	1.6157	1.6226	1.6296	1.6366	1.6438	1.6510
.93	1.6584	1.6658	1.6734	1.6811	1.6888	1.6967	1.7047	1.7129	1.7211	1.7295
.94	1.7380	1.7467	1.7555	1.7645	1.7736	1.7828	1.7923	1.8019	1.8117	1.8216
.95	1.8318	1.8421	1.8527	1.8635	1.8745	1.8857	1.8972	1.9090	1.9210	1.9333
.96	1.9459	1.9588	1.9721	1.9857	1.9996	2.0140	2.0287	2.0439	2.0595	2.0756
.97	2.0923	2.1095	2.1273	2.1457	2.1649	2.1847	2.2054	2.2269	2.2494	2.2729
.98	2.2976	2.3235	2.3507	2.3796	2.4101	2.4427	2.4774	2.5147	2.5550	2.5987
.99	2.6467	2.6996	2.7587	2.8257	2.9031	2.9945	3.1063	3.2504	3.4534	3.8002

Glossary

A. Sets

E_i	General notation for an element of a set.
X_i	An element which is a number (a score).
x_i	A deviation score; $X_i - \mu$.
z_i	A standard score; x_i/σ.
X_i'	A linearly transformed score; $X_i' = aX_i + b$.
(X_i, Y_i)	An element which is an ordered pair of numbers.
A, B, C	A set of sample points; an event.
U	The set of all possible sample points, called the universe set.
$\bar{A}$	The complement of set A; i.e., the set of all elements in U which do not belong to set A.
$\emptyset$	The null (empty) set; a set which has no elements.
$A \cup B$	The *union* of sets A and B, defined as the set of elements each of which belongs to A or to B or to both A and B.
$A \cap B$	The *intersection* of sets A and B, defined as the set of elements which each belong to *both* set A and set B.
$m(A)$	The measure of set A, defined here as the number of elements, n, which belong to set A.
$P(A)$	The probability of event A, defined as $m(A)/m(U)$.

B. Descriptive Statistics

Sample	Population	
S	S_*	General notation for any statistic; a function of a set of numbers.
S'	S_*'	A statistic based on transformed scores, X_i''s; for example, s'.
N	N_*	The number of elements in a set.
T	T_*	The sum of the scores within a set.

M	μ	The mean; the sum of scores divided by the number of scores.
SS	SS_*	The sum of the squared deviation scores in a set (sum of squares).
s^2	σ^2	The variance; the average squared deviation score for a set.
s	σ	The standard deviation; the positive square root of the variance of a set.
f_k	f_k	The number (frequency) of elements within a subset, k.
P_k	p_k	The proportion of scores in a set which belong to a subset, k.
T_k	–	Sum of the scores within a subset k.
GT	–	Sum of the scores in all partitions of a set; equal to T; called *the grand total.*
SS_{TOT}	–	Equal to the SS based on all N scores.
SS_{BG}	–	The between groups sum of squares.
SS_{WG}, SS_{S_s}	–	The sum of squares within groups (pooled); equal to the sum of SS values for all subsets.
$SS_A, SS_B,$ SS_T, SS_G	–	Sum of squares based on differences between means of the different levels of an independent variable, i.e., based on subsets formed with respect to a given independent variable.
$SS_{A\times B}$	–	A sum of squares reflecting the degree of interaction between independent variables A and B.
$SS_{S_s\times T}$	–	The sum of squares for subjects-by-trials interaction; the measure of correlated variation on the subject level.
$f_{o,k}$	–	Observed number of elements (frequency) in subset k.
f_{Ri}, f_{Cj}	–	The number of elements in row i (or column j); corresponds to the frequency of elements belonging to some level of one characteristic, ignoring a second characteristic.
r	ρ	The product moment correlation coefficient' based on paired scores, X_i's and Y_i's.
r_{pb}	ρ_{pb}	Point biserial r, a product moment correlation based on two variables, one of which has only two possible values (dichotomous).

ϕ	–	The phi coefficient; an r based on two variables, both of which are dichotomous.
Rho (ρ)	–	An r based on rank-order variables; Spearman's rho.
τ	–	Kendall's *tau*, a measure of relationship between pairs of ranks; a function of the proportion of inversions between scales.
Z	–	A logarithmic transformation of r, developed by Fisher.
U	–	The Mann-Whitney U; a statistic based on the sum of ranks found in one of two subsets of scores. Also used to denote the universe set, i.e., the collection of all possible sample points.

C. Estimators of Parameters

$\hat{S}_*$ The general symbol for a sample statistic used to estimate S_*.

$\hat{\mu}$ An estimate of μ, usually taken to be M.

$\hat{\sigma}^2$ An estimate of σ^2; equal to $SS/(N-1)$ for an unbiased estimate.

$\hat{\sigma}$ An estimate of σ.

MS_X The estimate of a population variance; SS_X/df_X, where X is a symbol indicating the SS used to obtain the estimate, and df is the number of degrees of freedom associated with SS_X; called *the mean square estimate.*

$\hat{\sigma}_S$ The estimated standard deviation of the hypothetical distribution of random sample values of a statistic; denoted as σ_S, where S may be any particular statistic; in particular, $\hat{\sigma}_M$ is called *the estimated standard error of the mean.*

$\hat{\sigma}_M$ The estimated standard error of the mean.

D. Statistics Based on Random Sampling Distributions

$RSD(S)$ Denotes the random sampling distribution of some statistic, S. Such a distribution gives the theoretical probability (or density) of all possible values of S from random samples of given size from a specified population. The mean and standard deviation of the distribution are denoted as μ_S and σ_S.

μ_M The mean, μ, of sample means, M's.

σ_M The standard deviation of the $RSD(M)$; called *the standard error of the mean.*

$\sigma_M{}^2$ The variance of the $RSD(M)$.

μ_{bin} The mean of the binomial distribution; equal to Np, where p is the population proportion of elements of type A, N is the sample size, and μ_{bin} is the average number of A type elements found in a random sample.

σ_{bin} — The standard deviation of the number of elements of type A found in random samples of size N; equal to $\sqrt{Npq}$.

$E(U)$ — The expected value of the Mann-Whitney U, i.e., the mean of the $RSD(U)$.

σ_U — The standard error of the Mann-Whitney U. Also see below.

$f_{e,k}$ — The mean (expected) frequency of elements in class k.

$f_{e,i,j}$ — The expected frequency in ith row, jth column cell.

σ_Z — The standard error of the Z transformation of r.

E. Test Statistics

z_S — The deviation of S from μ_S divided by σ_S; most useful when z_S is normally distributed.

z_M — The deviation of a sample mean from μ_M divided by σ_M; $(M - \mu_M)/\sigma_M$.

t — Student's t; the ratio of a normally distributed deviation score divided by an independently estimated standard deviation; for example, $(M - \mu_M)/\hat{\sigma}_M$, where the population is normal.

χ^2 — The sum of squared z-scores drawn at random from a normal distribution; for large N, equal to the sum of $[(f_{o,k} - f_{e,k})^2/f_{e,k}]$, for k classes.

F — The ratio of two independent estimates of σ^2, for a normal population of scores; MS_1/MS_2.

F. General Terms and Notations

α — Alpha, the probability of rejecting an hypothesis (H_0) when the hypothesis is actually true.

β — Beta, the probability of not rejecting an hypothesis (H_0) when the hypothesis is actually false. Also, the weight used on z_x for predicting a related z_y.

${}_nC_r$ — The number of combinations possible when selecting r elements, without replacement, from a set of n elements, equal to $n!/[(n - r)!\,r!]$.

Confidence interval — A range of values used to estimate a population parameter. The range is so developed that it has a known probability (confidence) of covering the value of the parameter.

δ — Delta, the number of degrees of freedom associated with a value of chi square.

! — The factorial sign. $X!$ is equal to $X(X - 1)(X - 2) \cdots (2)(1)$.

Homogeneity of variance	The situation in which all of a number of populations have equal variances.
H_0	A null hypothesis; the hypothesis that the population parameter or parameters have some specified value or values.
H_1	An alternative to H_0, usually the set of all parameter values not specified by H_0.
Level of confidence	The probability that a given test statistic (or one more unexpected) could have been obtained if H_0 were actually true.
μ_0	An assumed value of μ (used in H_0 testing).
NS	Not significant, refers to a statistical test.
Point estimate	An estimate of a parameter, given as a single number rather than a range.
Power	The probability (for a test) that H_0 will be rejected when it is actually false; equal to $1 - \beta$.
Question I	The probability of obtaining a random sample with a given value or range of values for S is required, given knowledge of the population.
Question II	The estimate of S_* is required, given some random sample from the population.
Question III	Requires that the probability of obtaining a test statistis (computed from samples) within some range of values is found, given certain assumptions about the population(s).
Region of rejection	A range of values for a test statistic which will allow rejection of the H_0 being tested.
r_r	A correlation used to describe the degree of *reliability* of a measuring instrument.
r_v	A correlation between the scores from a measuring instrument and the related true (criterion) scores which the instrument is supposed to measure; an index of *validity*.
Sample point	A description of one possible outcome for a random sample.
Sample space	The set of all *sample points* possible in a sampling experiment; defined here in terms of sample points which are mutually exclusive of one another and have equal likelihoods of occurrence.
σ_P^2	The variance of the scores on Y as predicted from the X variable; used in correlation and called the *predicted variance*.
$\sigma_{P_z}^2$	The *predicted variance*, measured in z-score terms; the variance of $\tilde{z}_y$, where $\tilde{z}_y = rz_x$.

σ_U	The *standard error* of *prediction;* equal to $\sqrt{\sigma_U{}^2}$.
$\sigma_U{}^2$	The *unpredicted variance,* i.e., the variance of errors in prediction; $s_y{}^2 \sqrt{1 - r^2}$.
$\sigma_{U_z}{}^2$	The *unpredicted variance,* measured in z-score terms, where an error in prediction is $z_y - \tilde{z}_y = z_y - rz_x$.
Type-I error	The rejection of H_0 when H_0 is actually true; the probability of this error is α.
Type-II error	The failue to reject H_0 when H_0 is actually false; the probability of such error is β.
Unbiased estimate	A *point estimate* of a parameter developed by a method of getting $\hat{S}_*$ with which the *average* $\hat{S}_*$, in the long run, is equal to the parameter.
$\tilde{Y}$	A predicted score on Y, where $\tilde{Y} = (X - M_x)(r)(s_y/s_x) + M_y$.
Z_H	The value of Fisher's Z transformation of r which corresponds to the hypothesized population correlation (ρ).
$\tilde{z}_y$	A predicted z-score on Y, equal to βz_x or rz_x.

Answers to Exercises

Chapter 1

1. (a) 49, (b) 56.
2. (a) Ratio, (b) nominal or ordinal (possibly), (c) ordinal or interval (possibly), (d) ordinal, (e) interval, (f) ordinal (probably).
3. (a) $1 - B^2/A^2$ (b) $\sum_{i=2}^{9} X_i^2$ (c) $\sum_{i=1}^{N} X_i^2 + 2c \sum_{i=1}^{N} X_i$

   (d) $(k + c) \sum_{j=1}^{N} X_j$ (e) $-\sum_{i=1}^{3} X_i - \sum_{i=7}^{100} X_i$
4. (a) $T = 80$, $N = 10$, (b) 77, (c) 762, (d) 0, (e) 122,

   (f) $\sum_{i=2}^{4} X_i - \sum_{i=8}^{10} X_i + \sum_{i=5}^{8} X_i^2$

Chapter 2

1. (a) Men–black, 2; men–blond; 1; men–brown, 4; men–red, 0. Women–black, 0; women–blond, 3; women–brown, 1; women–red, 1.
   (b) Blue–black, 1; blue–blond, 0; blue–brown, 2; blue–red, 0, brown–black, 1; brown–blond, 4; brown–brown, 3; brown–red, 0; green–black, 0; green–blond, 0; green–brown, 0; green–red, 1.
   (c) Men–blue–black, 1; men–blue–brown, 1; women–blue–brown, 1; men–brown–black, 1; men–brown–blond, 1; women–brown–blond, 3; men–brown–brown, 3; women–green–red, 1; all others, zero.
4. $M = 10.2$
5. $Mdn = 10.5$, mode = 11
6. (a) $M = 5$, (b) $SS = 230$, $s^2 = 11.5$, $s = 3.39$
7. $SS = 130$, $s^2 = 5.42$
8. $SS = 20$, $s = 2$; for both sets
9. $SS = 40$, $s = 2$; $SS = 4000$, $s = 20$

Chapter 3

1. $SS = 143.43$, $s^2 = 10.245$
2. $SS = 175.93$, $s^2 = 2.315$, ($M = 14.38$)
3. $M = 198.32$, $SS = 34{,}482$, $s^2 = 453.7$, $s = (14)\sqrt{2.315} = 21.3$
4. z's: $-.96$, $-.71$, $-.71$, $-.47$, $-.47$, $-.22$, $-.22$, $-.22$, .27, 1.00, 2.72

5. $z' = (X' - M')/s' = (aX + b - aM - b)/as = a(X - M)/as = z$

Chapter 4

2. (a) Question III, (b) Question I, (c) Question II, (d) Question III, (e) Question I, (f) Question II
3. $X' = X + 5$; $M' = M + 5$, so $T' = T + 20$. For $T' = 28$, $T = 28 - 20 = 8$; $P(8) = .0165$
4. $T_{Max} = 18$ so $T'_{Max} = 18 + 20 = 38$

Chapter 5

1. .1662
2. .3108
3. $\mu = 22.79$
4. .4857
5. 69.64 to 80.36, using $t_{.01}$
6. .1949
7. $z = -2.00$, reject
8. $z = -3.33$, $P = .005$ (approximately)
9. $z = 2.30$, probably not typical days, $P = .02$
10. $\hat{\sigma}^2 = 29.6/36 = .822$
11. $\hat{\sigma}_M = \sqrt{1.938} = 1.39$, $\hat{\mu} = 11.40 \pm (2.26)(1.39)$, $\hat{\mu} = 8.26$ to 14.54
12. $t = -2.95$, reject 15.5. $t = 4.24$, reject 5.5
13. $z_M = (2.375 - 4.5)/1.436 = -1.48$. Proportion below -1.48 is .0694. Table 4–2 shows .0715 below 2.375, using means of 2.25 and less.
14. Since $\hat{\mu}_M = \hat{\mu}$, an E is a deviation from its own mean. The average squared deviation ($\Sigma E^2/n$) is a variance. The notion of a linear transformation shows that $\sigma_M{}^2 = \sigma_{M-\mu}{}^2 = \sigma_E{}^2$.

Chapter 6

1.

	SS_{BG}	df_{BG}	df_{WG}
(a)	34.87	2	27
(b)	.3417	1	46
(c)	91.08	4	35
(d)	95.23	2	26

2. (a) $F(2,27) = 1.88$ *NS* (not significant)
   (b) $F(1,46) = 3.14$ *NS*
   (c) $F(4,35) = 0.64$ *NS*
   (d) $F(2,26) = 3.89$*
3. $F(2,12) = 9.02$†
4. $F(1,13) = 1.64$ *NS*
5. $SS_{BG} = \dfrac{T_1{}^2 + T_2{}^2}{n} - \dfrac{(GT)^2}{N} = \dfrac{2(T_1{}^2 + T_2{}^2)}{2n} - \dfrac{(T_1 + T_2)^2}{2n}$

   now expand the numerators, simplify and factor into $(T_1 - T_2)^2/N$

* $P < .05$. † $P < .01$.

6. (a) $F(2,33) = 3.19\ NS$
   (b) $\hat{\sigma}^2 = MS_{WG} = 28.94$
   (c) $\hat{\sigma}_M = \hat{\sigma}/\sqrt{12} = \sqrt{28.94/12} = 1.553, df = 33$
   $\hat{\mu}_1 = 37.00 \pm t_{.05}\,\hat{\sigma}_M$: 33.85 to 40.15
   $\hat{\mu}_2 = 41.92 \pm (2.03)(1.55)$: 38.77 to 45.07
   $\hat{\mu}_3 = 37.25 \pm 3.15$: 34.10 to 40.40
7. $F(1,18) = 2.50\ NS$, $F(1,38) = 5.28^*$, assuming $SS_{WG} = 1{,}152$, or 5.00^*, assuming $\hat{\sigma}_{WG}^2$ remains equal.

Chapter 7

1. (a) $SS_A = 831.37$, $df = 3$; $SS_B = 419.36$, $df = 2$; $SS_{A\times B} = 79.42$, $df = 6$
   (b) $SS_A = .544$, $df = 2$; $SS_B = .241$, $df = 2$; $SS_{A\times B} = 4.727$, $df = 4$
   (c) $SS_A = 20.417$, $df = 1$; $SS_B = .417$, $df = 1$; $SS_{A\times B} = 252.149$, $df = 1$
2. (a) $F_A(3,60) = 2.55\ NS$, $F_B(2,60) = 1.93\ NS$, $F_{A\times B}(6,60) = .12\ NS$
   (b) $F_A(2,81) = .35\ NS$, $F_B(2,81) = .16\ NS$, $F_{A\times B}(4,81) = 1.53\ NS$
   (c) $F_A(1,56) = 4.63^*$, $F_B(1,56) = .09\ NS$, $F_{A\times B}(1,56) = 57.16\dagger$
3. $SS_{TOT} = 155.867$, $SS_{WG} = 114.8$, $F_A(1,24) = 5.46^*$, $F_B(2,24) = 1.45\ NS$, $F_{A\times B}(2,24) = .11\ NS$
5. The one-variable case is appropriate, $F(1,18) = 3.08\ NS$, $SS_{TOT} = 180.95$
6. Use a two-variable case of the analysis of variance. Now the F_T for treatments is $F(1,16) = 4.74^*$ while $F_{Sex}(1,16) = 10.99\dagger$ and $F_{T\times Sex}(1,16) = .73\ NS$. The reduction in the MS error more than compensates for the reduction in df error, compared to exercise 5.

Chapter 8

1. (a) $SS_{BG} = 406.69$, $df = 1$; $SS_T = 205.33$, $df = 2$, $SS_{G\times T} = .23$, $df = 2$
   (b) $SS_{BG} = 24.67$, $df = 2$; $SS_T = 19.04$, $df = 4$, $SS_{G\times T} = 5.76$, $df = 8$
   (c) $SS_{BG} = 616.45$, $df = 3$; $SS_T = 96.80$, $df = 1$, $SS_{G\times T} = 25.30$, $df = 3$

2.

Source	*df*	*MS*	*F*
B.G.	2	81.00	9.72*
Ss	6	8.334	–
Trials	3	109.85	156.26†
$G \times T$	6	2.30	3.27*
$Ss \times T$	18	.703	–
Total	35	$SS = 568.00$	–

$^*P < .05$. $\dagger P < .01$.

3. $MS_{BG} = 324$, $MS_{WG} = 33.33$; $F(2,6) = 9.72^*$, the same as in the previous analysis.

$^*P < .05$. $\dagger P < .01$.

4.

Source	df	MS	F
B.G.	1	.05062	16.65†
Ss	6	.00304	–
Stim (trials)	1	.01210	11.52*
$G \times$ Stim	1	.00003	–
$Ss \times$ Stim	6	.00105	–
Total	15	SS = .08730	–

* $P < .05$. † $P < .01$.

5. Use a repeated-measures analysis of variance. The between-groups F tests for mean responding differences while the groups-by-trials F reflects differences in rate of conditioning. $F_{BG}(1,8) = 11.15$,* $F_T(4,32) = 20.74$,† $F_{G\times T}(4,32) = .21$ *NS*.
6. Since subject's scores are not ordered, a repeated-measures is not appropriate. A one-variable case of analysis of variance may be used if each subject is given a single score, e.g., the sum of his four estimates. This analysis gives $F(1,14) = 4.06$, which is not quite significant.

   Another approach is to examine the variability of scores. Each subject could be given a score, X, which is the SS of his four estimates (or the s^2 or s). These variability scores can then be used as the raw scores in a simple two-group analysis of variance. This approach does not yield significant differences either. From the data given, one cannot justify the rejection of the hypothesis that the independent variable has no effect on judgments.
7. $F(1,30) = t^2 = 5.52$; $T_1 = (16)(12.54)$, $T_2 = (16)(15.60)$, $GT = T_1 + T_2$.

$F = \dfrac{MS_{BG}}{MS_{WG}}$, $(MS_{WG})(5.52) = SS_{BG}/df_{BG}$. $MS_{WG} = 13.57$ (see 8–5.1 and 8–5.2)

Chapter 9

1. (a) 48, 18, 54

   (b) All six outcomes have expected frequencies of 50.

   (c) f_e's: $A = 858.8$, $B = 171.8$, $C = 21.5$, $D = 407.9$

   (d)

		Grade—f_e's				
		A	B	C	D	E
Sex	M	16.8	50.4	84.0	46.2	12.6
	W	14.4	43.2	72.0	39.6	10.8

* $P > .05$. † $P < .01$.

2. (a) χ^2 (2 *df*) = 11.08,† reject.
   (b) χ^2 (5 *df*) = 8.16 *NS*
   (c) χ^2 (3 *df*) = 225.5,† reject.
   (d) For the global hypothesis stated, two restrictions have been used; χ^2 (8 *df*) = 17.47* reject.
3. Using just the observed frequencies of grades, ignoring sex, χ^2 (4 *df*) = 11.53*; reject hypothesized curve. For the relationship between grades and sex, a contingency chi square must be used: χ^2 (4 *df*) = 8.48 *NS*.
4. Each hand dealt is considered an element and must be placed in some class. We have then 13 no-ace events and 15 hands with aces. The χ^2 test of $p_1 = .3038$, and $p_2 = 1 - p_1 = .6962$, corrected for continuity, χ^2 (1 *df*) = 2.693 *NS*.
5. A contingency chi square, corrected for continuity (since *df* = 1), shows χ^2 (1 *df*) = 5.68,* reject.
6. Equation (9–2.1) is appropriate here. The sum of the squared z-scores is interpreted as χ^2 with *df* = 15. The $\chi^2 = 11.09$, which does not cause rejection of the theory. Note that each z^2 is independent of the others, an unusual application of chi square.
7. Although each hit has been classified, a single subject has a variable score assigned to each class of hits. This feature makes chi square inappropriate. A repeated measures analysis of variance should be used, with the class of hits being the correlated (trials) variable.
8. This relational hypothesis is best tested with a contingency chi square. We must assume that such events are independent of one another within a given year and city.
9. (a) We may use the normal distribution for a test of the H_0. Since $\sigma_M = \sigma/\sqrt{N}$, $z_M = (112 - 100)/3 = 4.00$. The hypothesis is rejected.
   (b) The probability of an I.Q. being below 100 or above 100 is taken as .50. Using χ^2 we obtain $f_{e,1} = f_{e,2} = 12.5$. Correcting for continuity, we obtain χ^2 (1 *df*) = 7.84.†
10. A two-variable analysis of variance is appropriate. This results in: for colleges, $F(2,36) = 3.93$*; for time of testing, $F(2,36) = .76$ *NS*; and for the interaction, $F(4,36) = .47$ *NS*. One problem of interpretation arises from the apparent violation of random sampling. It is difficult to describe the populations of which the samples might be representative.

Chapter 10

1. (a) $12^3 = 1{,}728$, (b) $6^4 = 1{,}296$, (c) $6! = 720$, (d) $2^5 = 32$, (e) (52)(51)(50) = 132,600 (sequences) or ${}_{52}C_3 = 22{,}100$ (combinations)
2. .125
3. $m(U) = 100$ permutations, $m(A) = 4$, $P(A) = .04$
4. $m(U) = {}_6C_4$, $m(A) = (1)({}_3C_1)$, $P(A) = \frac{3}{15} = .20$
5. (a) $P(A \cup B) = P(A) + P(B) - 0 = (4)(3)/(9)(8) + (2)(1)/(9)(8) = .194$ (using permutations without replacement)

* $P < .05$. † $P < .01$.

(b) Using (a), we must include P(2 red socks), which is (3)(2)/(9)(8). $P(A \cup B \cup C) = P(A \cup B) + P(C) = .194 + .083 = .28$

6. $m(U) = {}_{26}C_5$, $m(A) = {}_{13}C_5$, $P(A) = .0196$
7. $m(U) = 2^7$ permutations, $m(A) = 1$, $P(A) = .0078$
8. (a) $m(U) = (10)(9)(8)$ permutations, $m(A) = 1$, $P(A) = .0014$ (b) $m(U) = {}_{10}C_3$, $m(A) = 1$, $P(A) = .0083$

   (c) Use permutations, $m(A) = 8 + 8$, ($X67$ or $67X$), $P(A) = .022$
9. $P(A \cup B) = P(A) + P(B) - P(A)P(B)$. $P(A) = P(B) = \frac{3}{36}$, using permutations with replacement. $P(A \cup B) = \frac{207}{1296} = .160$.
10. (a) Use binomial with $p = .5$. P(4 A's and 2 B's) $= {}_6C_4p^4q^2 = (15)(\frac{1}{2})^6 = .234$

    (b) Use $p = .25$. P(4 A's and 2 B's) $= {}_6C_4p^4q^2 = .033$

    (c) Let A = a spade, $p = .25$. P(0 A's and 6 B's) $= (\frac{3}{4})^6 = .178$
11. Use A = a six and expand $(p + q)^5$. $P(0) = .4019$, $P(1) = .4019$, $P(2) = .1608$, $P(3) = .0321$, $P(4) = .0032$, $P(5) = .0001$
12. Use the normal approximation to the binomial. $p = \frac{1}{6}$, mean $= np = 10$, $\sigma_{bi} = \sqrt{npq} = 2.886$

    $z = (15.5 - 10)/2.886 = 1.91$; $P(A) = .0281$, from normal curve table.
13. (a) May use normal approximation to the binomial or χ^2. $z = [19.5 - (.25)(50)]/\sigma_{bi} = 2.29$. There are .0110 of the cases above this z-score; for a two-tailed test of H_0, we include equally extreme values below the expected mean, hence the probability is .0220, sufficient for rejection. The χ^2 (1 df) $= 5.228$,* which also has a probability of about .02

    (b) Since expected number of events is less than 5, use the binomial expansion for $k = 0$, 1, and 2, and add the resulting probabilities. $P(\text{event}) = (\frac{3}{4})^{15} + 15(\frac{1}{4})(\frac{3}{4})^{14} + 105(\frac{1}{4})^2(\frac{3}{4})^{13}$.
14. First find $P(T)$, $T = 1$, 2, or 3. Sampling responses without replacement gives 6 permutations. The correct response is found twice in each position, hence, $P(1) = P(2) = P(3) = \frac{1}{3}$. Now we may use a χ^2 test of the observed frequencies: χ^2 (2 df) $= 4.22$ *NS*
15. Use analysis of variance, because trials for each animal cannot be assumed to be independent and values of p are not given. $F_{BG}(1,14) = 3.91$ *NS*

Chapter 11

1. (a) $r = .0833$ (b) $\sigma_{U_z}{}^2 = .993$
2. $r = .793$
3. (a) $z = 2.41$.* (b) $z = 1.92$ *NS*. (c) Both X and Y are assumed to be normally distributed in each case, and the N's are assumed to be large, which is not quite true of exercise-2 data.
4. $\sigma_U{}^2 = s_y{}^2(1 - r^2) = 5891.08$
5. $r_{pb} = .87$ or $-.87$ if women are given a score of 1
6. $\phi = .27$, χ^2 (1 df) $= 13.91$† (uncorrected) or 12.79,† corrected for continuity.

* $P < .05$. † $P < .01$.

7. Rho = −.976, tau = −.929
8. One approach would be to count the number of inversions between rankings for each person and then use analysis of variance. The number of inversions form a ratio scale; however, the range of scores is rather restricted. The $F(1,6) = 4.73^*$ and certainly supports the notion of a true effect of the independent variable; but some caution would be wise in view of the restriction in raw scores and their discontinuous nature.
9. Since the levels of the independent variable have not been held constant, the analysis of variance would have no within-groups error term. However, a correlation between weight and trials would be a meaningful approach to the problem, since the weights have been varied independently of subjects. The $r = -.750$, which gives a z of $-4.46\dagger$ for a test of H_0: $\rho = 0$. Even though the weights are not normally distributed, the very high value of the z allows a clear rejection of H_0. The different weights have caused a significant linear change (indirect relationship) in trials to learn.
10. Since the maximum validity is r_r^2, or .49, the test, at best, could predict only 24 per cent of the variance of criterion scores; see (11–7.3). He should change tests or alter the old one.

Chapter 12

1. (a) Using (12–1.1), we have $M_T = 58.7512 \pm .062 = 58.8132$ through 58.6892 or $M_T = 58.8$ (or 58.7).
   (b) Changing all scores by a factor of 10, $M_T' = 1{,}274.3650 \pm .0277 =$ 1,274.3927 through 1,274.3373, or $M_T' = 1{,}274.3$, and $M_T = 127.43$
   (c) $M_T' = 21.586 \pm .103 = 21.689$ through 21.483. $M_T' = 21$ and $M_T =$ .21 with the next digit subject to more than an error of ±1
2. Using (12–1.3) and the correction for the mean; approximately,

$$\begin{aligned} s_T{}^2 &= s^2 + .1 \pm 2\,\sigma_{s^2} = 144.1 \pm 2(24)/\sqrt{10(99)} \\ &= 144.1 \pm 1.53 \\ &= 145.6 \text{ through } 142.6 \end{aligned}$$

3. (a) Letting $\sigma_M = 1$, relative efficiency of M is $(1.253)^2 = 1.57$
   (b) $(79.37)/(49.14) = 1.62$
4. Use the normal approximation of the binomial distribution. We have $p = .134$, $\sigma_p = \sqrt{(pq)/500} = \sqrt{(.134)(.866)/500} = .01523$. Then, $\mu_p =$ $p \pm 1.96\,\sigma_p = .164$ through .104
5. If the assumptions given are true, we first find the value of M which would just yield significance: $(M_X - 100)/\sigma_M = -1.96$, $\sigma_M = 10/\sqrt{25} = 2$, and $M_X = 96.08$; since the sample means will be distributed normally around a μ_M of 90 with $\sigma_M = 2$, we wish the proportion of sample means less than M_X, all of which will lead to rejection of H_0. $z = (96.08 - 90)/2$ $= 3.04$; P (rejection) $= .9988$
6. The badly skewed scores and unequal n's suggest the Mann-Whitney U test rather than a variance test. The ranks in Group 1 are: 1.5, 1.5, 3.5, 3.5, 5, 6, and 8. $T_1 = 29$, $z = 3.12,\dagger$ which is significant even for small n's and a number of ties.

$^*P < .05$. $\quad \dagger P < .01$.

7. Since the χ^2 values are independent and the theory suggests which group should have the greater variation, we may use (12–3.5). $F(15,15) = 25.405/6.453 = 3.94^*$ (the v's are equal and cancel).

8. $\sigma_M = 16/\sqrt{4} = 8$. A mean deviation of 24 is then a z_M of 3.00. Equation (12–4.1) gives the maximum probability for such large deviations as $\frac{1}{9}$ or .111. This is most certainly a conservative figure because even an N of 4 will tend to make the distribution of means less irregular than the population.

$^* P < .01$.

Index